AF251307

Developments
in
Geochemistry

3

HELIUM ISOTOPES IN NATURE

DEVELOPMENTS IN GEOCHEMISTRY
Advisory Editor: W.S. Fyfe

1. W.S. FYFE, N.J. PRICE and A.B. THOMPSON
FLUIDS IN THE EARTH'S CRUST

2. P. HENDERSON (Editor)
RARE EARTH ELEMENT GEOCHEMISTRY

QD
181
H4
M36
13
DISCARDED

Developments
in
Geochemistry

3

HELIUM ISOTOPES IN NATURE

B.A. MAMYRIN AND L.N. TOLSTIKHIN

Geological Institute
Kola Department of USSR Academy of Sciences
Apatite 184200 (USSR)

ELSEVIER
Amsterdam — Oxford — New York — Tokyo 1984

ELSEVIER SCIENCE PUBLISHERS B.V.
1 Molenwerf
P.O. Box 211, 1000 AE Amsterdam, The Netherlands

Distributors for the United States and Canada:

ELSEVIER SCIENCE PUBLISHING COMPANY INC.
52, Vanderbilt Avenue
New York, N.Y. 10017

ISBN 0-444-42180-7 (Vol. 3)
ISBN 0-444-41635-8 (Series)

© Elsevier Science Publishers B.V., 1984
All rights reserved. No part of this publication may be reproduced, stored in a retrieval system or transmitted in any form or by any means, electronic, mechanical, photocopying, recording or otherwise, without the prior written permission of the publisher, Elsevier Science Publishers B.V., P.O. Box 220, 1000 AH Amsterdam, The Netherlands.

Printed in The Netherlands

PREFACE

During the last quarter of our century isotope cosmochemistry and geo-chemistry have made a great step forward. Continuous improvement of the sensitivity and accuracy of isotope analysis led to the introduction of mass spectrometry in space and terrestrial sciences. A considerable quantity of new experimental data, often of paramount importance, have been obtained. This has brought about a deeper understanding of many natural processes and served as a basis for new conceptions.

Isotopic investigations of noble gases have had their share in the recent achievements of isotopic methods, and several key problems of the origin and history of volatile elements have been resolved.

The isotopic ratios of noble gases are far more variable than those of other elements. This peculiarity is the result of a high noble gas depletion in spatial and terrestrial materials. Consequently, we are able to distinguish three major processes producing the noble gases in nature, each of them being responsible for a specific isotope composition of a gas. It has been established that nuclear fusion yields primordial noble gases; radioactive decay, fission and nuclear reactions stimulated by these processes are the sources of radiogenic gases; and the interaction of cosmic rays with matter produces spallogenic gases (Table 1).

Hence, the knowledge of the isotope composition of noble gases enables us to estimate the share of each genetic type in the observed mixture of gases. This aids in solving the genetic problem as well as in obtaining important information about the original conditions of the matter containing the gases,

TABLE 1

Isotope composition of primordial, radiogenic and spallogenic light noble gases

Gas	^{3}He/^{4}He	^{20}Ne/^{22}Ne	^{21}Ne/^{22}Ne	^{40}Ar/^{36}Ar	^{38}Ar/^{36}Ar
Primordial	$3 \cdot 10^{-4}$	12—13	0.03	10^{-4}	0.17—0.18
Radiogenic	$2 \cdot 10^{-8}$	0	0.3—1.0	10^7	1
Spallogenic	$2 \cdot 10^{-1}$	0.9	0.95	0.01	0.65

its age, its thermal history, etc. Often enough gases yield important information about matter that is inaccessible for direct observation.

The chemical inertness of noble gases is responsible for the most important features of their behaviour in nature as well as for the peculiarities of their elemental and isotopic analysis:

(1) The low contents of primordial noble gases (as compared with their cosmic abundance) enable one to define the occurrence of gases of other origin (radiogenic, spallogenic).

(2) In many cases it is possible to confine oneself to the discussion of physical processes alone without reference to the extremely complicated chemical ones. This, in turn, simplifies the models of natural processes and offers an opportunity for a more or less reliable quantitative approach to the interpretation of experimental data.

(3) The high volatility of noble gases makes it possible to study the migration processes as well as the structure of the matter through which the gases move.

(4) As a rule noble gases are easier to extract, purify and analyze mass-spectrometrically than other elements.

Although all noble gases are advantageous for the study of the origin of terrestrial matter, there is one gas whose properties are even more beneficial in this respect. We refer to helium — the only gas that escapes from the terrestrial atmosphere (as well as from the atmospheres of terrestrial planets), and the only gas that forms a flux rising from the earth's interior through the atmosphere into space. Due to continuous losses of helium, its atmospheric concentration is extremely low and there is practically no contamination of terrestrial rocks and fluids by it as we shall see below. This unique property of helium geochemistry is of great importance.

All this accounts for the unfailing interest of many researchers shown towards noble gases and for the scores of hundreds of papers that were devoted to the problems of noble gas isotope cosmochemistry and geochemistry in recent years.

In this "boundless sea of information" the isotopic geochemistry of helium has long been practically a "desert island"; a systematic exploration of this field began only in 1969. The main results obtained before this date were only few and may be outlined as follows.

Helium is the only element which has been initially discovered not on the earth, but in the sun; this famous discovery was made independently and simultaneously by D. Lockyer in England and P. Janssen in France in 1868.

Several decades later the heavy helium isotope, ^{4}He, was observed on the earth by W. Ramsay (1895) as a product of radioactive decay.

The second helium isotope, ^{3}He, was discovered in 1939 by L. Alvarez and R. Cornog (1939a) by means of a 60-inch cyclotron.

In 1948 L. Aldrich and A. Nier reported their first observations on helium isotope composition in some terrestrial gases and lithium minerals: it was established that in the latter the ^{3}He/^{4}He ratio reaches 10^{-5} — that is, two orders of magnitude higher than the same ratio in the gases. Simultaneously V. Khlopin and E. Gerling (1948) determined the ^{3}He/^{4}He ratio in uranium mineral as equal to $3 \cdot 10^{-10}$ — that is, two orders of magnitude lower than the ratio in the gases.

Interpreting these results, P. Morrison and J. Pine (1955) concluded that gas-well helium comes "neither from radioactive minerals as such, nor from the atmosphere, nor from preplanetary materials, but from a large mass of ordinary granite rock containing the usual diffuse amounts of trace elements . . ." According to Morrison and Pine, ^{4}He is produced by the α-decay directly, while ^{3}He is produced in nuclear reactions initiated by radioactive decay and fission. These authors also showed that the isotopic abundance of helium in radioactive and lithium minerals can be explained in the same way.

These works, based on all the data available at the time, laid the foundation of the conception of the earth's radiogenic helium. Its isotope composition being fairly homogeneous, radiogenic helium could be studied by measuring the ^{4}He content alone, and there seemed to be no need to resort to the isotope analysis of helium, which was extremely complicated at the time.

Investigations of helium on the "elemental level" were covered by several well-known reviews based on the conception of the radiogenic source of terrestrial helium as the only one (Gerling, 1957; Yakutseny, 1968; Moor and Esfandiary, 1971).

Nevertheless, it was known that the conception was in disagreement with some experimental results. In the earth's atmosphere the ^{3}He/^{4}He ratio was found to be $\sim 1 \cdot 10^{-6}$ — that is, two orders of magnitude higher than the ratio in terrestrial gases, which can be considered a source of atmospheric helium. At the same time there was no doubt about the genetic relationships between other volatile components of the gases and these components in the atmosphere. When the fact of helium escape from the atmosphere was established and the escape process was reconstructed in the most convincing way (Nicolet, 1957), it became clear that all the known sources of helium taken together might not produce the observed abundance of helium isotopes in the atmosphere.

The discovery of primordial noble gases in meteorites by Gerling and Levsky (1956) showed that the ratio of ^{3}He/^{4}He in helium of this type is about 10^{-4}, which is a hundred times higher than the same ratio in the atmosphere. Thus, a most powerful source of helium was found, which suggested that a similar source might exist in solid earth. If so, it would solve the mystery of the relationships between helium in the earth and in the atmosphere. Somewhat later Signer and Suess (1963) and Wasson (1969) came to the conclusion that primordial neon and heavier noble gases were outgassed into the atmosphere by the solid earth.

All these disconnected but important facts awaited quantitative systematization, which would lead to an explanation of the origin and distribution of helium isotopes in available terrestrial materials.

This work was initiated in 1967 by the authors of the present book. We worked out, and made extensive use of, the most fruitful up-to-date mass spectrometric method which is equal to the difficult task of a thorough isotope analysis of helium. Moreover, we employed modern equipment which guarantees the extraction of helium from all existing samples and its preparation for mass-spectrometric measurements.

The line pursued in this work has shown its worth in the realization of a wide program of isotope investigations of terrestrial helium; this included increasing the number of available $^3He/^4He$ values in various terrestrial samples by a factor of ca. 100 or even 1000. All this brought us nearer to the solution of the most vital problems of the geochemistry of volatiles.

An important contribution was also made by Canadian, American and, later, Japanese scientists, who carried out substantial and thorough investigations of the helium isotope abundance in sea water, basalts of oceanic crusts, rocks and gases of probable mantle origin.

As a result, a complete distribution of helium isotopes on the earth, in the ocean and the atmosphere has emerged and is presented on the following pages.

In recent years a new branch of isotope geochemistry has developed in relation to the previously referred to "desert island"; this has drawn the attention of many researchers from various parts of the world: the establishment of several important regularities of the isotope distribution of helium; the discovery of primordial helium in the earth's interior; the achievements in the exploration of helium escape from the earth's atmosphere; the introduction of earth degassing models based on helium isotopic data; the explanation of the nature and history of terrestrial volatile elements; the establishment of clear relationships between the concentrations of several elements in a rock and the isotope composition of its helium; the setting up of several isotopic criteria, important for applied geology and prospecting — this is only an incomplete list of the major results obtained in this new branch of isotope geochemistry.

This book is the first that aims to review the results so as to put them in the right perspective; it describes the origin, the history and the contemporary distribution of isotopes of helium, this most peculiar, even unique, volatile element. The book is mostly based on original experimental data obtained in the Soviet Union; however, Chapters 6 and 9 and some sections of other chapters summarize and interpret international publications; these are better known to the English-speaking reader. We believe that the book ought to include a detailed description of the apparatus and the mode of operations; their inefficiency had long been an obstacle in the way of the successful progress in helium isotope geochemistry.

Hence, the first part of the book, written by B. Mamyrin, gives a description of mass-spectrometric techniques and measurements of the helium isotope abundance. In this part methods of collecting various terrestrial samples are discussed as well as the apparatus for helium extraction, purification, volumetric measurements, etc.

In the second part of the book, written by I. Tolstikhin, recent data on the origin and distribution of helium isotopes in meteorites, in the earth's mantle, the crust and ocean, and in the atmosphere are discussed. It concludes with a model of the earth degassing and differentiation which is the logical outcome of the preceding analysis. The model shows a quantitative correspondence between the well-known data of the abundance of radioactive elements, primordial and radiogenic noble gases on the one hand and new data in the field of helium isotope geochemistry on the other.

We are deeply indebted to many well-known scientists in the Soviet Union and elsewhere: Academician A.P. Vinogradov; Professors E.K. Gerling, N.I. Ionov, L.K. Levsky, Yu.A. Shukolyukov, V.P. Yakutseny, E. Anders, S. Matsuo, M. Ozima, J.H. Reynolds, R.D. Russell, G.J. Wasserburg and others, who showed interest in the work and provided helpful advice and improvements. We also thank Doctors I. Azbel, G. Anufriev, E. Drubetskoy, I. Kamensky, L. Khabarin, V. Kononov, B. Polyak, E. Prasolov and B. Shustrov, who contributed experimental results, carried out calculations and took part in discussing various sections of the book. The authors are also grateful for the assistance of Doctors I. Alimova, B. Boltenkov, V. Gartmanov, B. Bogoluybov, V. Glebovskaya, A. Krylov and E. Matveeva, who helped in the preparation of the equipment, presented their samples for determination of the helium isotope composition and interpreted the results obtained.

CONTENTS

XIV

COLLECTION AND PREPARATION OF NATURAL HELIUM SAMPLES FOR MASS-SPECTROMETRIC ANALYSIS

Helium is widespread in nature — that is, it can be found in all natural gases, fluids and solids. It is obvious that different sampling techniques and methods of helium extraction for analysis should be used depending on the aggregate state of the materials sampled and their helium concentration.

However, there is one important requirement to be met in collecting these samples: one should by all means avoid air contamination. This is because many samples of terrestrial helium may have a ^{3}He/^{4}He ratio of about 10^{-8} or even 10^{-10}, whereas the atmospheric ratio amounts to $1.4 \cdot 10^{-6}$. Hence, even if 1% of air helium is contributed to the helium of the sample, it might lead to a 100% error in the measured isotopic ratio.

The error in measuring the ^{3}He and ^{4}He concentrations due to air contamination reflects the helium content in the sample; in the air the helium concentration is constant and equal to $5.24 \cdot 10^{-4}\%$.

Atmospheric helium may find its way into the sample in the course of its formation, or while being collected, stored or processed with the aim of extracting helium for isotope analysis. Besides, contamination may occur through leakage into ampoules used for the storage of helium.

Various techniques and rules for sampling, purification and storage of helium guaranteeing minimal changes in the original content of helium isotopes are discussed in the present chapter.

1.1. Vessels for collecting and long storage of natural helium samples

Stainless steel is the best material for ampoules and containers in which natural helium is to be collected and stored for a long time. However, a researcher seldom has at his disposal vessels with adequate metallic valves and, therefore, he uses glass ampoules or bottles.

Barer (1957) and Vostrov and Bol'shakov (1966) reported that some glasses and expecially quartz are penetrable to helium and their penetrability is temperature-dependent.

Let us assess the inaccuracy due to penetrability of glass during storage of helium in glass ampoules. The amount of helium, Q (cm^3 STP at standard temperature and pressure), penetrating through the glass wall under steady-state conditions during time t (s) may be represented as:

$$Q = k \, \Delta P \, t \, S/d \tag{1.1}$$

where k is the penetrability coefficient of the wall material, cm^3 mm cm^{-2} $Torr^{-1}$ s^{-1} (several values of k are listed in Table 1.1); ΔP is the constant difference of the partial pressures of helium on either side of the wall, Torr; S/d is the ratio of the area to the thickness of the wall, cm^2 mm^{-1} (Barer, 1957; Vostrov and Bol'shakov, 1966).

Since all glasses contain dissolved air helium, the leakage of helium into the ampoule starts immediately after it has been pumped out, and transitional processes are of no importance for the effects discussed here.

The leak-in ($P_0 > P_v$) or losses ($P_0 < P_v$) of helium isotopes depend on their partial pressures in the atmosphere (P_0) and in the ampoule or vessel (P_v). At sea level the partial pressures of helium isotopes are $P_{04} = 4 \cdot 10^{-3}$ Torr and $P_{03} = 5.6 \cdot 10^{-9}$ Torr for ^{4}He and ^{3}He, respectively.

The pressure of helium isotopes in an ampoule (vessel) is determined by its volume, the helium concentration in the sample and the amount of gas. The concentration and partial pressure of ^{4}He in terrestrial gases are as a rule higher than $\sim 10^{-4}\%$ and $\sim 10^{-3}$ Torr (the total gas pressure in the ampoule is assumed to be equal to ~ 760 Torr). This means that the partial pressure of ^{4}He in the ampoule is higher than the atmospheric pressure, and $P_0 < P_v$ is the only option.

TABLE 1.1

Permeability coefficients

Temperature ($^\circ$C)	Permeability, k (cm^3mm cm^{-2} $Torr^{-1}s^{-1}$)
Molybdenum glass	
25	$5 \cdot 10^{-14}$
200	$5 \cdot 10^{-12}$
600	$1.3 \cdot 10^{-10}$
Soda glass	
25	10^{-16}
200	10^{-13}
600	$2 \cdot 10^{-11}$
Quartz	
25	10^{-13}
600	$2 \cdot 10^{-10}$

Partial pressures of ^{3}He can be four to nine orders of magnitude lower and, hence, the lower limit reaches 10^{-12} Torr. Therefore, both $P_0 < P_v$ and $P_0 > P_v$ are possible for ^{3}He.

We will start by considering the case of helium loss ($P_v > P_0$). Let us assume that in eq. 1.1 $\Delta P \approx P_v$; this will slightly increase the estimated loss of helium. Then:

$$P_v = \frac{760}{V} Q_v \qquad (1.2)$$

where P_v is the partial pressure of helium in a vessel (ampule), Torr; V is its volume, cm^3; Q_v is the amount of helium in the vessel, cm^3 STP.

Hence, for our case eq. 1.1 can be re-written as:

$$-Q = 760 \frac{k\,S\,t}{V\,d} Q_v \qquad (1.3)$$

At $k = 5 \cdot 10^{-14}$ (molybdenum glass, $T = 25°C$), $S = 200$ cm^2, $d = 1$ mm, $V = 100$ cm^3, and $t = 2.6 \cdot 10^7$ s (ten months), the amount of helium leaked out from the sample may be found from eq. 1.3: $-Q \approx 2 \cdot 10^{-3} Q_v$.

Thus, normally the loss of helium isotope through glass walls of ampoules is neglibible, even when stored for about a year; it affects neither the amount of helium nor its isotopic ratio because the differences in the diffusion rates of ^{4}He and ^{3}He are small and, in this case, insignificant.

On the other hand, the leak-in of air ^{3}He, when the pressure of the isotopes in an ampoule is lower than the atmospheric pressure ($P_0 > P_v$), can strongly affect both the isotopic ratio and the ^{3}He content in a sample. Let us assume that terrestrial gas contains $10^{-3}\%$ of helium, with an isotopic ratio of ^{3}He/^{4}He equal to 10^{-9}. The gas has been kept in a molybdenum glass ampoule ($S = 200$ cm^2, $V = 100$ cm^3 and $d = 1$ mm) for a year. Then the amount of ^{3}He leaked in from the atmosphere may be determined from eq. 1.1 as $1.7 \cdot 10^{-12}$ cm^3, whereas the original amount of ^{3}He in the sample $Q_v = 10^{-12}$ cm^3. The use of soda glass (see Table 1.1) makes it possible to neglect the leak-in of air ^{3}He, even in this particular case.

All this means that sampling and long-term storage of natural helium in glass ampoules is possible. Occasional errors are only small (a few percent). However, preference should be given to ampoules of soda glass.

If glass apparatus is used for purification of helium from chemically active gases, one should beware of heated quartz elements of such apparatus. When the temperature of a quartz furnace or ampoule filled with titanium or calcium getters reaches $1200°C$ or $600°C$, the leak-in of atmospheric helium can considerably change the isotope composition of helium in a sample. For example, the ^{3}He leak-in through a quartz ampoule heated to $600°C$ for about one hour with a partial pressure of ^{4}He equal to the atmospheric pressure, causes an increase of several times the ^{3}He/^{4}He ratio for typical samples characterized by a ^{3}He/^{4}He ratio of $\sim n \cdot 10^{-8}$.

4

Examination of helium isotope leak-in through a heated glass neck of an ampoule, sealed off from a glass vacuum apparatus, shows a practically negligible contribution of air helium to helium of terrestrial samples in the ampoule.

1.2. Collecting of gaseous and fluid samples

Natural gases emerge as gaseous jets in rocks, wells, volcanoes or as bubbles in pools and water reservoirs; they can be dissolved in water, brine or oil. The following is some helpful advice in connection with the sampling technique: (a) the sampling system and ampoules for sampled gas or fluid should be thoroughly cleaned; and (b) any possibility of air entry during sampling, sealing the vessel and storage should be ruled out.

The effect of helium contamination by air is illustrated in Fig. 1.1 which shows curves of the ^{3}He/^{4}He ratio in the sample versus an air contaminant (in percent); this changes the ratio in the sample by 1%.

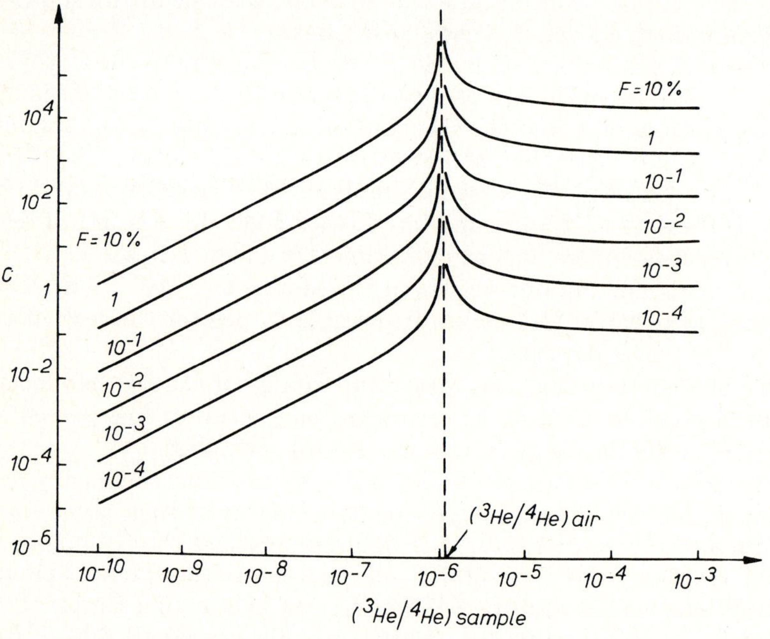

Fig. 1.1. Air contamination of a natural gas resulting in a 1% change of the ^{3}He/^{4}He ratio in the gas. C and F are the concentrations of air and helium in a gas, respectively, % volume (Mamyrin and Tolstikhin, 1981).

1.2.1. Sampling of atmospheric air

A simple and reliable technique is as follows: a bottle is filled with a sodium salt solution (which decreases the gas content in water), then the solution is removed just before the start of sample collecting, after which the bottle is plugged with a rubber stopper. To be sure, a water layer of about 0.5 cm is left above the stopper (necessary when gas is collected into bottles). The bottle is transported and stored bottom up.

1.2.2. Sampling of gas from wells and high-pressure lines

A steel vessel tested for a certain pressure is coupled with the well nipple. Cleaning takes place by letting gas into and out of the vessel several times. To reduce air contamination in the vessel to about 0.001%, the washing procedure is repeated five times at 5 atm.; at 2 atm. seventeen cycles are needed. The washing procedure can be performed by means of a cross fitted with nipples for coupling it with the hole, the vessel, the pressure gauge and the valve for gas release from the vessel when cleaned.

1.2.3. Sampling of gas from low-pressure sources

The sampling is done by displacing water (or a sodium salt solution) from a vessel by admitting gas through a pipe put into the vessel; the procedure is performed in an auxiliary liquid-containing vessel. At a very low gas pressure it can be injected by water flowing from the vessel.

In case of poor emission (water reservoirs, dry exposures), gas is collected by means of wide funnels; unfortunately, in this way atmospheric helium can easily get into the sample. If the helium concentration in a gas sample is below 10% and the ratio of $^3\mathrm{He}/^4\mathrm{He} < 10^{-7}$, it can lead to considerable errors.

1.2.4. Sampling of dissolved gases

As a rule in this case water itself is collected. The following techniques can be recommended:

(1) the so-called siphon technique, with water repeatedly changed in a vessel which is kept atmospheric air-proof;

(2) the vacuum technique, which consists of evacuating air from a vessel to obtain a vacuum sufficient to prevent undesirable contamination of a sample by atmospheric helium;

(3) the so-called piston technique which does not require cleaning (injector technique); it is suitable only in cases of a fairly high helium concentration and a high $^3\mathrm{He}/^4\mathrm{He}$ ratio (see Fig. 1.1).

Outgassing of a water sample can occur spontaneously or as a result of jerking or heating the vessel.

A detailed description of sampling techniques for gaseous and liquid samples is given in special manuals (Nesmelova, 1969; Bogolyubov et al., 1975).

1.2.5. Sampling of well gas dissolved in water

For this purpose special devices (samplers) are used. If properly designed they make it possible to collect water with dissolved gases from wells several kilometers deep as well as from deep water layers in seas and oceans. To decrease the effect of air contamination the cavity of such a sampler is pumped out or cleaned while the sampler is plunged into the water. Samplers are operated by a weight which is thrown along a wire attached to a sampler or by special catches which provide the sealing of the cavity the moment the sampler starts moving up. This latter system is only applicable for wells. A description of deep-water samplers can be found in Nesmelova (1969) and Bogolyubov et al. (1975). Tolstikhin and Kamensky (1970) showed a design of a well-water sampler with a device that cleans the cavity when it is sunk and provides a reliable sealing after sampling.

An interesting technique for sampling water with dissolved gas from ponds is reported by W.B. Clarke and Kugler (1973). Water from a certain level of the pond is pumped out through a thin long tube with a copper tube attached to its upper part. The latter can be hermetically closed at both ends. The system is continuously cleaned with water which eliminates air contamination before the sealing of a sample; it guarantees very clean sampling of helium dissolved in water.

1.3. Helium extraction from fluid microinclusions in minerals

Only a relatively small amount of gas is obtainable from microinclusions in minerals. Therefore, special measures should be taken to prevent air contamination of the gas extracted, viz. the instrument with a sample should be heated up and well pumped out prior to grinding; no additional loading of vacuum gaskets is allowed during the procedure, etc. Tolstikhin and Prasolov (1971) discussed in detail the technique of gas extraction from fluid microinclusions and gave a description of the devices.

Fig. 1.2 shows three types of devices for the comminution of mineral samples in vacuum; viz., a vacuum press with a bellows drive, a grinding device and an electromagnetic mill.

In the vacuum press (Fig. 1.2a) a sample is crushed between a steel mortar and a pestle operated by a screw press. The best application of the device is for extracting volatiles from relatively large inclusions.

The grinding device (Fig. 1.2c) is intended for attrition of samples containing small inclusions; here the mortar is rotated by a power-operating bellows. Samples weighing about 1 or 2 g are used.

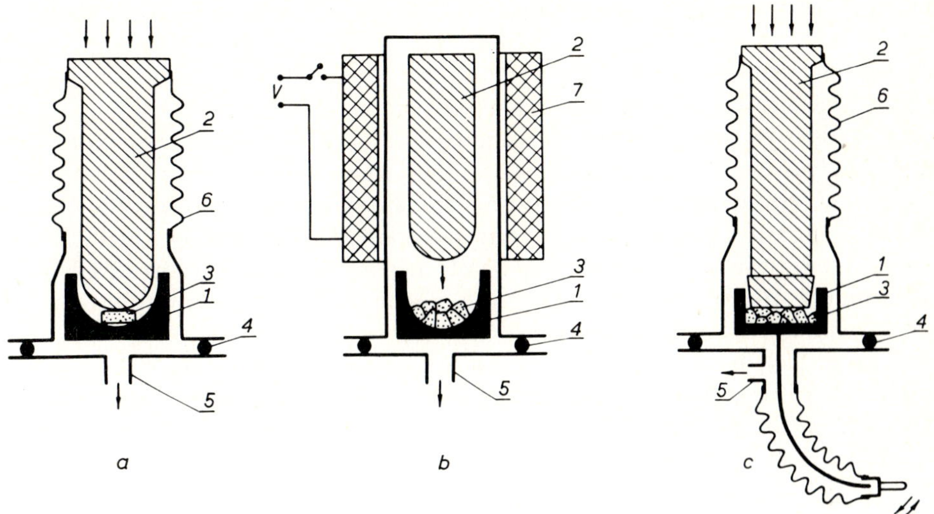

Fig. 1.2. Devices to grind solid samples in vacuum: vacuum press (a), electromagnetic mill (b) and rotating mortar (c) (Prasolov, 1972).
1 = mortar; *2* = pestle; *3* = sample; *4* = high-vacuum seal; *5* = nipple to analytical instrument; *6* = bellows; *7* = electromagnetic coil.

The high-vacuum electromagnetic mill (Fig. 1.2b) is of use for comminution of rocks and minerals when the sample weights amount to ~ 10 g. The pestle of magnetic steel weighing ~ 500 g is driven by a pulsed electromagnet. The pestle is lifted to a height of 200 mm and then dropped; its acceleration is provided by the same electromagnet. The period and acceleration can vary considerably.

Before the comminution both the device and the sample must undergo a prolonged heating up to $100-200°$ C under continuous pumping. This decreases the background of the instrument in such a way that, if released helium amounts to about 10^{-6} cm^3, the error in the subsequent determination of the ^{3}He/^{4}He ratio will not exceed 5% (Tolstikhin and Prasolov, 1971).

During the comminution of a mineral, helium released from microinclusions is normally accompanied by that from solid lattice. Because of the different origins of these types of helium, the interpretation of the results obtained for their mixture may lead to erroneous conclusions. However, if comminuting several sample weights of the mineral to various grain sizes, one can separate helium of microinclusions from that of solid lattice. The results of such an analysis can be combined with data of thermal extraction of helium from an undisturbed sample.

After extraction from microinclusions the gas is purified by a routine procedure (see section 1.5).

1.4. Helium extraction from solids

Helium is extracted from solid samples by means of heating and subsequent melting in a vacuum furnace. The main parameter of the apparatus used for this purpose is the "blank measurement" — that is, the amount of analyzed gas which is released without a sample in the instrument. The "blank measurement" or "background" of the system depends on the degree of contamination by trapped air, the so-called "memory effect" of a given isotope, the time and temperature of the experiment and — most of all — on preliminary preparation of the equipment, its heating and pumping out. It goes without saying that the leak-in of air helium must be prevented. The background of helium in modern metal instruments is about 10^5 and 10^9 atoms of ^{3}He and ^{4}He, respectively (Costa et al., 1975; Anufriev et al., 1977; Mamyrin and Khabarin, 1977; Shukolyukov et al., 1979).

The activation energy of a helium isotope (atom) varies depending on the type of lattice, the degree of its destruction, the number and the types of dislocations (defects) per gram (cm^3) of a sample, the origin of the helium isotopes in the sample, etc. Therefore, a complex relationship between the efficiency of outgassing (dQ/dt) and the temperature of the sample (T) is, as a rule, observed when the temperature rises gradually; one or more peaks of the dQ/dt value versus T are normally observed during the measurement.

For a total gas extraction by heating, two things are required: a temperature high enough to initiate the movement of atoms with the highest activation energy, and sufficient time for activated atoms to emerge at the surface of a sample. In fact, a temperature of $1100-1300°C$ is enough for complete helium extraction from the majority of minerals and rocks. Helium extraction lasting $10-15$ min is so complete that one hour of subsequent heating at $1000°C$ yields only a very small fraction of the original amount of helium in a sample. An almost complete removal of the surface gas film takes place at $200-220°C$, with less than 1% of helium released from a solid lattice.

A complete extraction of helium from a solid achieved by heating or melting enables determination of the total helium content ($cm^3\ g^{-1}$) in a sample and also measurement of the ^{3}He/^{4}He ratio. Moreover, it enables estimation of the diffusion parameters: the activation energy E and the frequence factor C for simple systems.

1.4.1. Determination of the diffusion parameters

In case of complex systems (various locations of atoms in a lattice and its defects, a mixture of different minerals in a rock, etc.) thermal extraction experiments enables determination of the parameters for each temperature fraction of the gas. The results of thermal extraction of helium isotopes together with data on the structure of the sample are indispensable for resolving the problem of the origin of helium in solids.

Gas diffusion from a solid follows Fick's law (Barer, 1957):

$$I = -D \text{ grad } Q*$$

$$\frac{dQ*}{dt} = D \text{ div grad } Q* \tag{1.4}$$

The first equation describes the gas flow I through the surface in a solid at a given gas concentration gradient (grad $Q*$) and the diffusion coefficient D. The second equation describes the distribution of concentration $Q*$ in each point of the medium depending on time. The flow, concentration and diffusion coefficient are given as g cm^{-2} s^{-1}, g cm^{-3} and cm^2 s^{-1}, respectively.

For a spherical solid with a radius r_0, diffusion coefficient D, and an originally uniform gas distribution ($Q_0* = $ constant), provided that the gas is released into an evacuated volume (at $r > r_0$, $Q* = 0$), the above eq. 1.4 may be reduced to a simpler one:

$$-\frac{dQ*}{dt} = C*Q* \tag{1.5}$$

where $C*$ is the frequency factor, viz. the relative rate of gas losses within a solid per second:

$$C* = \frac{\pi D}{r_0^2} \tag{1.6}$$

According to eq. 1.5 and 1.6, the gas losses from solids may be described by two parameters — that is, the diffusion coefficient D showing the properties of a lattice and diffusible atoms and the size of the solid, r_0. However, it was established that in natural samples of rocks and minerals characterized by a considerably altered structure, the frequency factor $C*$ does not depend on the size of the solid, and the gas release from solids of different sizes is, therefore, determined by the equation:

$$-\frac{dQ}{dt} = CQ \tag{1.7}$$

where Q is the amount of gas preserved in a sample at time t, and C is determined by the following empirical relation (Shewman, 1966):

$$C = C_0 \exp(-E/RT) \tag{1.8}$$

Here E is the activation energy of gas atoms necessary for the initiation of their movement in the lattice, T is the temperature of the solid, and R is the universal gas constant.

$$C_0 = \gamma \nu_0 \exp(\Delta S/R)$$

10

C_0 is the pre-exponential factor determined by: the coefficient γ which depends on defects in a solid; the Debye frequency of oscillations of an atom in the lattice of a solid, ν_0; and the entropy change of the system when atoms are activated in the lattice, ΔS.

Lack of dependence between the frequency factor and the size of the solid is the consequence of a great number of lattice dislocations and reflects a peculiar mechanism of "single-jump diffusion" of gas atoms into lattice defects as well as the subsequent annealing of defects described by Gerling (1939), Levsky (1963) and Morozova and Ashkinadze (1971).

Several methods have been proposed for the experimental determination of the diffusion parameters, such as the activation energy E and the frequency factor (or the pre-exponential factor C_0), for natural samples of rocks and minerals: (1) isothermal outgassing; (2) outgassing by step-wise heating; (3) dynamic outgassing by gradual heating; and (4) integral outgassing by gradual heating.

(1) In isothermal outgassing the amount of gas released is a function of time t at a constant heating temperature. The experiment is performed with two identical samples at heating temperatures T_1 and T_2. E and C_0 are calculated from the following experimental data. An exponent:

$$Q = Q_0 \exp(-Ct) \tag{1.9}$$

correspond to the solution of eq. 1.7. Here Q_0 is the total amount of gas in a sample determined at the end of the run and Q is the amount of gas in the sample at time t. The amount of gas released from the sample at time t is $\Delta Q = Q_0 - Q$, and hence:

$$\ln \frac{Q_0 - \Delta Q}{Q_0} = -Ct \tag{1.10}$$

The slopes of the experimental plots (eq. 1.10) for the outgassing temperatures T_1 and T_2 determine the angular coefficients C_1 and C_2:

$$C_1 = C_0 \exp(-E/RT_1)$$
$$C_2 = C_0 \exp(-E/RT_2)$$

hence it follows:

$$E = R \frac{T_1 T_2}{T_1 - T_2} \ln \left(\frac{C_1}{C_2}\right), \text{ and } C_0 = C_1 \exp(E/RT_1)$$

The plot of $\ln[(Q_0-Q)/Q]$ as a function of heating time t, which can be obtained from the isothermal outgassing, enables us to test whether the applied law (see eq. 1.7) is consistent with the real outgassing process or not. The deviation of this plot from the straight line implies either different diffusion mechanisms or a large amount of gas fractions with various diffusion parameters.

(2) In the process of outgassing by step-wise heating (Levsky, 1963) the temperature of the sample is raised step-by-step during a certain time interval, the temperature throughout one step being constant. At the end of a single heating interval, n, the amount of released gas, ΔQ_n, is measured. After completion of the whole run the total amount of gas contained in the sample, Q_0, is determined as a sum of all the fractions. E and C_0 are estimated with the help of experimental data according to eq. 1.7, its solution (eq. 1.9) and the formulae for the frequency factor (eq. 1.8).

For two steps of heating at T_1 and T_2 with heating time $\Delta t_1 = \Delta t_2$, the amounts of gas released, ΔQ_1 and ΔQ_2, and the total amount of gas in the sample, Q_0, we obtain:

$$\Delta Q_1 = Q_0 \left[1 - \exp\left(-C_1 \Delta t_1\right)\right]$$
$$\Delta Q_2 = (Q_0 - \Delta Q_1) \left[1 - \exp\left(-C_2 \Delta t_2\right)\right]$$

whence:

$$\ln \frac{Q_0 - \Delta Q_1}{Q_0} = -C_0 \Delta t_1 \exp\left(-E/RT_1\right)$$

$$\ln \frac{Q_0 - \Delta Q_1 - \Delta Q_2}{Q_0 - \Delta Q_1} = -C_0 \Delta t_2 \exp\left(-E/RT_2\right) \tag{1.11}$$

From eq. 1.11 we find C_0 and E as follows:

$$E = R \frac{T_1 T_2}{T_1 - T_2} \ln \left[\frac{\ln\left(1 - \dfrac{\Delta Q_1}{Q_0}\right)}{\ln\left(1 - \dfrac{\Delta Q_2}{Q_0 - \Delta Q_1}\right)} \right]$$

When several gas fractions with different activation energy values occur, a large number of steps enables us to estimate E and C_0 for each fraction by using two adjacent steps (n and $n + 1$); the calculations are similar to those given above.
In this case:

12

$$E = R \frac{T_n T_{n+1}}{T_n - T_{n+1}} \ln \left[\frac{\ln \left(1 - \dfrac{\Delta Q_n}{Q_f - \Sigma \Delta Q_{n-1}}\right)}{\ln \left(1 - \dfrac{\Delta Q_n}{Q_f - \Sigma \Delta Q_{n-1} - \Delta Q_n}\right)} \right]$$

where Q_f is the total amount of gas in a fraction with a given activation energy; $\Sigma \Delta Q_{n-1}$ is the amount of gas which belongs to the fraction Q_f and has been released in the course of previous steps (before step n); ΔQ_n and ΔQ_{n-1} are the amounts of gas released during steps n and $n + 1$, at heating temperatures T_n and T_{n+1}.

The pre-exponential factor C_0 is found from the formulae similar to eq. 1.8:

$$\ln \left(1 - \frac{\Delta Q_n}{Q_f - \Sigma \Delta Q_{n-1}}\right) = -C_0 \Delta t_n \exp \left(-E/RT_n\right)$$

where $\Delta t_n = \Delta t_{n+1}$, the time of heating at temperatures T_n and T_{n+1}, respectively.

Step-wise outgassing enables determination of the diffusion parameters by using one sample only, which improves the reliability of the results obtained. However, if there are several activation energy values of a gas in a sample studied, this technique will be time-consuming and less sensitive.

(3) Under gradual outgassing the temperature of a sample increases linearly, $T = T_0 + \alpha t$. The diffusion parameters are determined in two runs with two sample weights of one sample. Each run includes heating of the sample at the rate of temperature increase α_1 (or α_2) and measuring of T_1 (or T_2), the temperature corresponding to the maximum rate of gas release. This latter parameter is found by measuring the amount of a gas either in a vessel which is being pumped out at constant velocity or that released from the sample over certain time intervals.

The parameters E and C_0 are calculated from eqs. 1.7 and 1.8:

$$-\frac{dQ}{dt} = C_0 Q(t) \exp \left[-E/RT(t)\right]$$

Since $d^2Q/dt^2 = 0$ at a point where the outgassing curve reaches an extremum, and assuming that $T = T_0 + \alpha t$, we find from eqs. 1.7 and 1.8:

$$\frac{E\alpha_1}{C_0 R T_1^2} = \exp \left(-E/RT_1\right)$$

$$\frac{E\alpha_2}{C_0 R T_2^{\,2}} = \exp\left(-E/RT_2\right)$$

From these C_0 and E are inferred:

$$E = R\,\frac{T_1 T_2}{T_1 - T_2}\left(\ln\frac{\alpha_1}{\alpha_2} + 2\ln\frac{T_2}{T_1}\right)$$

The outgassing curve enables us to distinguish the gas fractions having different activation energy values with the best resolution and to estimate the diffusion parameters for each fraction.

(4) The parameters E and C_0 can also be found by the integral outgassing method, which is similar to the dynamic one: both methods involve the linear temperature increase ($T = T_0 + \alpha t$), but in this method the gas released is continuously accumulated in the analyzer of a mass spectrometer, the valves between the analyzer and vacuum pumps being closed. The maximum of outgassing rate is estimated by means of a graphic differentiation of the integral outgassing curve. In addition to some difficulties of interpreting the experimental results, the method of integral outgassing has to face the vacuum problem because the mass spectrometer must operate throughout the complete heating time (several hours) without being pumped.

The use of both the dynamic and the integral outgassing techniques involves a special device to provide a linear temperature increase.

1.4.2. Sample heating techniques

Various heating techniques are used for vacuum thermal extraction of gases from solid samples; the principal ones are shown in Fig. 1.3.

(1) The simplest equipment consists of a crucible made of high-temperature steel which is heated by an external resistor furnace (Fig. 1.3a). The red-hot wall of the crucible comes into contact with atmospheric air, which limits the maximal temperature of the heating to about 1400°C.

(2) Another method is the use of a vacuum container which includes a crucible and an electric heater. This device is used to increase the heating temperature (Fig. 1.3b). In such furnaces the temperature may reach 1500–2000° depending on the type of the heater and the crucible (Mamyrin and Khabarin, 1977). These furnaces are liable to a high sorption capacity of thermal and outer water-cooled screens, which inevitably increases the background of the system.

(3) Probably the most advanced method is the double-vacuum furnace (Fig. 1.3c). A molybdenum seamless crucible is coupled (through a sample loading section) with a high-vacuum apparatus for noble gas purification. In

14

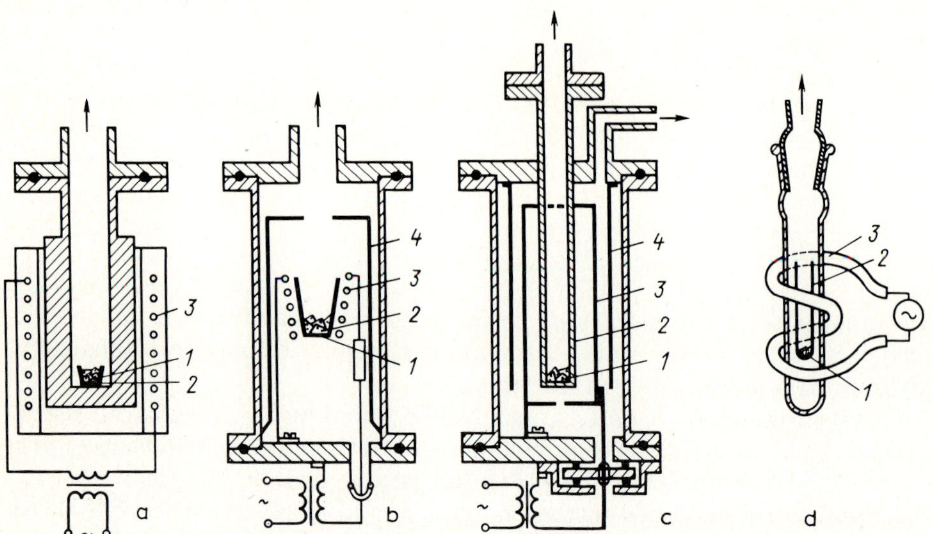

Fig. 1.3. Devices to heat solid samples in vacuum (Mamyrin and Tolstikhin, 1981). a. A high-temperature steel container heated by an outer resistor furnace. b. A crucible heated by a resistor furnace, both the crucible and the furnace arranged in a vacuum container. c. Double-vacuum system where a molybdenum container with a sample is heated by means of a resistor furnace, isolated from the atmosphere in a special vacuum chamber. d. Molybdenum crucible in a glass vacuum tube heated by a radiofrequency generator. *1* = sample; *2* = crucible or container with a sample; *3* = resistor furnace (in Fig. 1.3d radiofrequency generator); *4* = heat screen. Arrow leads to an analytical instrument.

addition, the crucible is placed (through a vacuum seal) into a separately pumped system where the outer surface of the crucible is heated in a tantalum resistor furnace.

(4) Radiofrequency generators are also used. The induction heating warms up the metal crucibles to a temperature of about 2000°C (Fig. 1.3d). The heating system of such apparatus is characterized by a low background (Ashkinadze et al., 1976). However, such systems are very expensive, complicated and cumbersome; the electric power of the radiofrequency generator which is used for such purposes is no less than 10 kW whereas its efficiency is low: only 15—20% of the power is really utilized for the heating of the crucible.

The temperature of a sample can be measured by a thermocouple (Pt/Pt-Rh or W/W-Re) and pyrometrically. Unfortunately, at high temperatures and in contact with the crucible the thermocouples soon become defective. Therefore, it is better to attach thermocouples to the system in places of lower temperature with a subsequent calibration of a sample temperature; naturally it decreases the accuracy of the measurement. The thermocouple can be preserved by means of a special manipulator which attaches the thermocouple to the crucible only when it is necessary to measure the temperature.

When the temperature of a crucible is measured pyrometrically the system is provided with glass windows with movable shutters and screens preventing dust collecting on the windows.

To decrease the background pre-heating and pumping should precede the gas extraction from a sample. This procedure requires thermo-isolation of the sample loading system from the crucible and their pre-heating at different temperatures. The dropping of samples into the crucible is accomplished by various drives (magnetic or bellows) transferring movements into vacuum (Anufriev et al., 1977; Mamyrin and Khabarin, 1977).

Attention should be paid to the preparation of representative samples because some procedures can lead to additional and unexpected errors.

The grinding of a rock to obtain a fraction of a required size can result in losses of some amount of helium due to its emanation from newly exposed grain surfaces as well as through losses of some small-size fractions after screening; the fractions can contain accessory minerals, rich in helium. On the other hand, a capture of air helium can occur due to sorption, dislocation movements and fissure development when minerals are affected mechanically (Klyavin et al., 1976). These effects decrease the accuracy of measurements and depend on the grain size — that is, the smaller the grain size, the lower is the accuracy. Vacuum grinding without subsequent screening appears to be the most effective method.

A considerable error may come from employing a metal foil for encapsulating of the powdered samples. Investigations carried out by Alimova et al. (1970) and Mamyrin et al. (1978) showed that aluminium, nickel and copper foils, even when made of very pure metals (by partial melting), can contain helium with an unusually high $^3He/^4He$ ratio reaching 10^{-1}, the ^3He content amounting to 10^9-10^{10} atoms/g^{-1}. The cause of this effect is unknown and we cannot, therefore, suggest a testing procedure which proves that excess ^3He is absent. At the same time even a small amount of ^3He resulting from metal foil which accompanies a sample can lead to serious errors in the results. The pre-heating of the foil in vacuum to a temperature exceeding that required for sample heating actually eliminates contamination.

It is necessary to keep in mind a possible contamination of the foil by ^3He when helium isotopes are studied in space by both the foil technique (Geiss et al., 1970) and the implantation of helium ions into metal plates (Alimova et al., 1966; Boltenkov, 1973).

In such experiments relatively large areas of foil are used, so it is difficult to test the distribution of excess ^3He over the foil. A most irregular pattern can occur: over an area of several hundred square centimeters the total amount of ^3He can be accumulated in several points covering an area of 1 mm^2 (Mamyrin et al., 1978).

1.5. Purification of helium from accompanying gases

The concentration of helium in natural gases and gases extracted from liquid and solid samples is usually very low, reachting 10^{-4} to $10^{-5}\%$ vol. and even less. The concentration of ^{4}He in the richest samples amounts to several percent, while ^{3}He even in these exceptional cases does not exceed $10^{-8}\%$.

Therefore, it is always necessary to separate helium from accompanying gases prior to mass-spectrometric isotope analysis. The most objectable among these gases, as will be shown in Chapter 3, are hydrogen and hydrogen-bearing components, because they can preclude mass-spectrometric measurements of ^{3}He amounts.

According to their physical and chemical properties accompanying gases may be subdivided as follows: water vapors, chemically active gases, heavy inert gases (argon inclusive) and, finally, neon. Various techniques are used to separate these gases from helium.

Water vapors can be removed by adsorption in a trap cooled by liquid nitrogen.

Chemically active gases are trapped by absorbents, such as a heated calcium or titanium sponge.

Traps with metal absorbents are made in the form of high-temperature steel cylinders (about 30 mm in diameter) with Ni-Cr outer heaters. The cylinders are filled with a 10—20 g calcium or titanium sponge. The operating conditions of metal absorbents are shown in Table 1.2. Absorption of chemically active gases takes place mainly during the first stage at a high temperature, while the second stage, characterized by a lower temperature, decreases the hydrogen residual pressure, which is important for the ^{3}He measurement.

Traps filled with activated charcoal and cooled to the boiling point of liquid nitrogen are used to absorb heavy noble gases. Such a trap may be designed as a U-shaped tube or cold finger. A fairly adequate design of a metal pass-through trap is shown in Fig. 1.4. Under the gas absorption regime the trap is cooled by liquid nitrogen and then heated by a Ni-Cr furnace to about 300°C under a regeneration regime.

Helium can be separated from neon by helium diffusion through a thin

TABLE 1.2

Operating conditions of metal absorbents

Absorbent	Regeneration regime	First absorption stage	Second absorption stage
Calcium	800°C	600°C	300°C
Titanium sponge	1100—1200°C	800°C	400—500°C

heated quartz partition or freezing neon on surfaces cooled below the boiling point of neon. The possible procedures of Ne separation from He are discussed elsewhere (Ermolin, 1957; Mamyrin et al., 1970a; see also section 2.3). It is clear that the separation should be carried out when the share of Ne is large enough. Small contributions of Ne do not interfere with mass spectrometric measurements of helium isotopes (see section 3.4).

A complete procedure of He purification from associated gases consists of the following operations: (1) drying of the gas by means of P_2O_5 or a freezer; (2) absorption of chemically active gases by means of calcium or titanium traps; (3) separation of argon and heavy noble gases in a charcoal trap cooled to liquid nitrogen temperature; and (4) separation of neon by freezing on the surface cooled by liquid helium.

In most cases when separation of chemically active gases from heavy noble gases is not compulsory one can use a simpler procedure:

(1) Absorption of major accompanying gases by means of a cold finger filled with charcoal having a large sorbing surface; the trap is cooled by liquid nitrogen.

(2) After the isolation of the finger the second charcoal trap is connected with the furnace for final purification.

In such a way, the purification procedure is completed, and neon is not separated from the helium sample; its amount is controlled through mass-spectrometric analysis.

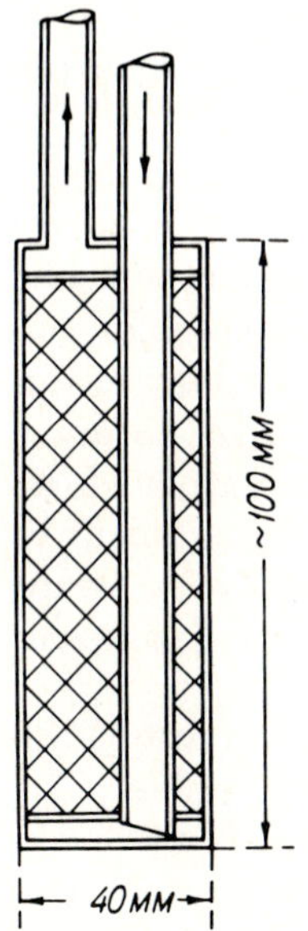

Fig. 1.4. Steel trap with activated charcoal.

1.6. A modern all-metal system for extraction and purification of helium

Recently several metal extraction systems with a background of ^{4}He approaching 10^{10} atoms were reported.

Costa et al. (1975) described the construction and the method of operation of an all-metal system for extraction and purification of noble gases. The system is capable of handling up to twenty samples in batches of about 15 g total weight. The typical background is about $4 \cdot 10^{10}$ atoms for ^{4}He and $\sim 10^{11}$ atoms for ^{40}Ar$_{atm}$.

Mamyrin and Khabarin (1977) reported on a simple furnace with a sample loading system. A big crucible and a powerful heater enable one to use sample weights of several tens of grams. It takes about 10 min to heat and outgas a rock or mineral sample. The background is about 10^{10} atoms and 10^6 atoms of ^{4}He and ^{3}He, respectively.

Shukolyukov et al. (1979) described a double-vacuum furnace characterized by a heating temperature of $2000°$C. The background of ^{4}He in their system amounts to $\sim 10^{10}$ atoms.

By way of example, let us consider in more detail a compact all-metal system for gas extraction and helium purification from accompanying gases which shows a lower background of ^{4}He, about 10^9 atoms, and ^{3}He, about $3 \cdot 10^5$ atoms (Anufriev et al., 1977). Fig. 1.5 illustrates the vacuum and electric circuits of the assembly. A description of the elements is given in the figure caption. The equipment was intended to attain the lowest background and minimize the working volume, which is important for studies of small amounts of helium.

The entire system is made of stainless steel and permits pre-heating for outgassing up to $400°$C; its volume is about 250 cm^3; the overall sizes are $410 \times 220 \times 60$ mm. The furnace for sample heating is designed as a high-temperature steel finger. Its sizes are: outer diameter 30 mm; wall thickness in the heating section 2 mm; wall thickness of the circular groove, which is turned for better thermal isolation of the lower heated part of the finger, 0.5 mm. This groove and heavy flanges protect samples in the loading section from heating by the operating furnace. The furnace is heated by an outer Ni-Cr 0.8 mm diameter resistor coil; the maximal heating temperature of the samples in crucible *11* is about $1450°$C; the temperature rises from ambient to $1200°$C for 10 min. The temperature can be raised automatically by means of a temperature rise-rate computer program to study the kinetics of gas losses and determine the diffusion parameters.

The sample loading section permits successive drops of ten and more samples, their number depending on the number of cells in a special holder. The holder is operated by an outer electromagnet. Window *8* enables us to observe the sample dropping and determine the temperature of the crucible pyrometrically.

Auxilliary volume *22* is used for a short-time storage of the gas extracted as well as for preparation or storage of standards of noble gases. The small size of the system enables us to couple it with a mass-spectrometric analyzer by means of a short pipe; it also decreases the background and improves the sensitivity of measurements.

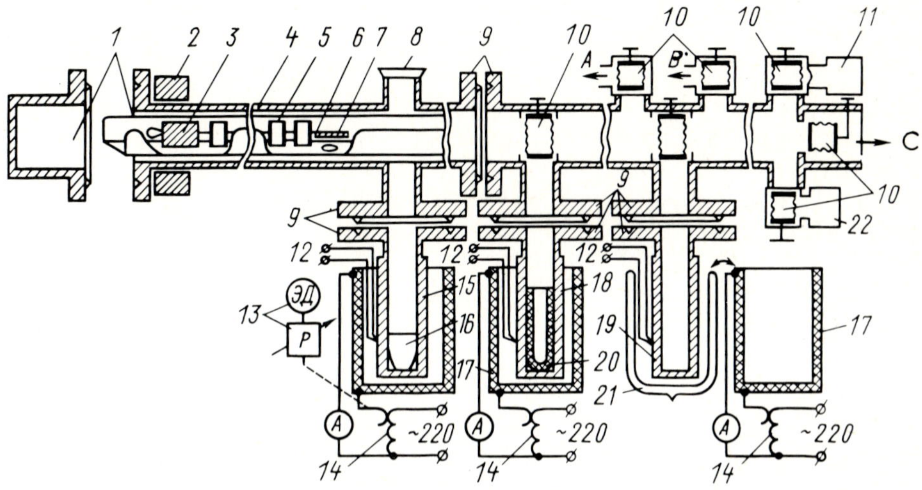

Fig. 1.5. All-metal vacuum system to extract and purify noble gases (Anufriev et al., 1979).

1 = sample loading section (volume V = 87 cm³); 2 = electromagnet; 3 = magnetic pusher; 4 = housing tube; 5 = movable sample holder; 6 = heat isolation screen; 7 = window screen; 8 = window; 9 = high-vacuum joints; 10 = high vacuum valves; 11 = high vacuum ion gauge; 12 = thermocouples; 13 = automatic electric drive of moving-coil voltage regulator; 14 = voltage regulators; 15 = furnace to heat a sample (V = 40 cm³); 16 = crucible; 17 = resistor heaters; 18 = titanium getter (V = 40 cm³); 19 = charcoal fingers; 20 = steel cap for titanium getter; 21 = Dewar flask; 22 = additional vessel. Arrows A, B and C indicate high-vacuum and fore pumps and mass spectrometer, respectively.

Chapter 2

MEASUREMENT OF HELIUM CONTENTS IN SAMPLES; HELIUM STANDARDS WITH A GIVEN ISOTOPE RATIO

The measurement of the absolute amount and concentration of helium in samples is carried out in installations combining elements necessary for extraction, purification and volume determination. One of such installations is described below (see section 2.1).

Mass-spectrometric measurements of the absolute amount of helium as well as the helium/neon and the helium isotope ratios are required in standards.

Sections 2.2 and 2.4 deal with the techniques and equipment for preparing standards with an accuracy suitable for most measurements of the ^{4}He concentrations and ^{3}He/^{4}He ratios in terrestrial samples.

2.1. Measurement of helium contents in terrestrial gases

The helium contents in natural gas samples or gases extracted from liquids and solids can be measured by means of the volumetric method or with the help of the mass-spectrometric technique.

When the helium concentration is determined by means of the volumetric method, the total gas amount is calculated first through measuring its pressure in a calibrated volume. Then the helium/neon mixture is purified from chemically active and heavy inert gases and the amount of the mixture is measured. Neon is not usually separated because its contribution as a rule is smaller than that of helium.

Now let us consider the entire procedure of volumetric measurements shown by the system (Fig. 2.1) designed by Kamensky (1970) and Tolstikhin (1975a).

After opening ampoule *2* the sample gas enters manometer *7* through trap *5* in which water vapors are sorbed. The measuring burette of the manometer is a tube 3 mm in diameter with three spheric volumes measuring (from top to bottom) 2.5, 10 and 30 cm^3, respectively. Such a design allows us to measure gas over a wide range — that is, from 0.1 to 25 norm. cm^3 with a relative error of less than ± 0.5%.

After the gas has been measured by manometer *7*, it is purified from chem-

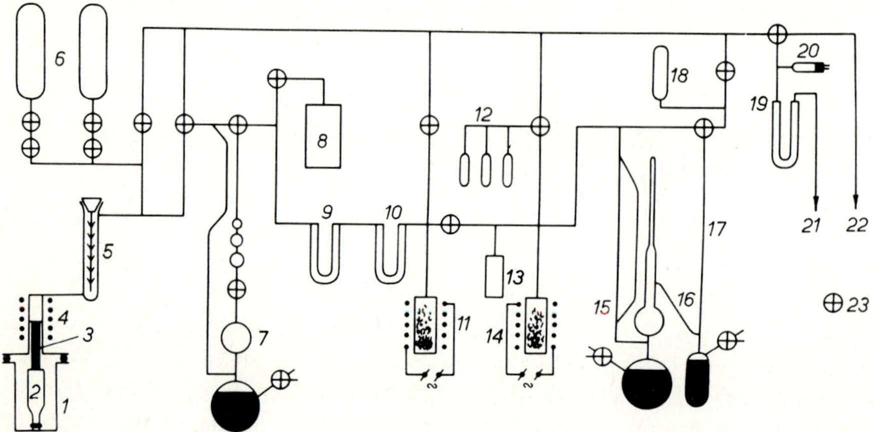

Fig. 2.1. Vacuum system to measure helium and argon concentrations in natural gases (Kamensky, 1970).
1= device to break glass vessels; *2* = vessel with a sample; *3* = breaker; *4* = electromagnet for breaker lifting; *5* = trap for desiccation of a gas; *6* = additional vessels for gas storage; *7* = McLeod gauge; *8* = additional volume; *9* and *10* = cold trap and cold charcoal trap, respectively; *11* = quartz ampoule with titanium getter and outer heating furnace; *12* = ampoules with activated charcoal for argon; *13* = cold finger; *14* = quartz ampoule with calcium getter and outer heating furnace; *15* = McLeod gauge; *16* = mercury valve; *17* = glass tube; *18* = ampoule for helium; *19* = cold trap; *20* = pressure gauge; *21* and *22* = high-vacuum and fore pumps, respectively.

ically active and heavy noble gases in the pre-cleaning charcoal trap *10*, while helium and neon are being pumped out by a McLeod pump, *15*. The U-shaped mercury valve *16* with an additional mercury volume enables collecting almost all the gas from the system in tube *17*. Then the pre-cleaning purification trap is cut off and the helium/neon mixture is completely purified by means of the second charcoal trap *13* and a hot calcium getter, *14*.

The amount of the pure helium/neon mixture is measured by manometer *15*. To improve the accuracy of measurements the lower and upper parts of the McLeod indication capillary tube measures 2 and 0.5 mm in diameter, respectively. This allows us to measure various amounts of a noble gas ranging from $1 \cdot 10^{-3}$ to 0.25 cm^3 with a relative error of less than ± 0.5%. For small amounts of gas, for example ca. 10^{-5} cm^3, the error increases to 20%.

The system as a whole (with a correction for the background) enables us to measure concentrations of the helium/neon mixture in a gas sample from 10% to 0.01%, with an error of less than 1%; for low helium contents of about 10^{-3}% the error increases to 3%.

Usually neon constitutes a small proportion of the terrestrial helium/neon mixture, and is, therefore, completely neglected in the measurements. The mass spectrometer provides a more accurate determination of each component of the mixture. If necessary, neon may be separated by means of a trap cooled by liquid helium prior to measuring the helium content (see section 2.3).

When the absolute helium amount in a sample is very low and difficult to measure by the volumetric method, it can be estimated from the height of the ^{4}He peak in the mass spectrum. In this case it is required that the mass spectrometer be calibrated by standards, and its sensitivity be high and constant with a linear dependence of the output ion current on the pressure. Modern instruments allow us to determine ^{4}He in a sample with an absolute error of about 1% when the ^{4}He atoms amount to 10^{11}.

In cases of small amounts of helium in samples and if high accuracy of measurement is required, the isotope dilution method can be used with the ^{3}He monoisotope as a tracer (Damon and Kulp, 1957; Costa et al., 1975).

2.2. Preparation of standard mixtures of ^{3}He and ^{4}He

^{3}He/^{4}He ratios measured by the mass-spectrometric technique are at variance with the actual ratios in a sample because of the great difference in abundance of ^{3}He and ^{4}He in terrestrial helium and a large difference between the ^{3}He and ^{4}He masses. Therefore, to get accurate results by mass-spectrometric analysis, it is necessary to use standard mixtures of helium isotopes with ^{3}He/^{4}He ratios which are known beforehand. This is especially important for two-beam instruments with different sensitivities in the beam paths (see Chapter 5).

Standard mixtures of helium isotopes were obtained by Mamyrin et al. (1970a) to measure the isotopic ratio of atmospheric helium, ^{3}He/^{4}He = $1.4 \cdot 10^{-6}$ (see section 2.3). However, the apparatus and method of operation which were worked out and described by the authors enable one to prepare various mixtures of ^{3}He and ^{4}He, the ratio ranging from 10^{-8} to 10^{-4}.

To prepare the standard mixture, Mamyrin et al. (1970a) used the ^{3}He monoisotope which contains less than 0.1% of contaminants and less than 0.01% of ^{4}He. Well helium (He$_w$) was used as the ^{4}He monoisotope; the ^{3}He/^{4}He ratio in this helium was equal to $(2.0 \pm 0.1) \cdot 10^{-8}$, the share of contaminants not exceeding 0.01%.

Mixtures of ^{3}He/^{4}He of ca. 10^{-6} are prepared in two ways: first the gases are mixed to obtain a ^{3}He/^{4}He ratio of ca. 10^{-4} and then, during the second stage, the ratio of about 10^{-6} is reached. Two different techniques, one after another, were applied in the first stage.

(1) First stage, first technique: a glass ampoule and a vessel were coupled with two different vacuum systems, pumped out to a pressure of $\sim 10^{-3}$ Torr and filled with known amounts of ^{3}He and He$_w$, respectively (Table 2.1). Then they were coupled with a third system pumped out to the same pressure. Finally ampoule *1* and vessel *2* were opened (Fig. 2.2).

TABLE 2.1

Initial ^{3}He and He$_w$ quantities mixed at the first stage and isotope ratios obtained[1]

^{3}He			He$_w$			^{3}He/^{4}He
V (cm³)	P (Torr)	T (°C)	V (cm³)	P (Torr)	T (°C)	
First mixing						
0.653 ± 0.001	102.2 ± 0.2	23.0 ± 0.2	1522 ± 0.5	696 ± 0.5	21.0 ± 0.2	0.626 · 10⁻⁴
Second mixing						
0.722 ± 0.001	99.7 ± 0.2	21.8 ± 0.2	1975 ± 0.5	702 ± 0.5	19.6 ± 0.2	0.515 · 10⁻⁴

[1] The ampoule volume for ^{3}He was determined by weighing of mercury filling the ampoule; the vessel volume for He$_w$ was determined by weighing distilled water; the pressure and temperature were measured by a mercury gauge and a mercury thermometer, respectively.

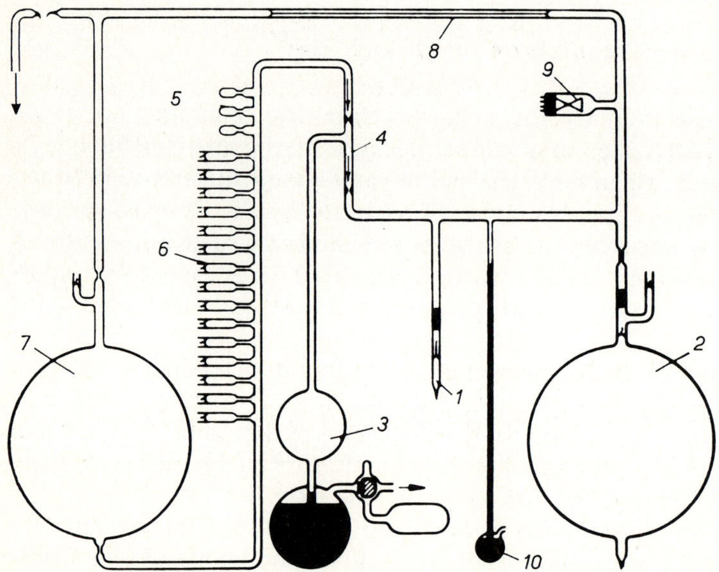

Fig. 2.2. Circulating system to mix ^{3}He and He$_w$ in proportions of about 10^{-4} (Mamyrin et al., 1970b).
1 = ampoule (V = 0.653 cm³) with ^{3}He; *2* = vessel (V = 1.522 cm³) with He$_w$; *3* = mercury pump; *4* = slide valves; *5* = set of control ampoules (V = 0.3 cm³); *6* = set of ampoules (V = 2 cm³) to prepare mixtures; *7* = additional vessel (V = 2000 cm³); *8* = glass capillary with inner diameter of 0.08 cm; *9* = thermocouple gauge; *10* = mercury gauge.

Mercury pump *3* and valves *4* allowed gases to circulate: it was shown that after 200 cycles of mixing the isotope ratio in sealed-off control ampoules was constant. The mixture in ampoules *6* (Fig. 2.2) was further used to prepare standard mixtures at the second stage (3).

(2) First stage, second technique: results obtained by the first technique

were compared with those obtained by means of another, modified technique.

The ampoule with ^{3}He and the vessel with He$_w$ were sealed within the same system with a set of ampoules, a thermocouple and a mercury manometer. The system was pumped out to a pressure of 10^{-3} Torr and the pump was sealed off, then the ampoule and the vessel were opened. To determine the time required for a complete isotope mixing the ampoules were sealed off one by one with considerable time intervals, and subjected to isotope analysis. When the ^{3}He/^{4}He mixture in the ampoules became constant and the time of a total mixing had been found, a similar system was arranged in order to get working mixtures. The initial ^{3}He and He$_w$ amounts used in this experiment are shown in Table 2.1. After a 58-day mixing, the ampoules were sealed off and used for the second stage.

(3) Second stage: preparation of mixtures with ^{3}He/^{4}He ratios of $\approx 10^{-6}$. At the second stage He$_w$ and the two mixtures prepared during the first stage were used (see Table 2.1). The mixing was carried out in a high-vacuum apparatus as shown in Fig. 2.3; the apparatus was pumped out to a pressure of 10^{-7} Torr. Ampoule 2 containing a ^{3}He/^{4}He mixture of $\approx 10^{-4}$ and ampoule 1 containing He$_w$ (both about 2 cm^3 in volume and filled with gases under a pressure of $\approx$ 700 Torr) were coupled with expansion vessels 16 and 17 of which the volumes measured ≈ 2000 and ≈ 200 cm^3, respectively. Then the gases were transferred from the ampoules into the vessels and kept there.

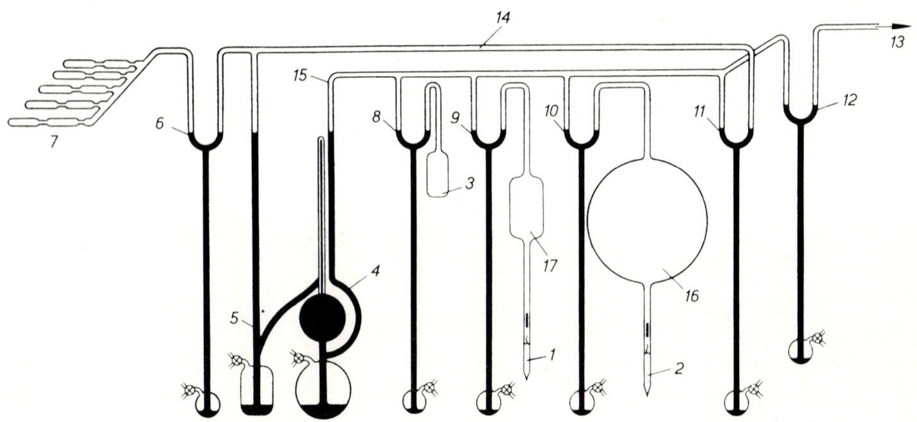

Fig. 2.3. High-vacuum system to prepare helium standards with ^{3}He/^{4}He values of about 10^{-6} (Mamyrin et al., 1970b).
1 = ampoule (V = 2 cm^3) with He$_w$; 2 = ampoule (V = 2 cm^3) with ^{3}He/^{4}He mixtures of $0.626 \cdot 10^{-4}$ or $0.515 \cdot 10^{-4}$, obtained at the first stage of mixing (see Fig. 2.2); 3 = ampoule (V = 70 cm^3); 4 = McLeod gauge; 5, 6, 8, 9, 10, 11 and 12 = mercury valves; 7 = set of ampoules (V = 4 cm^3) to prepare helium standards; arrow 13 indicates high-vacuum pump; 14 and 15 = glass tubes (V = 7 cm^3); 16 = vessel (V = 2000 cm^3); 17 = vessel (V = 200 cm^3).

26

He$_w$ was admitted into McLeod gauge *4* with a capillary of 0.5 mm in diameter, measured and then transferred through valve *5* into tube *14*; the remaining He$_w$ was measured and pumped out. Then the mixture from vessel *16* was introduced into the McLeod gauge and measured; to reduce the pressure of the mixture in tube *15* a small volume *3* and valve *8* were used. The gas that remained in tube *15* was pumped out, the mercury levels in the McLeod gauge and in valves *5* and *11* were lowered and the gas mixing went on for 20 min. Further, the gas was admitted through valve *6* into ampoules *7* and they were flame-sealed. Each ampoule contained about 10^{-3} cm^3 of the gas.

The isotope ratio of the final standard mixture was calculated from the following formula:

$$\left(\frac{^3He}{^4He}\right)_{mix} = \frac{He_1 \left(\frac{^3He}{^4He}\right)_1 + He_w \left(\frac{^3He}{^4He}\right)_w}{He_1 + He_w} \qquad (2.1)$$

were He$_1$ is the amount of helium mixed at the first stages (1) or (2); (^{3}He/^{4}He)$_1$ is the isotope ratio in this mixture (see Table 2.1); He$_w$ is the amount of He$_w$ taken for mixing; (^{3}He/^{4}He)$_w$ is the isotope ratio in the well helium.

With the help of the above techniques we obtained a set of mixtures with the following (^{3}He/^{4}He)$_{mix}$ ratios: $0.665 \cdot 10^{-6}$, $0.803 \cdot 10^{-6}$, $1.02 \cdot 10^{-6}$, $1.26 \cdot 10^{-6}$, $1.77 \cdot 10^{-6}$, $3.78 \cdot 10^{-6}$, $4.08 \cdot 10^{-6}$ (group 1).

To prevent a possible systematic error the capillary in the McLeod gauge was replaced by another one (0.53 mm in diameter) and standard mixtures, with (^{3}He/^{4}He)$_{mix}$ ratios of $1.01 \cdot 10^{-6}$, $1.40 \cdot 10^{-6}$ and $1.99 \cdot 10^{-6}$ (group 2) were prepared.

The standard mixtures with isotope ratios of $\sim 10^{-4}$ and $\sim 10^{-6}$ were used in numerous measurements and allowed us to determine a precise helium isotope ratio in the atmosphere (see section 2.3).

To prepare a mixture of ^{3}He and He$_w$ with a ^{3}He/^{4}He ratio from 10^{-4} to 10^{-6}, He$_w$ can be used; to provide a further decrease of the ^{3}He/^{4}He ratio in the mixture, helium characterized by an extremely low ratio is required. Such helium can be obtained by extraction from uranium minerals, by separation of ^{3}He from well helium in diffusion columns or by implantation of ^{4}He into metal sponges in electromagnetic separators, etc.

2.3. Helium isotope composition in the lower atmosphere

A precise determination of the ^{3}He/^{4}He ratio in the air is very important for helium isotope studies, geochemical and aeronomical calculations, etc.

Moreover, air helium is used as a standard for mass-spectrometric measurements.

Before 1970 the value of the ^{3}He/^{4}He ratio in the air was not known for certain (Alvarez and Cornog, 1939a, b; Aldrich and Nier, 1946, 1948; Fairbank et al., 1947), neither was it clear if the ratio is constant in different layers of the atmosphere.

A precise determination of the helium isotopic ratio in the atmosphere was achieved at the A.F. Ioffe Physico-Technical Institute, Leningrad (Mamyrin et al., 1970a, b). It was found through a comparison of the isotope ratio of purified air helium with that of standard mixtures of helium isotopes (see section 2.2). Apart from a precise determination of the ^{3}He/^{4}He ratio for the air of Leningrad, it was established that the ratio was constant for the air collected from various parts of the lower atmosphere.

The atmospheric helium standards (He$_{st}$) were prepared in the following way:

A mixture of He, Ne and N$_2$ (37, 40 and 23%, respectively) was collected from the dephlegmator under a pressure of 6 atm.; then nitrogen was removed by sorption on charcoal traps cooled by liquid nitrogen, and the residual gas contained 57% of helium and 43% of neon.

Helium was separated from neon by freezing the latter at liquid helium

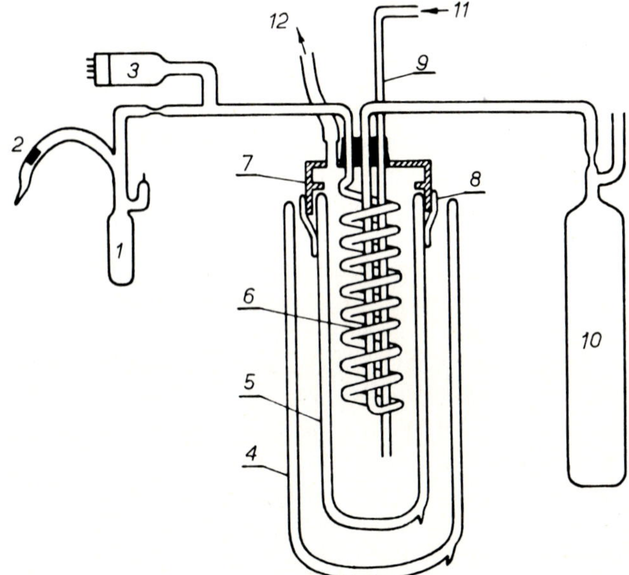

Fig. 2.4. Glass system to separate (He + Ne)$_{atm}$ mixture from Ne (Mamyrin et al., 1970b). *1* = vessel (V = 150 cm^3); *2* = soft iron weight; *3* = thermocouple gauge; *4* and *5* = Dewar vessels; *6* = helix; *7* = metal cover; *8* = rubber seal; *9* = pipe for leak-in of liquid helium; *10* = vessel (V = 1000 cm^3); arrow *11* shows the leak-in of liquid helium; arrow *12* shows the escape of evaporated helium.

temperature. A glass system (Fig. 2.4) was used for separation; it was pumped out to a pressure of 10^{-3} Torr and flame-sealed off from the fore pump. Then Dewar vessels *4* and *5* were put under glass helix *6*, the outer vessel was filled with liquid nitrogen, and one hour later liquid helium was introduced into the inner vessel. The ampoule containing the neon/helium mixture was opened with a small weight (*2*) and the gas leaked through helix *6* primarily cooled by liquid helium into a 1000-cm^3 vessel (*10*). Forty-five minutes later this vessel was submerged in liquid nitrogen and after another 15 min it was flame-sealed off from the system. From vessel *10*, He_{st} (He = 99.9%; Ne $\leqslant$ 0.1%) was distributed in 10-cm^3 glass ampoules.

Isotope analysis of all the helium samples was performed by means of a magnetic resonance mass spectrometer at the A.F. Ioffe Physico-Technical Institute (see Chapter 5). To determine the $^3He/^4He$ absolute value in pure air helium (He_{st}) nine runs were carried out, the air helium of each run being successively compared with standard mixtures. The absolute value of the $^3He/^4He$ ratio in He_{st} was calculated from the following formula:

$$\left(\frac{^3He}{^4He}\right)_{st} = \frac{\left(\dfrac{^3He}{^4He}\right)_{st}^{meas.} \left(\dfrac{^3He}{^4He}\right)_{mix}}{\left(\dfrac{^3He}{^4He}\right)_{mix}^{meas.}} \tag{2.2}$$

where subscripts "st" and "mix" denote the $^3He/^4He$ ratio in the air and in the mixtures (see section 2.2), respectively, and "meas." denotes measured values of the ratio. The calculations were carried out for successive measurements of $(^3He/^4He)_{st}^{meas.}$ and $(^3He/^4He)_{mix}^{meas.}$ values; the results are given in Table 2.2. The resultant isotope ratio for He_{st}, $(^3He/^4He)_{st}$, was $(1.399 \pm 0.013) \cdot 10^{-6}$. The error is a mean-square error for the mean calculated on the grounds of mean values of nine $(^3He/^4He)_{st}$ measurements.

To confirm the constancy of the $^3He/^4He$ value in the lower atmosphere air samples were collected in different areas of the USSR and the ratios in these samples were compared with the $(^3He/^4He)_{st}$ ratio. Table 2.3 shows that, within the error of a single measurement, the $^3He/^4He$ ratio may be considered constant and equal to $1.4 \cdot 10^{-6}$.

W.B. Clarke et al. (1976) reported the $^3He/^4He$ ratio for air helium in Ontario (Canada), which was determined by the technique developed by Mamyrin et al. (1970a); their value $(1.384 \pm 0.006) \cdot 10^{-6}$ within the experimental error coincides with that determined in Leningrad.

At present $(^3He/^4He)_{atm} = 1.39 \pm 0.01$ may be considered a reliable value for the entire lower atmosphere.

TABLE 2.2

^{3}He/^{4}He ratios determined in He$_{st}$

Set No.	Number of measurements	Absolute values of (^{3}He/^{4}He) ($\times 10^6$)
1	8	1.440
2	8	1.430
3	4	1.317
4	9	1.361
5	11	1.420
6	12	1.375
7	11	1.428
8	11	1.386
9	12	1.393
Average value		(1.399 ± 0.013)

TABLE 2.3

^{3}He/^{4}He ratio in air samples taken in different regions of the USSR

Sample characteristics		Number of measurements	Absolute value of (^{3}He/^{4}He) ($\times 10^6$)
Locality	Altitude above sea level (m)		
Standard He$_{st}$ Leningrad	10	2 1	1.40 1.40 ± 0.07
Standard He$_{st}$ Leningrad	10	2 2	1.40 1.40 ± 0.06
Standard He$_{st}$ Sochi	10 000	1 2	1.40 1.40 ± 0.07
Standard He$_{st}$ Sukhumi Sochi	0 10 000	5 2 2	1.40 1.43 ± 0.06 1.40 ± 0.06
Standard He$_{st}$ Sukhumi	0	4 2	1.40 1.40 ± 0.04

2.4. Universal mass-spectrometer leak-in system for helium isotope studies

Among the peculiarities of helium isotope studies there is the necessity to measure ^{3}He/^{4}He ratios whose values are very wide apart ranging from $\sim 10^{-10}$ to 10^{-1}. This implies the use of several standards as well as the preparation of required isotopic mixtures directly in the leak-in system.

A "static" regime of measurements (with no pumping of the analyzer, see Chapter 3), when a sample is admitted into the analyzer in small measured amounts, provides a high sensitivity of measurements. To obtain this amount the leak-in system must be able to separate a small proportion from the whole amount and use it for analysis. High-vacuum should be maintained in the leak-in system because the accuracy of the results (see Chapter 3) depends on the background of the hydrogen- and carbon-bearing components. These considerations allow us to formulate the requirements of a universal leak-in system for helium isotope studies.

The vacuum circuit of such a system (Alekseichuk et al., 1979) is shown in Fig. 2.5; its elements are described in the caption. The system has the following characteristics:

(1) Two vacuum ampoule-breakers (each holding fifteen glass ampoules) coupled with independent vacuum pumping lines are able to carry out a great number of analyses one after another; the nipples of the system enable us to use any additional ampoule-breakers.

(2) Six 1-l stainless steel vessels can be pumped out and commutated in-

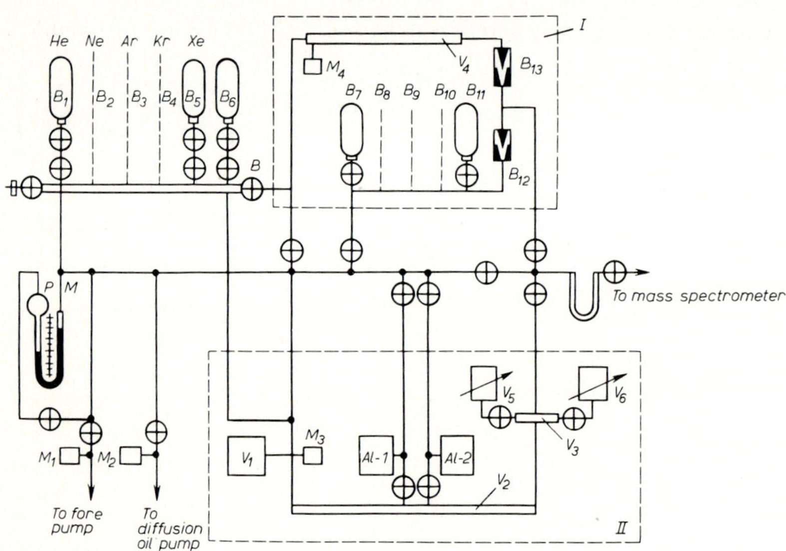

Fig. 2.5. Schematic diagram of the system for noble gas leak-in of mass spectrometer MRMS MI 9302 (Anufriev et al., 1979).
I = leak-in system to analyze large amounts of gas, while it is continuously flowing through the analyzer; II = leak-in system to analyze a small amount of gas when measurements are performed without pumping out the gas from the analyzer; B = vacuum valves; $B1$—$B6$ = vessels with standard gases; $B7$—$B11$ = vessels with study samples; $B12$ and $B13$ = needle vacuum valves; $V1$—$V4$ = permanent-volume vessels and tubes; $V5$ and $V6$ = variable (bellow) vessels; PM = mercury pressure gauge; $M1$—$M4$ = pressure gauges; Al-1 and Al-2 = ampoule breakers.

dependently; a set of standard isotopic mixtures or air noble gases can be stored in these vessels and used for measurements of a noble gas isotope composition.

(3) The division of the initial gas amount into a given proportion by a special system; the initial volume of the system is about 2 cm^3.

(4) The inital pressure of a gas is determined by a mercury gauge; the gauge performs measurements in the range of 0—250 Torr, the uncertainty of indication is 0.2 Torr.

(5) To provide the static mode of operation the system is supplied with a valve admitting a measured amount of a sample or standard into the analyzer.

(6) When the mass spectrometer operates in the dynamic mode (with the gas flowing through the ion source and pumped out from the chamber) some needle valves are coupled with vessel $B/4$ (standard gases) and with a set of ampoules; the latter are supplied with gate valves coupled to the system by threaded nipples.

(7) All the items of the vacuum system as well as gas-commutation tubes are made of stainless steel; metal gaskets are used in all seals.

(8) The leak-in system as a whole can be heated up to $300°C$; any given level of temperature within this limit is maintained automatically.

(9) The system is pumped out by a diffusion oil pump coupled with a shutter trap without warm walls; a residual pressure in the system is $\sim 10^{-8}$ Torr.

The leak-in system (Anufriev et al., 1977) is supplied with a furnace for the extraction of noble gases from solid samples and a system for their purification (see section 1.6).

Chapter 3

PECULIAR FEATURES OF MASS-SPECTROMETRIC ISOTOPE ANALYSES OF TERRESTRIAL HELIUM

The main peculiarity of the distribution of helium isotopes in nature is the extremely high range of the ^{3}He/^{4}He ratio, namely from 1 to 10^{-10}. The measurement of very low isotope ratios, such as 10^{-6} to 10^{-10}, which are typical of most terrestrial samples, is a very difficult problem. A mass spectrometer for isotope analysis of helium must combine various and sometimes incompatible characteristics such as a high resolution and great sensitivity, a good shape of the ion peak, etc. Some aspects of the problem are discussed below.

3.1. Resolution of a mass spectrometer

The measurement of the ^{3}He$^+$ ion beam when the ^{3}He/^{4}He ratio and, consequently, the ^{3}He amount in the chamber of the mass spectrometer are very low, requires not only high sensitivity of the instrument but also a high resolution power. If we assume that the upper bound of helium pressure in the instrument is 10^{-6} to 10^{-5} Torr and the ^{3}He/^{4}He ratio is equal to 10^{-9} then the ^{3}He partial pressure is 10^{-15} to 10^{-14} Torr. Hence, only 10^5—10^6 ^{3}He atoms are contained in the chamber of the analyzer with a volume of several litres.

The typical pressure of residual gas in the instrument is about 10^{-7}—10^{-8} Torr; then the pressure of H_2, one of the major components of the gas, is 10^7—10^8 times higher than that of ^{3}He. Consequently, the intensity of the HD ion peak (the mass of HD is practically equal to that of ^{3}He) is 10^3—10^4 times higher than that of the ^{3}He peak[1]. The intensity of the H_3 peak is also great as compared with that of ^{3}He. Thus, it is obvious that the ^{3}He measurement can be performed only when the ^{3}He$^+$ ions are separated from the HD$^+$ and H_3^+ ions. H_3^+, HD$^+$ and ^{3}He$^+$ ion peaks constitute the mass multiplet: $\Delta M(\text{HD} - {}^3\text{He}) = 5.9 \cdot 10^{-3}$ amu; $\Delta M(\text{H}_3 - {}^3\text{He}) = 7.45 \cdot 10^{-3}$ amu; $\Delta M(\text{H}_3 - \text{HD}) = 1.55 \cdot 10^{-3}$ amu.

---

[1] The ratio of D/H = $1.56 \cdot 10^{-4}$ is typical of natural hydrogen.

To resolve reliably the multiplet the resolution power of the mass spectrometer should be of the order of 1000 at the 10% level of the peak height.

The ^{4}He abundance in terrestrial materials is known to be many orders of magnitude higher than that of ^{3}He, and the multiplet peaks do not interfere with the measuring of the ^{4}He peak. However, if the ratio of ^{3}He/^{4}He approaches 10^{-1} (spallogenic helium in meteorites) and the total helium amount is low, the multiplet peaks at mass 4 must be resolved.

The ^{12}C^{3+} peak is the most intense one in the background mass spectrum at mass 4 when ions are produced by the electron impact. The mass difference is $\Delta M(^4\text{He}^+ - {}^{12}\text{C}^{3+}) = 2.6 \cdot 10^{-3}$ amu. An adequate resolution in this case is about 1600.

Consequently, the resolution of the mass sepctrometer for isotopic analysis of various samples of natural helium should be about $2 \cdot 10^3$ — that is, essentially higher than the resolution required for the separation of ^{3}He$^+$ from ^{4}He$^+$ ions.

Unfortunately, a high resolution itself is not enough for a successful measurement of the ^{3}He peak when the ^{3}He/^{4}He ratios are equal to 10^{-6} and less. Steps should be taken to eliminate the background of composite peaks.

3.2. Background of ^{4}He scattered ions and its suppression in the measurements of very low ^{3}He/^{4}He ratios

The "tail" of a very intense ^{4}He peak produced by ion scattering is a major component of the background.

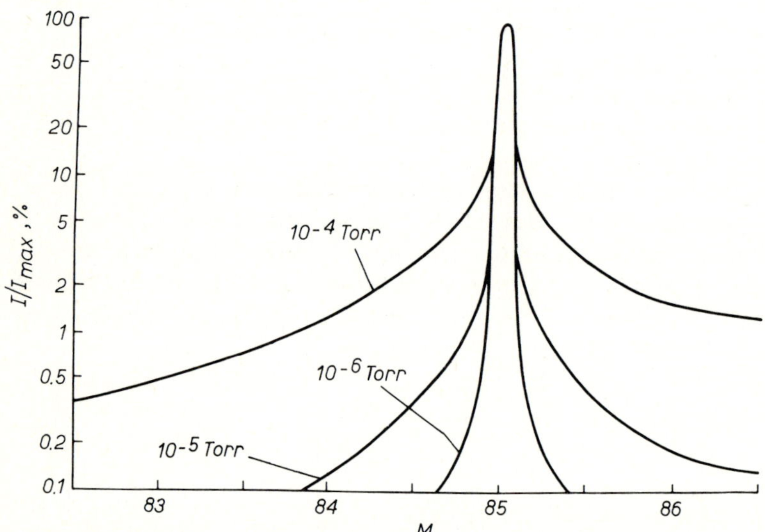

Fig. 3.1. Effect of gas pressure in the analyzer on the shape of a mass peak.

There are several causes of ion scattering within a wide range of angles: collisions of ions with gas atoms and molecules in the analyzer, their interaction with the chamber walls and elements of the analyzer units as well as with slit edges, bounding the ion beam along the ion trajectory. When slits are wide enough (thus no narrow channels), collisions of ions with neutral atoms contribute a large share of scattering ions, even if the total gas pressure in the analyzer is only 10^{-7}–10^{-8} Torr. The relationships between the gas pressure in the analyzer and the shape of the ion peak is illustrated in Fig. 3.1.

The scattered ion current entering the exit slit of the one-stage instrument is affected not only by the pressure (P) but also by dispersion (D), the geometric parameters of the analyzer and the relative shift of the mass peak ($\Delta M/M$).

Investigations of atom collisions (Coffey et al., 1969; Afrosimov et al., 1971, 1975) show that the value $\sigma\gamma^2$ slightly depends on $E\gamma$ (σ is the cross-section of scattering at an angle γ, and E is the ion energy). Depending on the types of combinations of scattering ions and neutral particles, $\sigma\gamma^2$ ranges within a factor of 2 to 5 when $E\gamma$ varies from 0.1 to 10 keV degrees. Hence, to estimate the scattered ion current one can assume that:

$$\sigma \approx \frac{\text{const}}{\gamma^2} \tag{3.1}$$

The scattered ion current, ΔI_{scat}, is proportional to the beam current, I_0, the pressure of the scattering gas, P, the scattering cross-section, σ, and the solid angle, F/l_1^2, at which the area of the exit slit is visible from the point of scattering. Hence, taking into account eq. 3.1 we can write:

$$I_{\text{scat}} \sim I_0 P\, \frac{F}{\gamma^2 l_1^2}\, \Delta l$$

where Δl is the element of the ion beam length in the instrument.

Fig. 3.2 shows that the scattering angle γ may be taken to be equal to $\Delta x/l_1$. As Δx is proportional to the mass peak shift, $\Delta M/M$, and the dispersion of the instrument, D, the scattered ion current passing through the entrance slit is:

$$I_{\text{scat}} \sim I_0\, \frac{PF\Delta l}{D^2\,(\Delta M/M)^2} \tag{3.2}$$

Introducing the constant A determined by dimensions and some non-specified factors, we derive for the scattered ion current for the total length of the ion beam:

36

$$I_{scat} = A \int_0^l \frac{PF}{D^2(\Delta M/M)^2} I_0 \, dl$$

Assuming that I_0 = const along the trajectory of the ion beam we obtain:

$$\frac{I_{scat}}{I_0} \approx A \frac{PlF}{D^2(\Delta M/M)^2} \tag{3.3}$$

Constant A may be found from experimental data. If all the geometric values are given in millimeters then the average A is about $6 \cdot 10^{-4}$ for a one-stage magnetic analyzer, and eq. 3.3 can be rewritten as:

$$\frac{I_{scat}}{I_0} \approx 10^{-3} \frac{PlF}{D^2(\Delta M/M)^2} \tag{3.4}$$

where P is the pressure in the analyzer, Torr; l is the length of the ion beam in the instrument, mm; D is the dispersion, mm per 1% of mass change; $\Delta M/M$ is the relative mass shift of the scattered ion beam centre from that of the slit; F is the area of the exit slit, mm^2.

Assuming that $P = 10^{-6}$ Torr, $D = 1$ mm, $F = 0.5$ mm^2, $l = 450$ mm, $\Delta M/M = \frac{1}{4}$, we find from eq. 3.4:

$$\frac{I_{scat}}{I_0} = 3.6 \cdot 10^{-6}$$

It is noteworthy that eq. 3.4 is valid only for ions scattered due to their collisions with neutral particles; it fits with the scattered ion current at high pres-

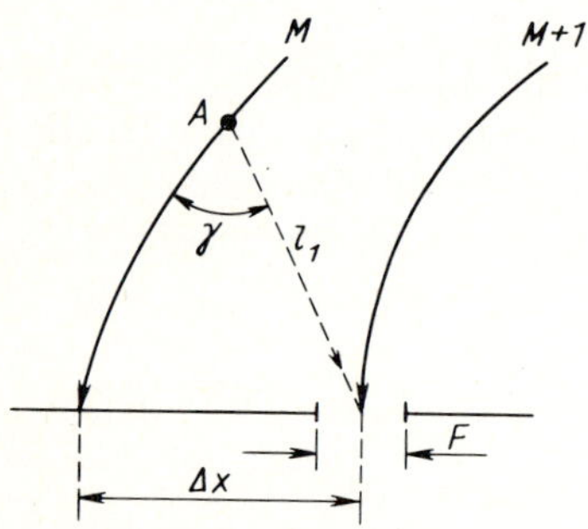

Fig. 3.2. Effect of dispersion on a scattering angle causing the entry of scattered ions of an adjacent mass peak into the exit slit of a mass spectrometer.
A = point of scattering; F = cross-section of the exit slit; Δx = distance between centers of adjacent peaks (proportional to dispersion); γ = scattering angle.

sure of the gas in the instrument. When vacuum is of the order of 10^{-8} Torr and $\Delta M/M = \frac{1}{4}$ (helium), then the scattered ion current in the one-stage instrument is directed mainly by collisions with chamber walls or slit edges (Aleksandrov et al., 1974) and by some processes in the ion source, in particular a direct passage of the ion current from the vicinity of the cathode. In this case calculation of eq. 3.4 will lead to erroneous (underestimated) results.

In fact, the pressure in the analyzer is usually not less than 10^{-7}–10^{-8} Torr; it is determined mainly by the sample leak-in and, in cases of the highest sensitivity, can reach 10^{-6}–10^{-5} Torr. Thus, one-stage ion separation in a constant magnetic field leads to a significant contribution ($\sim 10^{-5}$) of a scattered ion current to the intense peak of ^{4}He at mass 3, thus preventing measurements of ^{3}He/^{4}He ratios equal to 10^{-6} and less.

We have not discussed how the non-elastic processes affect the scattered ion current in the collision of ions with neutral particles. In principle, such processes result in the formation of scattered ions of a lower energy so that their trajectories in the magnetic field will perfectly coincide with those of authentic ions. Consequently, these scattered ions cannot be distinguished from those under analysis. However, they are very few as compared with the elastically scattered ions, and even with all the non-elastically scattered ions, because usually ions whose energy and angle are changed due to scattering cannot enter the exit slit. Moreover, the low energy ions can be easily eliminated if a simple system inducing a retarding electric field is placed in front of the detector.

The most efficient technique to suppress the background of elastically scattered ions of the adjacent peak is the two- and more-stage separation of ions. The processes occuring in this case are shown in Fig. 3.3. When the instrument is adjusted to a low intensity peak (to detect ions of mass M_1), some scattered ions of the intense mass M_2 pass through the exit slit of the first stage (S_1) into the second stage. The proportion of these fly-through ions of mass M_2 to the total number of ions amounts to 10^{-4}–10^{-6} when the ratio of $(M_1 - M_2)/M_1$ is not less than 0.1. At the second stage of separation the

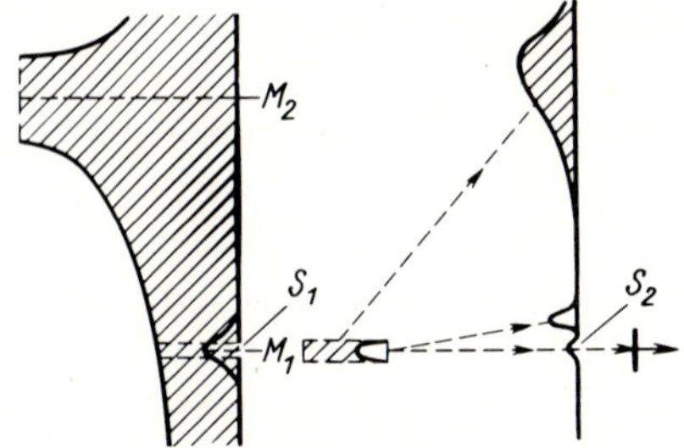

Fig. 3.3. Schematic diagram of ion separation in a two-stage mass spectrometer (Mamyrin and Tolstikhin, 1981).
M_1 = small measured mass peak; M_2 = adjacent large mass peak; S_1 = exit slit of the first stage; S_2 = exit slit of the second stage.

same proportions (10^{-4}–10^{-6}) are observed between scattered ions which pass through exit slip S_2 (and are detected together with ions of mass M_1) and those flown through slit S_1 (see Fig. 3.3).

It is obvious that the scattered ion current of mass M_2 passing through exit slit S_2 will amount to 10^{-8}–10^{-12} of the current coming out of the source. Naturally, the three-stage system will be even more efficient; however, a low sensitivity and complexity of design of such systems limit their practical use.

Any separating system may be used as the first stage of the two-stage mass spectrometer because a very low resolving power (10—20) is required in the case of helium analysis. The second stage should separate multiplets of mass 3 and 4 and should have a resolution of about $2 \cdot 10^3$.

In general, the background of scattered ions for a two-stage instrument can be estimated from eq. 3.4, but real values of parameters of the first and the second stages might be taken into consideration. In particular, the effective dispersion in the second stage for the proportion of scattered ions which are cut out by the exit slit of the first stage, can be lower than that for analyzed ions.

The side-effects become very important in such systems, since the output background current in the second stage can be about 10^{-11}–10^{-10} times of the maximum ion current in the first stage (that is, as high as the total current of the given ions in the first stage), even if a negligible damage in screening occurs or if secondary electrons are intercepted by ions in the vicinity of the exit slit.

3.3. Background of scattered HD$^+$ and H$_3^+$ ions

The second background component which interferes with measuring the ^{3}He peak is the "tall" of the HD and H$_3$ multiplet peaks. These peaks were shown to be several orders more intense than the ^{3}He peak. The multiplet ions together with the ^{3}He ions pass completely into the second stage of the double-stage mass spectrometer (see Fig. 3.3) and the presence of the first stage does not affect the intensity of the current multiplet ions entering slit S_2. It is noteworthy that the mass difference between the ^{3}He peak and the intense multiplet peaks ($\Delta M \approx 0.01$ amu) is much smaller than the corresponding difference for the ^{4}He peak ($\Delta M = 1$ amu). The current intensity of the scattered hydrogen ions may account for 10^{-2} of the total intensity of the multiplet peak and be much higher than that of the ^{3}He peak. If resolution is high in the first stage and low in the second, there will be no gain since scattered ions of the multiplet peak pass through slit S_1 (see Figure 3.3) and cannot be separated from the ^{3}He ions. It is technically difficult (though possible) to make a system characterized by a high resolution of both stages. However, a more complex design will inevitably decrease the sensitivity of the entire system.

To explain what happens when the $\Delta M/M$ values are low, we will discuss in more detail the shape of a mass spectrum peak of the double mass spectrometer. Let us assume that the resolution of the first stage, R_1, is much lower than that of the second stage, R_2, and that the ion density in the cross-section of the ion beam is constant.

Fig. 3.4 shows the peak shapes of the double mass spectrometer for different cases. The dashed line represents an ideal shape of the mass spectrum peak when broadening of the ion beam caused by aberrations is negligible — that is, the spread of initial energy of the ions and the initial angles in the ion source tend to be zero ($\Delta U \approx 0$ and $\Delta \alpha \approx 0$, respectively). Moreover, it is assumed that the magnetic field is homogeneous ($\Delta H \approx 0$) and the contribution of scattered ions is very low. In cases of aberrations, such as $\Delta U \neq 0$, $\Delta \alpha \neq 0$, $\Delta H \neq 0$, we observe "tails" of peaks strictly limited by the maximum values of ΔU, $\Delta \alpha$, and ΔH.

The shape of curve 2 corresponds to the case when both aberrations and ion scattering take place ($\Delta U \neq 0$, $\Delta \alpha \neq 0$, $\Delta H \neq 0$), assuming the first stage absent; tails of scattered ions trail behind at a distance of several mass units. Finally, the shape of curve 3 corresponds to a real two-stage system with an aberration broadening, scattering of ions and presence of the first stage; the

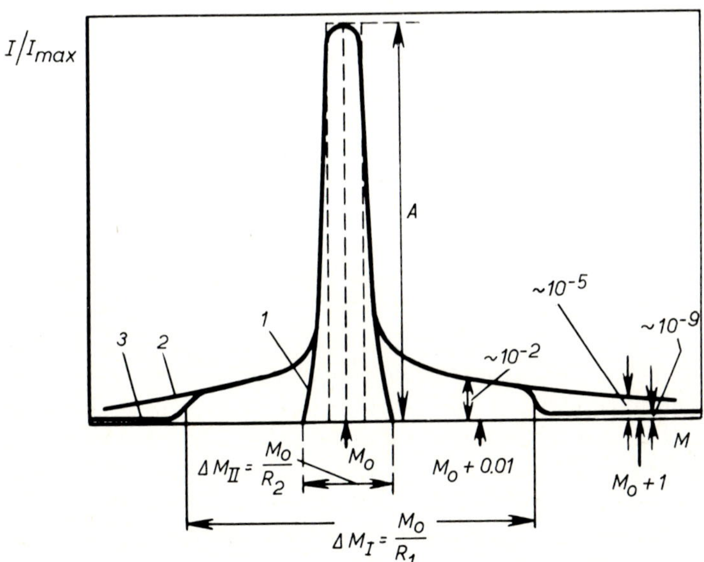

Fig. 3.4. Shape of the mass spectrum peak in a two-stage mass spectrometer (Mamyrin and Tolstikhin, 1981).
A = height of the peak, taken as 1; 1 = shape of the peak in the presence of aberrations and the absence of scattering; 2 = the same, in the presence of aberrations and scattering, but the first stage of the mass spectrometer is assumed to be absent; 3 = the same for a real two-stage system. For other explanations see text.

latter sharply decreases the scattered ion background. The impact of the first stage begins to have effect at a distance of one order of $M_0/2R_1$ from the centre of a peak, where R_1 is the resolution of the first stage (at a level close to the peak base $R_1 = M_0/\Delta M_1$). The position of the adjacent peak ($M_0 + 1$) and that of multiplet peaks ($M_0 + \Delta M$) are marked by arrows. Fig. 3.4 shows that special measures should be taken to suppress the background of scattered ions of a given peak within the area of neighbouring multiplet peaks, because the first, low resolution stage has no effect in this case.

The problem of suppressing multiplet peak "tails" can be solved by increasing the dispersion of the second stage. In fact, it follows from eq. 3.4 that:

$$\frac{I_{\text{scat}}}{I_0} \sim \frac{1}{D^2}$$

at a given resolution of the instrument and at a distance of $\Delta M/M$ from the peak from which ions are being scattered.

A high dispersion (at a given resolution) allows us to increase the sensitivity because of the use of a wide source and outlet slits as well as because of a high electron current in the source.

A dispersion $D \approx 40$ mm per 1% of $\Delta M/M$ is required to obtain $I_{\text{scat}}/I_0 \sim 10^{-4}$ at the distance between the $^3\text{He}^+$ peak and hydrogen multiplet peaks of $\Delta M/M \approx 2\ 10^{-3}$; at the same time the resolution of the second stage, $R_{10\%}$, and the pressure in the analyzer, P, might be about 10^3 and 10^{-6} Torr, respectively. In practice, it is hardly possible to achieve these parameters in a static magnet mass spectrometer; the radius of the ion trajectory reaches 4 m (!).

Unfortunately, an increase in dispersion always leads to an increase in the ion trajectory in the magnetic field of ordinary static instruments.

3.4. "Memory" and other factors affecting the measurements of ^3He and ^4He abundance in terrestrial samples

Some quantities of ^3He and ^4He are always found in an analyzer's chamber without any leak-in of the sample or standard helium. This is the third background component which also interferes with the measurements of helium isotopic abundance. This component is caused by air helium penetration through small leakages of some parts of the instrument; another source is the escape of helium from parts of the instrument and its desorption from surfaces.

Penetration of air helium into a mass spectrometer is much lower than in systems for extraction of helium from natural samples and its purification (see Chapter 1), because during the analyses vacuum connections and glass elements (lead-throughs) are not heated. However, some elements of the ion source undergoing intense heating might have consequences.

Considerable amounts of helium isotopes can appear in the analyzer due to the "memory" effect. It is widely accepted that owing to its inertness and high mobility helium can be easily pumped out from a heated system. However, it was proved experimentally that there exists a strong ^{3}He and ^{4}He "memory" effect. This effect is caused by helium penetration into parts of the instrument in the case of a considerable helium pressure and its reverse outgassing when the pressure in the system drops abruptly.

A high increase of partial helium pressure occurs when the instrument is opened and the inner surfaces come into contact with atmospheric air or when it is necessary to measure samples containing helium isotopes markedly different in proportion. For example, after the leak-in of helium in a metal chamber up to a pressure of 10^{-5}—10^{-6} Torr, the latter can be decreased by a factor of 10^3—10^4 by pumping out the chamber without heating. To decrease the pressure by a factor of 10^5—10^6 the leak-in should be followed by heating and prolonged pumping out of the chamber. However, it proved impossible to decrease a ^{3}He pressure of 10^{-5} Torr after leak-in by a factor of 10^9, though heating and evacuating the system continued for several months.

Helium is introduced into the instrumental units in various ways, the following being the most frequent: the implantation of accelerated ions, helium penetration into quartz and glass elements, the sorption on rough surfaces, the penetration into fractures and capillary passages of tightly pressed elements.

A scale of these processes is illustrated by the number of ions passing through a circuit of an ionization gauge. The total ion current of the gauge is 10^{-4}—10^{-8}A. During the analyses a partial component of the helium ion current may reach 10^{-7}—10^{-8}A — that is, 10^{11} to 10^{12} ions per second. Hence, if the gauge operates for one hour up to 10^{15} atoms (!) can be accumulated in its electrodes. If after a sharp decrease of the helium pressure the gauge is neither heated nor evacuated, it preserves the helium and a high background will remain in the instrument for a long time.

The contribution of helium in the residual gas of the instrument affects measurements of small samples taken from unique specimens such as lunar soil, cosmic dust, crystal thin-sections in diffusion studies and the like. In these cases major difficulties may arise in the determination of ^{4}He because its content in the air is 10^6 times higher than that of ^{3}He, and the ^{4}He residual pressure is generally much higher than that of ^{3}He. Accordingly, the sensitivity threshold of ^{3}He measurements is much higher than that of ^{4}He; at present it is impossible to measure the ^{4}He/^{3}He ratios if they are as low as the ^{3}He/^{4}He ones — that is, about 10^{-10}.

Three other factors which affect the results of mass-spectrometric measurements of the ^{3}He/^{4}He ratios should be taken into account:

(1) The first one is the technological difficulty of separating helium from neon since the latter is also not sorbed on surfaces cooled by liquid nitrogen. The volume charge in the ionization chamber of the ion source increases

when the neon pressure exceeds that of helium in the analyzer. A sharp difference between the ^{3}He and ^{4}He masses and the ensuing difference of their residence time, as well as their recombination probabilities, result in changes of the ^{3}He and ^{4}He source efficiency, which, in turn, affects the measured ^{3}He/^{4}He ratio. The problem is that, even when one uses standard samples there is no guarantee of high accuracy of measurement if the neon content in the standard differs from that in the sample.

Kamensky (1970) and Boltenkov (1973) showed that, if the neon pressure in a sample is several times higher than that of helium, the ^{3}He/^{4}He ratio is changed by several tens of percents.

The only way to neutralize the negative effect of neon is to reduce its pressure below that of helium (or if the total gas pressure in the source is less than $\sim 5 \cdot 10^{-7}$ Torr).

(2) The second factor is also related to the sharp difference between the atom masses of helium isotopes. In fact, the fractionation effect occurs when helium isotopes pass through small holes and thin tubes. This may be demonstrated by the following example: some amount of helium from a 3-cm^3 vessel was admitted into the analyzer several times, then measured and pumped out. Helium was admitted through a 4-mm tube and a narrow gap in the valve (~ 0.1 mm). The first measurement yielded a ^{3}He/^{4}He ratio of $8 \cdot 10^{-8}$. When this procedure was repeated five times and almost all the helium was withdrawn from the vessel the ^{3}He/^{4}He ratio decreased to $5.5 \cdot 10^{-8}$.

Mamyrin et al. (1970a) observed the process of stabilization of the isotope ratio when two types of helium distinguished by ^{3}He/^{4}He ratios were mixed in a glass tube with a set of ampoules (see Fig. 2.3). Despite the fact that the diameters of the tubes and the orifices in the system were only a matter of some millimeters, the ^{3}He/^{4}He ratios became constant after more than 15 h of mixing. In one hour the isotope ratio in the control ampoules differed from the final value by one order of magnitude (Fig. 3.5).

Obviously, the effect of isotope fractionation should be taken into consideration when a sample is admitted into the mass spectrometer or else when standards are prepared by mixing various types of helium. To prevent fractionation the volume to be filled and that containing helium has to stay coupled long enough. The same requirement should be met when helium is being pumped through small holes and thin tubes.

(3) The third factor is connected with errors in measurements of ^{3}He/^{4}He ratios resulting from the contribution of tritium. Tritium is practically inseparable from ^{3}He because a resolution of the order of 200 000 is required for such a separation; hence the superposition of ^{3}H$^+$ and ^{3}He$^+$ ion peaks is inevitable. As a rule, helium is purified from other volatiles (^{3}H among them) prior to mass-spectrometric measurements; but even in this case the analyses can be erroneous. Some amount of helium-3, ^{3}He$_{rad}$, is produced owing to the ^{3}H β-decay. The half-life is only 12.3 years and, consequently, a certain amount of ^{3}He$_{rad}$ can be accumulated soon, even though the total amount of

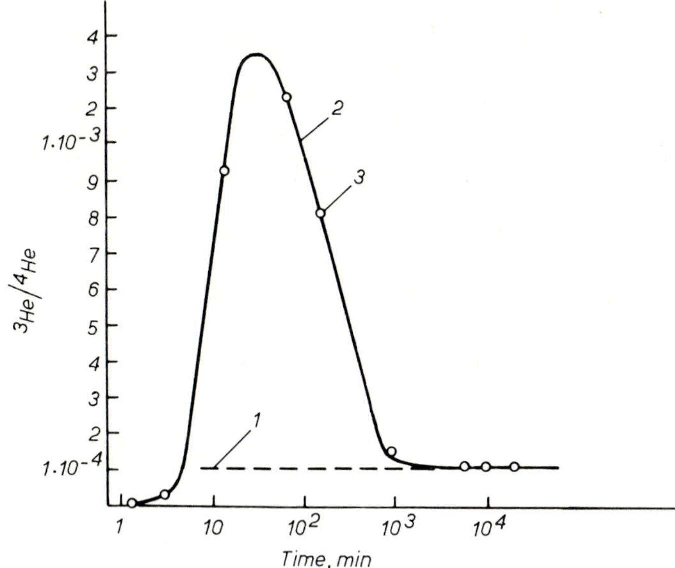

Fig. 3.5. ^{3}He/^{4}He ratio as a function of time of ^{3}He and He$_w$ mixing (Mamyrin et al., 1970b).
1 = expected ^{3}He/^{4}He ratio; 2 = plot of the measured ^{3}He/^{4}He ratio versus time (in min); 3 = measured ^{3}He/^{4}He ratios in control ampoules.

^{3}H is not very large. For instance, 10^{11} atoms of tritium produce 10^7 atoms of ^{3}He$_{rad}$ in 24 h, which in many cases leads to 100% error.

3.5. Two-beam mode of operation and static mode of pumping a mass spectrometer

(1) A two-beam mode of operation, which means that the ^{3}He and ^{4}He output ion currents enter two collectors of the mass spectrometer simultaneously, enables one to achieve maximum sensitivity and accuracy of helium isotope analysis. The re-adjustment of an instrument characterized by the one-beam mode with its switching over from the ^{3}He current to the ^{4}He current leads to a waste of time during which a useful signal could be measured (and, hence, to a less accurate analysis). Moreover it also leads to a greater uncertainty due to inaccurate reproducibility of the analyzer operation conditions and the necessity for a drastic change in the gain of the detecting system. The difference between the ^{3}He and ^{4}He ion currents may be as high as 10^{10}, so it is impossible to transmit them using the same gain and indication without a considerable readjustment.

The usage of standards is inevitable in the two-beam mode of operation because the channels for the ^{3}He and ^{4}He measurements differ greatly. However, to reduce a systematic error in mass-spectrometric measurements, it is preferable to use standards such as air noble gases or artificial mixtures of isotopes whatever instrument is employed.

In addition to high accuracy and sensitivity the two-beam mode of operation allows a decrease in the time of analysis since no re-adjustment of the measuring system is needed.

(2) Sensitivity is an important parameter of the mass spectrometer especially for determining low ^{3}He/^{4}He ratios, reaching 10^{-9}—10^{-10}.

As mentioned in section 3.1, the ^{3}He partial pressure can be as low as 10^{-11}—10^{-15} Torr and, consequently, a sensitivity of the order of 10^{-3}—10^{-4} A Torr^{-1} is required to obtain a measurable ion current of 10^{-18} A at the instrument outlet (or at the inlet of the secondary electron multiplier). At a resolving power of about $2 \cdot 10^3$ such a sensitivity can be achieved only under severe conditions in the ion source and with fairly wide slits (that is, at a high instrumental dispersion).

In some cases the amount of ^{3}He in a sample is only 10^5—10^6 atoms and so the ordinary gas leakage into the source with continuous pumping does not allow us to run measurements. In fact, it will take about one second to pump out a sample containing 10^6 atoms of ^{3}He with a partial pressure of 10^{-14} Torr in a 1-l chamber and a pumping speed of $1\,\mathrm{l\,s^{-1}}$. Therefore, very small gas quantities can be analyzed only in the so-called "static" pumping mode whereby a gas is admitted into the analyzer without any pumping out (Mamyrin et al., 1972a; Anufriev et al., 1979). In this way the output ion current will not be constant because the analyzed gas is pumped out by the ion source, trapped by sorption on the analyzer surface, etc. However, this mode enables us to decrease the pressure fairly slowly and makes high accurate measurements possible.

The so-called "quasistatic" pumping mode is advantageous for helium isotope analysis; in this case diffusion or any other type of pump which removes all gases (helium inclusive) is disconnected and a sorption pump is coupled with the mass-spectrometer chamber. A sprayed Ti mirror cooled by liquid nitrogen can serve as such a pump. If the area of the mirror is about $100\,\mathrm{cm}^2$, a residual pressure of 10^{-9}—10^{-10} Torr can be maintained in the mass-spectrometer chamber. Unfortunately, the pump does not remove neon, often present in helium samples.

3.6. Requirements for a mass spectrometer used for natural helium isotope measurements

Summarizing the peculiar features of helium isotope analysis we can list the following requirements which a mass spectrometer should meet:

(1) The resolving power must be no less than 1600 at 10% level of a peak height.

(2) In the case of the static pumping mode the sensitivity threshold must be no less than 10^6 atoms of ^{3}He in a chamber that is equal to its partial pressure of about 10^{-14} Torr.

(3) The contribution of ^{4}He$^+$ scattered ions in the output current of the ^{3}He$^+$ ions must not exceed 10^{-10} of the total ^{4}He$^+$ current.

(4) The contribution of H$_3^+$ and HD$^+$ scattered ions in the output current of ^{3}He$^+$ ions must be lower than 10^{-4} of the total current of the hydrogen multiplet.

(5) The design should provide a low "memory" effect which permits working without heating when measured isotope ratios vary within several (3—4) orders of magnitude.

(6) Leakage of atmospheric helium must be prevented and a static mode of operation must be ensured.

(7) Simultaneous measurement of the ^{3}He$^+$ and ^{4}He$^+$ ion currents must be provided by a two-beam operation mode.

(8) Taking into account the extremely wide variation of the ^{3}He/^{4}He ratio and error of 2% to 10% is allowed for a single analysis (except for special cases).

Some additional requirements of the total system applied for isotopic analysis of terrestrial helium:

(1) A fairly high production is desirable because often a large number of analyses is needed to provide a reliable interpretation of the natural relationships.

(2) A two-beam mode of operation calls for the use of standards and, thus, the leak-in system has to provide a set of standards, preparation of mixtures with a given ^{3}He/^{4}He ratio from the standards, leak-in of the standards (or the mixtures) and samples into the chamber of the analyzer under a static and a dynamic mode of evacuating.

(3) The design of the leak-in system and the mode of operation should provide a considerable difference in masses of helium isotopes.

The above requirements show that the most complicated problem is to ensure simultaneously many different objectives which are at times counteractive.

Chapter 4

ISOTOPIC ANALYSIS OF NATURAL HELIUM BY A STATIC MAGNETIC MASS SPECTROMETER

A resolving power sufficient for helium isotope analysis, namely $R_{10\%} \simeq$ 1000 (the mass multiplet resolution of mass 3) and $R_{10\%} \simeq 2000$ (the mass multiplet resolution of mass 4) can be achieved in a small magnetic analyzer with a radius of the ion trajectory in the magnetic field equal to 150—200 mm and a width of the ion source and the analyzer slits of about 50 μm. Such slits enable one to obtain an output ion current of 10^{-9}A, which is sufficient for the determination of ion current ratios of 10^8—10^9.

Unfortunately, in practice low ^{3}He/^{4}He ratios cannot be measured in one separation stage due to background of scattered ions of the ^{4}He$^+$ intensive peak and the HD$^+$ and H$_3^+$ multiplet peaks. A fairly low background of scattered ions is achieved in various "tandem" mass-spectrometer systems consisting of two or more stages: separation magnetic and focusing electrostatic at the same time. A possible application of such systems for the measurements of low ^{3}He/^{4}He ratios is discussed in the following chapter.

4.1. One-stage magnetic static mass spectrometer

Fig. 4.1 shows typical schemes of one-stage magnetic static mass spectrometers, in which a 60°, 90° and 180° deflection of ions in the magnetic field

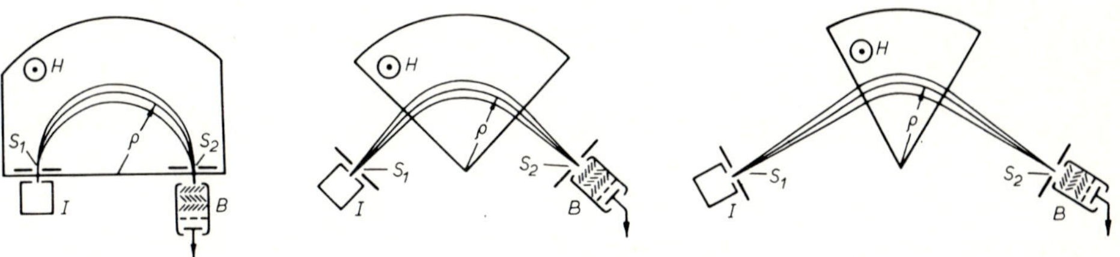

Fig. 4.1. Diagrams of magnetic static mass spectrometers with 180°, 90° and 60° ion deflections in the magnetic field.
H = magnetic field; I = ion source; S_1 and S_2 = entrance and exit slits of the analyzer, respectively; B = secondary electron multiplier; ρ = radius of the ion trajectory in the magnetic field.

is realized. No matter what the angle of the deflection, the dispersion D of such instruments (in mm per 1% of the ion mass change) is determined by radius ρ of the ion trajectory (mm) in the magnetic field:

$$D = \rho/100 \tag{4.1}$$

The resolution at the peak base is:

$$R = \frac{100D}{S_1 + S_2 + \Delta l} = \frac{\rho}{S_1 + S_2 + \Delta l} \tag{4.2}$$

where ρ is the radius; S_1 and S_2 are the widths of entrance and exit slits in mm; Δl is the ion beam aberration broadening depending on the energy and the angle spread of ions.

Employing pole pieces of a particular type as well as improved parameters of the ion source and power supply, it is possible to decrease Δl and obtain a resolving power $R_{10\%} \simeq 1600$ for $D = 2$ mm per 1% of the mass change when $S_1 = S_2 = 50$ μm and $\rho = 200$ mm.

High sensitivity and the possibility to use a small amount of gas for the analysis may be provided by a static pumping mode of the instrument. This equalizes the pressure of the analyzed gas in the ion source and in the analyzer. The required gas pressure in the analyzer can be found in the following way.

The rise of pressure (P) increases the $^3\text{He}^+$ ion current in proportion to P, while the $^4\text{He}^+$ current of scattered ions increases in proportion to P^2. This follows from eq. 3.4 where $I_0 = I(^4\text{He}) \sim P$. Consequently, at a low P, the ratio of the $^3\text{He}^+$ ion current to the scattered ion current, I_{scat}, might be high. However, a too sharp decrease of gas pressure in the chamber is undesirable because of the residual pressure providing a useless current of $^3\text{He}^+$ ions, $I(^3\text{He})$, and a scattered ion current, I_{scat}. Moreover, the $I(^3\text{He})$ current, also decreasing simultaneously with the pressure, should be higher than the noise current of the detecting system, reduced to the inlet of the secondary electron multiplier.

Let us assume that:

$$I(^3\text{He})_{\text{min}} = 10^{-18}\text{A} = 2\,I_{\text{scat}} \tag{4.3}$$

Then taking into account that in eq. 3.4:

$$I_0 = I(^4\text{He}) = S_i P \tag{4.4}$$

where S_i is the sensitivity of the instrument, A Torr^{-1}; and P is the ^4He pressure, Torr; we find from eq. 3.4, 4.3 and 4.4:

$$P_{\text{opt}} = 2 \cdot 10^{-8} D \frac{\Delta M}{M} \sqrt{\frac{1}{S_i Fl}} \tag{4.5}$$

For $R_{10\%} \simeq 1600$, $D = 2$, $\Delta M/M = \frac{1}{4}$, $S_i \simeq 10^{-5}$ A Torr^{-1}, $F = 0.3$ mm^2, and $l \simeq 10^3$ mm, we obtain $P_{\text{opt}} \simeq 2 \cdot 10^{-7}$ Torr and $[I(^3He)/I(^4He)]_{\min} \simeq 3 \cdot 10^{-7}$.

From section 3.3 we know that the current of the (HD + H$_3$) hydrogen peaks can reach $10^4 \cdot I(^3He)$ and that the scattered ion current of the hydrogen multiplet peaks should be taken into account. When $\Delta M/M \simeq 6 \cdot 10^{-3}/3$ and the other values are the same as in the above example, we find from eq. 3.4:

$$\frac{I_{\text{scat}}}{I_0} \, (HD + H_3) \simeq 3 \cdot 10^{-3}$$

Hence, the (HD + H$_3$) scattered ion current in the centre of the ^{3}He peak will be:

$$I(HD + H_3)_{\text{scat}} = 3 \cdot 10^{-3} \cdot 10^4 \cdot I(^3He)_{\min} = 30 \cdot I(^3He)_{\min}$$

Thus, the minimum measureable ^{3}He/^{4}He ratio is limited by the HD$^+$ and H$_3^+$ scattered ion current. Allowing for the ^{4}He scattered ion current which is not related to the scattering of gas molecules, the lowest obtainable limit of the ^{3}He/^{4}He ratio will be in the range of 10^{-5}—10^{-6}, if one magnetic separation stage is used.

In principle, the helium isotope ratio can be determined with more precision in a one-stage instrument provided its dispersion is high enough and if a special technique of removing hydrogen from the vacuum system is employed and additional purification of the investigated helium is achieved. Without the static operation mode we can use the difference in pressure between the ion source and the analyzer, which increases the ^{3}He ion yield but reduces the scattering effect of ^{4}He, HD and H$_3$ ions. However, all this leads to a waste of time and limits the scope of the possible application of the described technique.

By means of the above techniques, Aldrich and Nier (1948) managed to measure for the first time the isotope composition of terrestrial helium employing a one-stage sector mass spectrometer. But their instrument required up to 1 norm.cm^3 of helium for the analysis which is 10^5 times more than the amount used in mass spectrometer 9302 (see Chapter 5); sometimes Aldrich and Nier had to use about 0.5 kg of the mineral for one determination! Even so, it was impossible to measure ^{3}He/^{4}He ratios less than 10^{-7}.

One-stage instruments with a non-uniform magnetic field (Alekseevsky et al., 1966) will not be discussed here as in such system an increase in dispersion in accompanied by extremely great "tails" of the mass spectrum peaks, which makes it impossible to measure helium isotope ratios less than 10^{-6}.

Recently mass-spectrometric systems with prism optics have been developed by V.M. Kel'man and his colleagues (Kel'man et al., 1972, 1976, 1979). These systems enable increasing the dispersion as compared to sector magnetic mass spectrometers (with similar overall dimensions). However, to get an increase in dispersion sufficient to eliminate the scattered ion current of the hydrogen multiplet, a design of mass spectrometers with prism optics must, by necessity, be extremely complex. As was shown in section 3.2, two-stage mass spectrometers are required to measure all possible varieties of the ^{3}He/^{4}He ratios in terrestrial helium. Nevertheless, analyzers with prism optics are the most adequate type among static systems of mass spectrometers for natural helium isotope studies.

4.2. Multi-stage static mass spectrometers

Fig. 4.2 shows different systems of two-stage mass spectrometers in which the separating and angle focusing magnetic stage is coupled with the electrostatic energy focusing stage.

The importance of the second stage in the two-stage magnetic system was

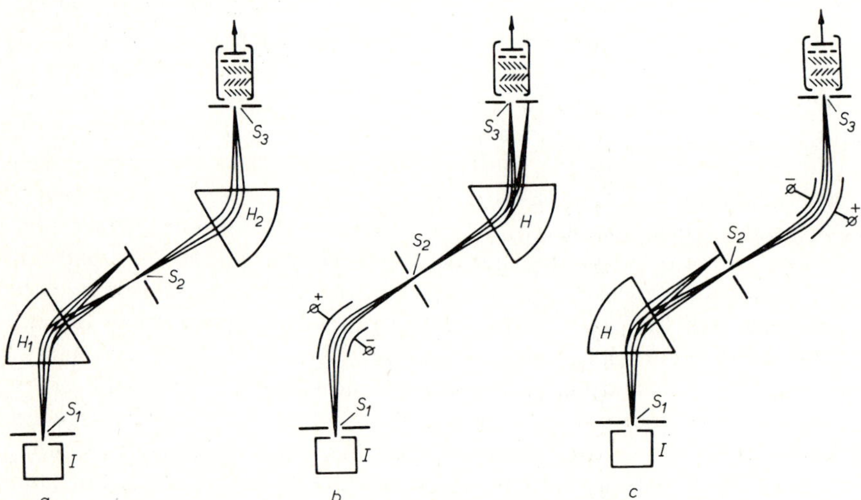

Fig. 4.2. Diagrams of two-stage static mass spectrometers.
a = two magnetic stage instrument; b = joint of electrostatic and magnetic stages; c = joint of magnetic and electrostatic stages. H_1 and H_2 = magnetic fields of the first and the second stages, respectively; S_3 = exit slit of the second stage. For other symbols see Fig. 4.1.

discussed in section 3.2 (see Fig. 4.2a). The $^4\mathrm{He}^+$, HD and H_3^+ scattered ion current entering slit S_2 was estimated for one magnetic stage in section 4.1. It is more difficult to calculate a further drop in the current (at the outlet S_3) because the ions scattered in the first stage and "proper" ions enter the second stage at different angles. Therefore, the effect of the secondary scattered ions passing through the exit slit of the second stage may be somewhat stronger than that defined by eq. 3.4. Examples provided in the next section show that a possible difference between the estimated and experimental values of I_{scat}/I_0 for the second stage of the two-stage mass spectrometer might reach one order of magnitude.

It is even more difficult to suppress scattered ions of the hydrogen peak in the two-stage system. This system may be advantageous when two magnetic stages, characterized by a resolution sufficient for the separation of multiplet peaks from the $^3\mathrm{He}$ (or $^4\mathrm{He}$) peaks, are used — that is, $R_{10\%} \simeq 1600$ (see section 3.1). Such a system is very complex and expensive. The second approach is to use a low resolution stage (~ 20) but a fairly high dispersion in the second stage. This is not easy to realize within the bounds of the static system, however.

An electrostatic analyzer of the Hughes—Rozhansky condensor type as the second stage of the "tandem" system is less efficient than the magnetic one. An electrostatic analyzer in front of the magnetic one (see Fig. 4.2b) will not, in fact, affect the $^4\mathrm{He}$ scattered ions entering slit S_3 as it does not separate the $^3\mathrm{He}$ and the $^4\mathrm{He}$ ion beams. The improved energy focusing of ions enables us to increase the sensitivity of an instrument at a given resolution which is equivalent to a reduction of the pressure and, hence, to a suppression of the scattering ion current. However, the effect will be only slight because the optimal pressure is proportional to $\sqrt{(1/S_i)}$ (see eq. 4.5).

Arrangement of the electrostatic stage after the magnetic one (Fig. 4.2c) appears to produce a better effect. The $^4\mathrm{He}$ ions scattered in the first magnetic stage enter slit S_2 at angles different from those of $^3\mathrm{He}$ ions; this makes their passage through the electrostatic stage as well as focusing and entering slit S_3 difficult. However, the electrostatic stage does not provide mass dispersion, so the separation of scattered ions from $^3\mathrm{He}$ ions becomes less efficient than it would be were there a second magnetic stage. According to the published data the addition of an electrostatic stage, as shown in Fig. 4.2c, can decrease the quantifiable $^3\mathrm{He}/^4\mathrm{He}$ ratio by one to two orders of magnitude as against one magnetic stage.

Instruments in which two separation stages are made in a single magnetic step form a separate class of two-stage mass spectrometers. Among them are instruments with a trochoidal ion trajectory in the crossed magnetic and electric fields, and instruments where a 360° ion deflection occurs in the magnetic field when the circular orbit radius changes step-wise (Fig. 4.3).

Kuz'min (1976) carried out a most exhaustive investigation of a trochoidal instrument and tested the system experimentally with a 360° ion deflection

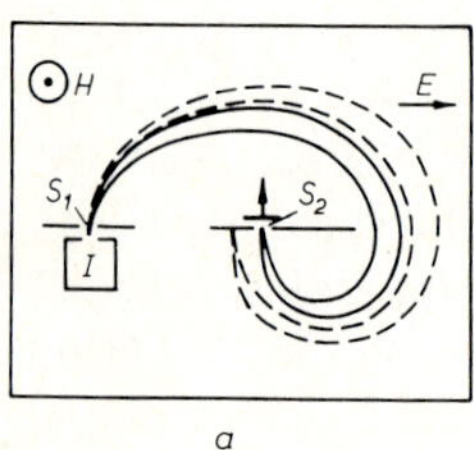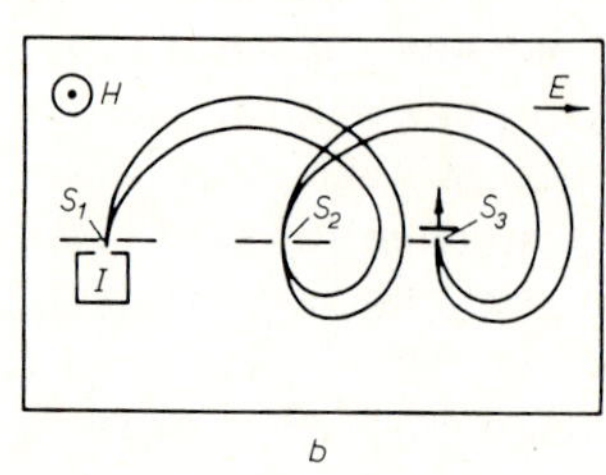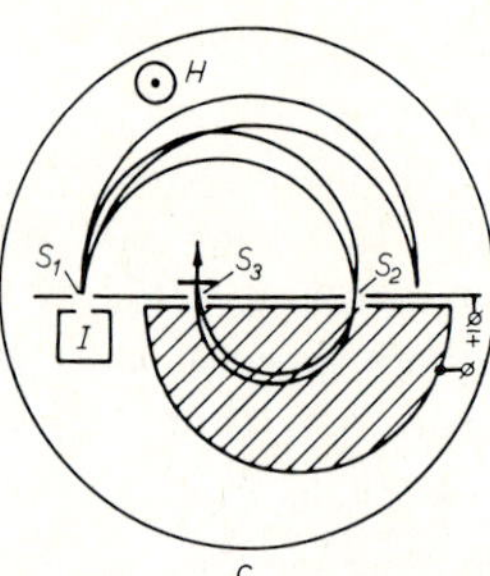

Fig. 4.3. Diagrams of two-stage mass spectrometers with one magnet. E = electrostatic field. For other symbols see Fig. 4.1 and 4.2.

(Fig. 4.3a) and 720° ion deflection (Fig. 4.3b). The trochoidal instruments with a 360° ion deflection (single-looped devices) are better than one-stage magnetic mass spectrometers because a lower proportion of the scattered ion current of 10^{-4}—10^{-5} at $\Delta M/M \simeq 1/40$ can be achieved, the resolving power being about 2000.

Two-loop instruments (with a 720° ion deflection) suppress the scattered ion current less efficiently than do instruments with two magnetic stages. The addition of a second loop (Kuz'min, 1976) decreases the current only by a factor of 10—20 as scattered ions directly enter slit S_3 without passing S_2; this is impossible in two-stage instruments.

An important problem of trochoidal instruments is the ion beam lead out to the electron multiplier, although this can be overcome (Hall et al., 1959). However, these instruments can resolve only some particular problems in helium isotope studies. It is very difficult to suppress the hydrocarbon background when ^{3}He is measured in such two-stage systems.

Another two-stage system with a 360° ion deflection with orbit radius change after 180° deflection (see Fig. 4.3c) was proposed and studied experimentally by Ionov et al. (1961) and Ionov and Karataev (1962). Despite good results obtained in potassium isotope separation (the proportion of scattered ions reached 10^{-7} at $\Delta M/M \simeq 1/40$) we can hardly recommend such apparatus as a universal instrument for helium isotope analysis. In order to obtain a resolution power of 1000 to 2000, this mass spectrometer requires a radius of the ion orbit of no less than 200 mm which corresponds to a diameter of pole pieces of about 500 mm. The HD and H_3 scattered ion current in such a system cannot be reduced to the required level since the resolution of one of the stages remains less than 1600. Also it is difficult to obtain a high dispersion. The lead out of the ion beam from the magnetic field to the secondary electron multiplier also presents some difficulties. Nevertheless, the mass spectrometer developed by Ionov and co-workers (1961) appears to be adequate for a successful solution of some problems of helium isotope analysis. One should bear in mind that a fairly low magnetic field is required for the separation of helium isotopes and permanent magnets provide the necessary magnetic field despite the large size of the pole pieces.

Systems consisting of three and more stages seem to be excessive because the separation of multiplets and the reduction of the ^{4}He scattered ion current to 10^{-9} can be achieved in a two-stage system. A "tandem" of two magnetic instruments, characterized by a resolution $R_{10\%} \simeq 1600$ (for the resolution of multiplet of mass 4) or $R_{10\%} \simeq 800$ (for the resolution of multiplet at mass 3) and a two-stage reduction of the HD and H_3 scattered ion currents is the best static system for isotope analyses of terrestrial helium.

4.3. Results obtained by means of magnetic static mass spectrometers

An ultimate suppression of the scattered ion current was reported for many static instruments.

Ehrenberg (1953) presented data for a one-stage sector $60°$ magnetic instrument with an ion trajectory radius in the magnetic field of about 200 mm. This instrument has a high sensitivity (the output currents $\sim 10^{-11}$A) and a resolution of ~ 1000 which provides the separation of multiplets of mass 3. The scattered ion background at $P \sim 10^{-6}$ Torr, $\Delta M/M \simeq 1/85$ is $I_{\text{scat}}/I_0 \approx 6 \cdot 10^{-4}$. According to eq. 3.4, this instrument may have $I_{\text{scat}}/I_0 \approx 10^{-6}$ at $\Delta M/M = 1/4$. The instrument dispersion of 2 mm per 1% of mass can reduce the share of scattered ions of hydrogen peaks to:

$$\frac{I(\text{HD})}{I(^3\text{He})} \sim 10^{-2}.$$

Gall' (1969) described an instrument with a magnetic stage ($90°$ sector magnet. ρ = 500 mm) followed by an energy focusing $60°$ electrostatic stage (ρ = 250 mm). At $P \simeq 10^{-7}$ Torr, $\Delta M/M \simeq 4 \cdot 10^{-3}$ and $R_{10\%} \simeq 1200$, the level of scattered ions, I_{scat}/I_0, is about 10^{-7}. In helium isotope analysis ($\Delta M/M = 1/4$) the ^{4}He scattered ion current can be decreased to 10^{-8}; however, this current would not be completely suppressed, and therefore the measured ^{3}He/^{4}He ratio can hardly be less than 10^{-7}.

Wilson (1963) discussed a complicated "tandem" with two similar magnetic stages ($90°$ sector magnets, ρ = 380 mm). The resolving power of each stage, $R_{10\%}$, is about 1200, the total length of the ion path within the instrument is about 3 m. At $P \simeq 2 \cdot 10^{-7}$ Torr and $\Delta M/M \simeq 5 \cdot 10^{-3}$, the proportion of scattered ion current, I_{scat}/I_0, equals 10^{-7}. In such a system at $\Delta M/M = 1/4$ the proportion of ^{4}He scattered ion current decreases to $I_{\text{scat}}/I(^4\text{He}) \sim 10^{-9}$ and that of HD and H_3 decreases to a required level of $I_{\text{scat}}/I(\text{HD} + H_3) \sim 10^{-4}$, because both stages have a resolution sufficient for separation of HD^+ and H_3^+ ions from $^3He^+$ ions.

White and Forman (1967) gave a description of a huge instrument consisting of two magnetic and two electrostatic stages placed in front of and behind the magnetic "tandem". The radii of the ion trajectory in the magnetic and electric stages are about 50 cm, and the total length of the trajectory is about 5 m. The proportion of the scattered ion current is $I_{scat}/I_0 = 10^{-7}$ at $P \approx 10^{-7}$ Torr and $\Delta M/M \approx 5 \cdot 10^{-3}$; hence, at $\Delta M/M = 1/4$ a value of $I_{scat}/I_0 \sim 10^{-10}$ can be reached in this instrument. The resolution of each magnetic stage being $R_{10\%} \approx 1500$, the proportion of the scattered hydrogen ion current in such a system decreases to $\sim 10^{-4}$ which provides measurements of low ^{3}He amounts. The instrument provides a z-focusing of the ion beam, and output currents appear to be sufficient to determine the ^{3}He peak at ^{3}He/^{4}He $\approx 10^{-9}$.

Despite the isotopic sensitivity obtained in the instrument it can hardly be considered as ideally suited for the measurement of the helium isotope composition due to its complex design, comparatively low absolute sensitivity resulting from the huge volume of the analyzer, and difficulties in providing the static pumping mode of operation.

Thus, we may conclude that none of the known static mass spectrometers (of reasonable size and complexity) can meet all the requirements of an instrument that can serve as a universal mass spectrometer for measurements of the isotope composition of natural helium. This conclusion stimulated the design of a special instrument for natural helium studies. It was developed on the basis of the magnetic resonance mass spectrometer at the A.F. Ioffe Physico-Technical Institute of the USSR Academy of Sciences. This system is discussed in Chapter 5.

MAGNETIC RESONANCE MASS SPECTROMETER (MRMS)

The most difficult cases of ^{3}He/^{4}He measurements require that the mass-spectrometric analyzer be of minimum volume and provide a maximum possible dispersion: the smaller the volume of the analyzer, the greater is its absolute sensitivity, the less prominent is the effect of ion scattering which is proportional to the total ion trajectory in the instrument (see eq. 3.4). The small volume of the analyzer is also beneficial for its vacuum.

A great dispersion has the following advantages: (1) it helps to obtain a high resolution, (2) it widens the slits bounding the ion beam, thus increasing the sensitivity at a given resolution, and (3) it reduces the background of scattered ions (see eq. 3.4). However, as dispersion in a static instrument is determined by the ion radius in the magnetic field, any gains in size are counter-productive with respect to dispersion. This differs from dynamic mass spectrometers of which dispersion depends on geometrical as well as time-of-flight relationships.

Let us turn to the process of mass dispersion illustrated in Fig. 5.1. We

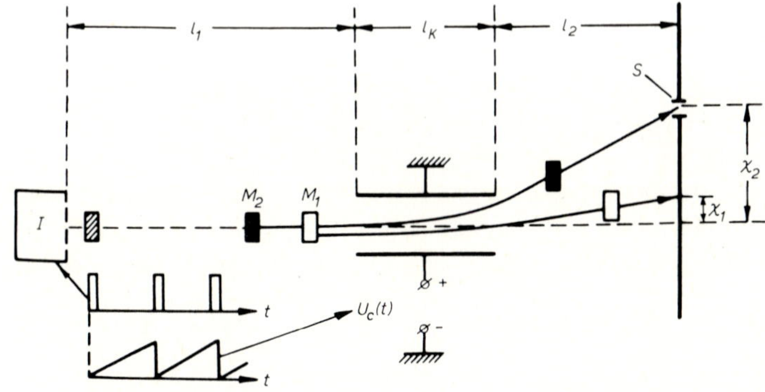

Fig. 5.1. Schematic diagram of a dynamic system which enables increasing dispersion independently of geometrical dimensions.
I = pulse ion source; l_1, l_2 and l_k = geometrical dimensions of the analyzer; M_1 and M_2 = ion packets of different mass; t = time; $U_c(t)$ = saw-tooth voltage of the capacitor; S = exit slit; x_1 and x_2 = deflections of the packets from the axis of the instrument.

assume that all ions leaving the source are characterized by equal energies while their velocities differ only with respect to the ratio of mass:charge:

$$V = \sqrt{\frac{2eU}{m}} \tag{5.1}$$

where e and m are the charge and the mass of the ion, respectively; U is the electric potential accelerating the ion with zero initial velocity. In the course of its flight, a packet of ions of different masses derived from the pulsed source is divided into different packets, each containing identical ions. Passing through the deflection system whose electrodes are supplied by saw-tooth voltage, $U_c(t)$, the ions of masses M_1 and M_2 are deflected in the plane of exit slit S at distances x_1 and x_2. The ion shift depends on: (1) the geometric relationships in the system (l_1, l_2, l_t,); (2) the time needed by ions of masses M_1 and M_2 for entry into the deflection system; and (3) the rate of the voltage curve, $U_c(t)$. If this rate is increased by a factor of 10 the dispersion should increase by the same factor with unchanged geometric sizes of the system.

Investigations of numerous schemes of dynamic mass spectrometers (Blaut, 1965; Mamyrin, 1966) show that the MRMS may be considered a promising instrument for isotope analyses of natural helium. The analytical parameters of the mass spectrometer have been described earlier (Mamyrin and Shustrov, 1957, 1962). The most efficient modification of the MRMS characterized by a compensation mode of operation and a sinusoidal voltage supply of the modulator (Shustrov, 1960; Mamyrin, 1966; Mamyrin et al., 1969b, 1972a, b) is discussed below.

5.1. Design and operating principles of the MRMS

According to the underlying operating principle the MRMS is a time-of-flight mass spectrometer, in which ions of different mass:charge ratios (M/e) are separated due to the difference in time of their flight along a certain path in a space with a constant electric potential distribution (in particular, in the field-free space). This principle is realized by a sinusoidal voltage supply of the modulator which cuts off ion packets from the ion beam continuously leaving the source; naturally, these packets contain all kinds of ions which have passed through the modulator slit. Since all ions in the source are accelerated by the same potential, their velocities (eq. 5.1) are proportional to $\sqrt{e/m}$, and the initial packet which has passed a certain path is divided into a number of packets each containing ions of the same mass:charge ratio.

The resolution of such systems depends on two factors: the initial time (or geometric) thickness of the packets and their final thickness after passing through the drift space. The sensitivity of the system is also determined by two factors — that is, the number of ions in the initial packet and the utilization factor, namely, the ratio of the number of ions in the ion packets at the detector entrance to all ions of a given kind formed in the source.

The above four factors are successfully combined in the MRMS designed at the A.F. Ioffe Physico-Technical Institute of the USSR Academy of Sciences, Leningrad.

Let us consider the operation of the MRMS analyzer (Fig. 5.2). The flat vacuum chamber is located in the magnet gap, the uniform magnetic field H being perpendicular to the figure plane.

The ions are formed in the conventional source with an electron impact and come out continuously through slit S_1. Owing to the magnetic field they are separated as in an ordinary 180° type mass spectrometer. The resolution of the thus formed first stage (low resolution stage) is sufficient for the separation of ions whose masses differ by one amu over the operational range; the dispersion of this stage, D_1, is $\rho_1/100$.

Ions with slightly different M/e (mass multiplets) ratios, not separated at the first stage, enter modulator slit S_2 (the second dynamic separation stage, or high resolution stage). A high frequency sinusoidal voltage is applied to the modulator's middle electrode and outer grounded electrodes. The voltage period, T_g, is much shorter than the flight time of the ions in circular orbit in the magnetic field — that is, the cyclotron period:

$$T_c = 651.21 \frac{M}{eH} \tag{5.2}$$

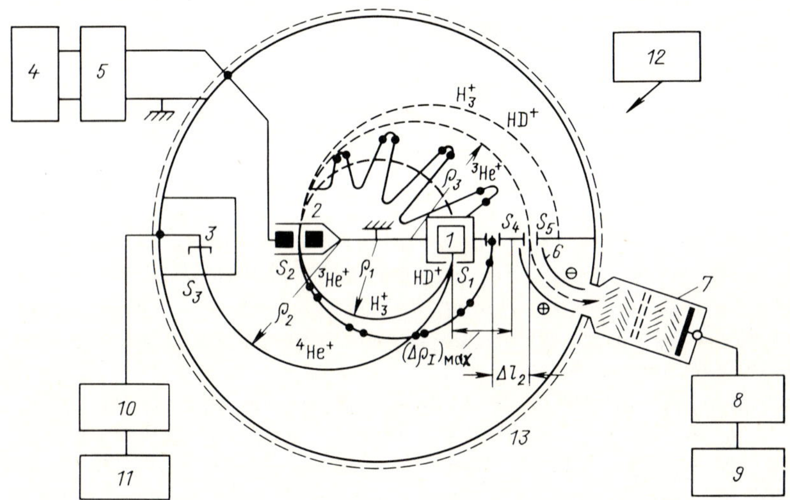

Fig. 5.2. Diagram of the analyzer of the magnetic resonance mass spectrometer (Mamyrin and Tolstikhin, 1981).
1 = ion source; *2* = modulator; *3* = ⁴He ion collector; *4* = high frequency oscillator; *5* = sinusoidal voltage amplifier; *6* = lead-out deflection capacitor; *7* = secondary electronic multiplier; *8* and *10* = D.C. amplifiers; *9* and *11* = recorders; *12* = ion source power supply; *13* = boundary of magnetic field. For other symbols see the text.

where T_c is the ion cyclotron period, μsec; M is the ion mass, amu; e is the ion charge (for a single charged ion $e = 1$); and H is the magnetic field intensity, Oe.

The size of the modulator, the ion velocity and the period of h.f. voltage, T_g, are chosen in such a way that, within two gaps of the modulator, an ion may obtain an acceleration of the same sign and almost equal magnitude — that is, when the ion passes through the field-free space, the h.f. voltage of the middle electrode changes by $T_g/2$. Since the change of the ion energy in the modulator is less than its total kinetic energy, an increment of the ion energy and, hence, that of the circular orbit radius, approximately follows the sine law according to the applied h.f. voltage.

An example of the ion flow from the modulator to the rear side of the source is illustrated by a wavy line in Fig. 5.2. During each h.f. voltage period slit S_4 cuts off two ion packets shown by dots. The time when a packet leaves the modulator is called drift start. Ions with different mass:charge ratios that start simultaneously are separated at their second entry into the modulator in accordance with the flight time of a complete revolution in the magnetic field.

The time of one complete revolution of ions in the uniform magnetic field, T_c, is dependent neither on the ion energy nor on the exit angle (Kel'man and Yavor, 1968), and the ions of types $(M/e)_1$ and $(M/e)_2$ re-enter the modulator at different times, T_{c1} and T_{c2}, precisely defined by formula 5.2.

Slit S_4 cuts off rather long packets so that each packet contains some 10 or 15% of the total ion number leaving the source. This provides a high utilization factor of the ion current. However, ions characterized by a small mass difference can move within the bounds of these long packets prior to their second passage through the modulator. In order to substantially increase the resolution and separate such ions, the MRMS employs the so-called "compensation" mode of operation (Fig. 5.3).

Suppose ions a and b are the first and last items in a packet cut off by slit S_4 (at the first acceleration in the modulator a and b are the most and the least accelerated ions, respectively). The corresponding changes of orbit radius at the first acceleration are shown in Fig. 5.3 by $\Delta\rho_{aI}$ and $\Delta\rho_{bI}$ for ions a and b, respectively; $\Delta\rho_{aII}$ and $\Delta\rho_{bII}$ are the orbit radius increments for ions a and b at the second acceleration.

If a and b are of the same kind their cyclotron periods, T_c, will be equal. Hence, for the time parameters shown in Fig. 5.3, the total increments of the orbit radii after two accelerations will be the same for these ions and the entire packet will actually move along the orbit of the same radius. A slight broadening of the packet (in radius) is caused by the width of the source slit and aberrations due to a certain diversity in the magnetic field and some irregularity in the compensation process. However, in practice the aberrations are small and their total effect is considerably less than the width of the source slit.

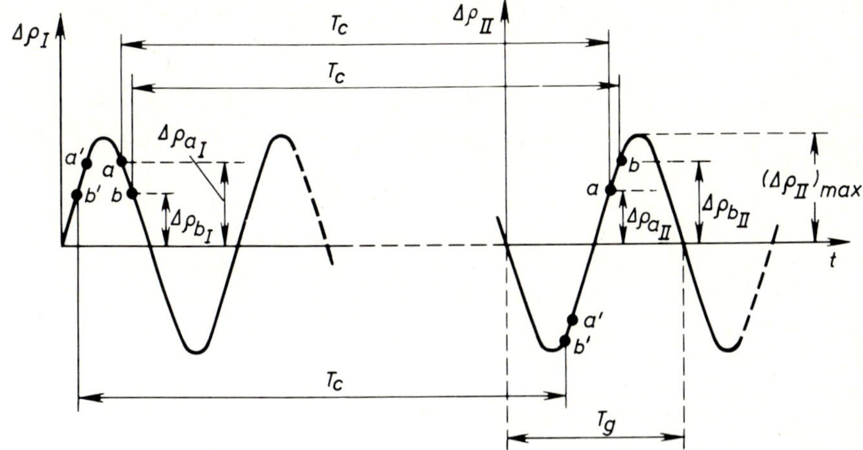

Fig. 5.3. Compensation of ion energy spread in the modulator of the MRMS (see text).

Ions a′—b′ cut off by slit S_4 (see Fig. 5.2) together with ions a—b have the same M/e ratio and the same cyclotron period. Therefore, in the course of their second passage (Fig. 5.3) they will acquire a total acceleration different from that of ions a—b and will miss the exit slit. The same holds for the ions whose $(M/e)_1$ value is slightly different from the M/e ratio of the ions a—b, because they possess a different cyclotron period, $(T_c)_1 \neq T_c$. Fig. 5.3 shows that such ions, starting simultaneously with the a—b ions, acquire increments of the orbit radii similar to those of the a—b ions at the moment of their first passage through the modulator. However, their cyclotron periods are not equal to T_c and having flown around the drift orbit they acquire such a second acceleration in the modulator that the total increments of their orbit radii are not equal to those of ions a—b. Due to this separation stage the ions characterized by $(M/e)_1$ will not enter slit S_5 (see Fig. 5.2).

In Fig. 5.2 ions a—b with the ratio M/e stand for the ${}^3\text{He}^+$ ions; the system is adjusted to the resonance for these ions and they pass through exit slit S_5; ions with $(M/e)_1$ stand for H_3^+ and HD^+.

A deflection system *6* schematically drawn in Fig. 5.2 serves as a highly efficient lead-out of the ion beam from the magnetic field. Subsequent employment of an electron multiplier enables measuring an output current as low as 10^{-18} or 10^{-19} A.

From Fig. 5.3 it is evident that if the period of h.f. voltage applied to the modulator varies over a wide range whereas the mass of the ions and, consequently, their cyclotron period are preserved constant, the output ion current will yield a complete sequence of peaks (Fig. 5.4).

When the T_c/T_g ratio is similar to that presented in Fig. 5.3, only ions a—b enter the exit slit and produce peak *1* since the total increment of their orbits,

60

$\Delta\rho_1 + \Delta\rho_2$, is equal to the distance between slits S_1 and S_5 (see Fig. 5.2). If we decrease T_g (T_c being preserved constant) then ions a—b acquire a larger increment of orbit radius and deflect from slit S_5. On the other hand, ions of another packet, a'—b', reach slit S_5 and make peak 2; however, this peak is considerably broadened due to the spread of the ion orbits because the compensation mechanism does not operate in this case. A further decrease in T_g leads to an increase in the orbit radius of the a'—b' ions. They leave slit S_5 after which the a—b ions, that have shifted along the falling slope of the sinusoidal voltage curve T_g, again reach slit S_5 (see Fig. 5.4). Peak 3, repeatedly produced by the a—b ions, is also broadened as the compensation mechanism is not in operation. Peaks 2 and 3 can overlap and form one broad peak. The following decrease in T_g results in peak 4 produced by the a'—b' ions; now they are accelerated in the modulator at the moment when voltage curve T_g is falling; a situation similar to the one that produced peak 1 (see Fig. 5.4). In this case the compensation mechanism operates properly and the output peak 4 can be measured. Further changes in T_g lead to another couple of peaks, etc.

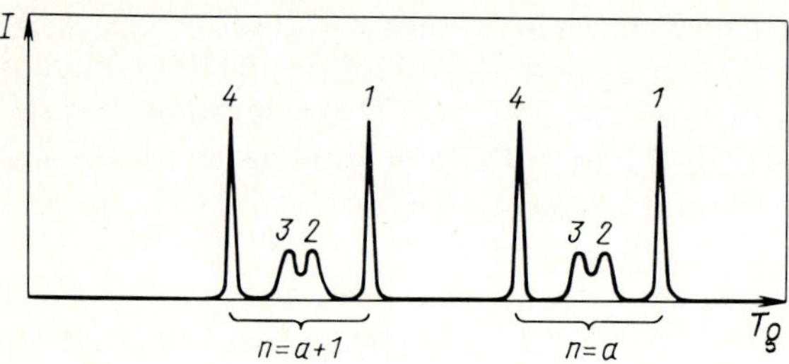

Fig. 5.4. Peaks of the MRMS output ion current, with the mass-spectrum scanning parameter T_g changing considerably greater than the mass multiplet width. 1 and 4 = compensation peaks; 2 and 3 = resonance peaks.

The identification of a mass spectrum produced by such a mechanism is simple because the peak positions result from significant changes of the h.f. frequency.

The entire multiplet spectrum of a given mass is located near the compensation peaks (peak 1 or 4 in Fig. 5.4); there are no ion peaks with a different mass number because they deflect from the modulator slit due to a sufficient resolution of the first 180° stage.

The width of ion packets in front of the exit slit (in radial direction) is small and approximately similar to that of the source slit width despite a wide slit S_4. The instrument's dispersion — the shifting of packets along the radius in the exit slit plane when the ion mass changes — can be very high since the shift of the packet centre can be independently increased by increasing the steepness of the h.f. voltage curve at a given amplitude. This, in turn, requires an increase in frequency.

In practice, it is easy to obtain a dispersion of several tens of millimeters per 1% of mass change at an ion trajectory radius of 50—60 mm — that is, two orders of magnitude higher than the dispersion of static instruments with similar ion trajectory radii in the magnetic field.

The resolution of such an instrument characterized by a high dispersion and a good focusing of the source slit image can reach $2 \cdot 10^3$; moreover, wide slits enable one to increase the sensitivity of the system and simplify its assembly and adjustment.

It is noteworthy that the initial energy and angle spread of the ions in the source are less important in the MRMS as compared to static mass spectrometers. When the source's slit width in the MRMS is about 1—2 mm, the normally observed beam aberration broadenings, $\rho\alpha^2$ and $\rho(\Delta U/U)$, do not actually affect the resolution. This enables using high currents of ionizing electrons reaching several mA.

Wide slits, a relatively short length of the ion flight and a high efficiency of the ion source ensure a major rise of ion currents and, despite the losses caused by cutting off the ion packets from a continuous beam (the utilizing factor is ~ 0.15), the instrument has a sensitivity as high as 10^{-4}—10^{-3} A Torr^{-1} combined with a resolution of $2 \cdot 10^3$.

Because of a number of repeated suppressions, the ^{4}He scattered ions do not actually reach slit S_5 when the dynamic compensation system is adjusted for detection of the ^{3}He ions. This is the most important feature of the MRMS modification shown in Fig. 5.2 and discussed in more detail below.

5.2. Ways to ensure the analytical parameters

The requirements that a mass spectrometer used for measuring the ^{3}He/^{4}He ratio is expected to meet were listed in section 3.6. It can be shown that all these requirements can be met when an instrument is based on the MRMS circuit.

The dispersion and resolution of the MRMS at a sinusoidal h.f. voltage applied to the modulator and the compensating mode of operation can be estimated from the following formulas (Mamyrin, 1966):

$$D = \frac{2\pi}{100}\, n(2\Delta\rho_{\mathrm{II}})_{\max} \sqrt{1 - \frac{\Delta l_2}{2(\Delta\rho_{\mathrm{II}})_{\max}}} \tag{5.3}$$

$$R = \frac{D\,100}{S_1 + S_5 + \Delta l_{\mathrm{aber}}} \tag{5.4}$$

where $n = T_c/T_g$ is the harmonic number (the number of packet pairs which are on the circular orbit) at the same time; Δl_2 is the distance between slits S_4 and S_5 (see Fig. 5.2); $(\Delta\rho_{\mathrm{II}})_{\max}$ is the maximum increment in the ion

62

orbit radius produced by the second acceleration in the modulator (see Fig. 5.3); S_1 and S_5 are the widths of the source and exit slits, respectively; and Δl_{aber} is the total radial broadening of the ion packet (usually $\Delta l_{\text{aber}} \ll S_1 + S_5$).

The highest dispersion and resolution are obtained at the optimum ratio of $\Delta l_2 / 2(\Delta \rho_{\text{II}})_{\text{max}} \simeq 0.5$. In this case:

$$D \simeq 0.1 \, n \Delta l_2 \tag{5.5}$$

$$R_{10\%} \simeq \frac{10 \, n \Delta l_2}{S_1 + S_3 + \Delta l_{\text{aber}}} \tag{5.6}$$

At the radius of ions emitted from the source $\rho_1 = 50$ mm, $\Delta l_2 = 20$ mm and $n = 40$, we find from eq. 5.3 a dispersion, D, of 80 mm per 1% of mass change. At such a dispersion and $R_{10\%} = 2 \cdot 10^3$, we obtain from eq. 5.4: $S_1 + S_5 + l_{\text{aber}} = 6$ mm. When l_{aber} is small (see section 5.1) the resolution required for helium isotope analyses is attained when the slits are wide enough (about 2 mm). This provides a high sensitivity along with a high dispersion and resolution.

Let us evaluate the maximum sensitivity of the MRMS. The maximum sensitivity, S_i, of any mass spectrometer is determined by the source efficiency, S_s, and the utilizing factor of the analyzer, k_i, that is:

$$S_i = k_i S_s \tag{5.7}$$

For the MRMS, $S_s \simeq 2 \cdot 10^{-2}$ A Torr^{-1} and k_i may be determined as follows: slit S_4 cuts off ion packets containing about 10% of the ions leaving the source. The current decreases also 10-fold due to the vertical divergence of the beam and, finally, the current losses amount to about 50% due to the beam lead-out from the magnetic field. Thus, $k_i \simeq 5 \cdot 10^{-3}$. The maximum sensitivity of the MRMS at such parameters is $S_i \approx 10^{-4}$ A Torr^{-1}.

At the lowest measured current, $I_{\text{min}} = 3 \cdot 10^{-19}$ A, the minimum partial sensitivity will be $P_{\text{min}} = I_{\text{min}}/S_i \approx 3 \cdot 10^{-15}$ Torr. If the analyzer's volume is about 2 l, such a partial sensitivity corresponds to an absolute sensitivity $P_{\text{abs}} \approx 3 \cdot 10^5$ ^{3}He atoms.

Such a sensitivity of the MRMS, together with a maximum pressure $P_{\text{max}} \approx 2 \cdot 10^{-5}$ Torr, enable us to obtain a minimum measurable ^{3}He/^{4}He ratio (isotopic sensitivity) of:

$$\left(\frac{^3\text{He}}{^4\text{He}}\right)_{\text{min}} = \left(\frac{I_{\text{min}}}{I_{\text{max}}}\right)_{\text{out}} = \frac{I_{\text{min}\,\text{out}}}{S_i P_{\text{max}}} \simeq 3 \cdot 10^{-10}$$

The validity of the estimates was confirmed experimentally (Mamyrin et al., 1972a).

A decrease in losses due to vertical divergence of the beam and its lead-out from the magnetic field, a higher source efficiency and employment of storage detectors (ion counting) can further increase the MRMS sensitivity (and its isotopic sensitivity).

Reduction of the scattered ion background, which actually determines the final isotopic sensitivity, is a parameter as important as resolution and sensitivity.

The plotting of possible trajectories of scattered ions (see Fig. 5.2) shows that none of the single scattered ^{4}He ions that have passed through the modulator slit are able to enter exit slit S_5 due to their insufficient energy, even though they have been accelerated in the modulator. Consequently, when the dynamic stage is adjusted to the ^{3}He measurement, the ^{4}He ions can enter slit S_5 only in the case of double scattering. However, not all of the rescattered ions that have entered slit S_5 can reach the multiplier input because the lead-out system is adjusted to the transmission of authentic high-energy ions. Thus, the MRMS scheme is similar to a "tandem" of three magnetic stages; this is confirmed experimentally by the peak shape (Khabarin, 1975; Fig. 5.5).

In fact, at the highest possible sensitivity it is difficult to detect the ^{4}He scattered ion current on the dynamic stage outlet, when it is adjusted to the multiplet peaks of mass 3, even with gas pressure P in the analyzer being $\sim 10^{-5}$ Torr.

To estimate the HD and H_3 scattered ion currents in the MRMS one should take into account the dynamic separation stage and the separation in the lead-out system 6 (see Fig. 5.2).

As ^{3}He, HD and H_3 masses are similar, the scattered ion current through exit slit S_5 can be calculated from eq. 3.4 with the value of dispersion taken

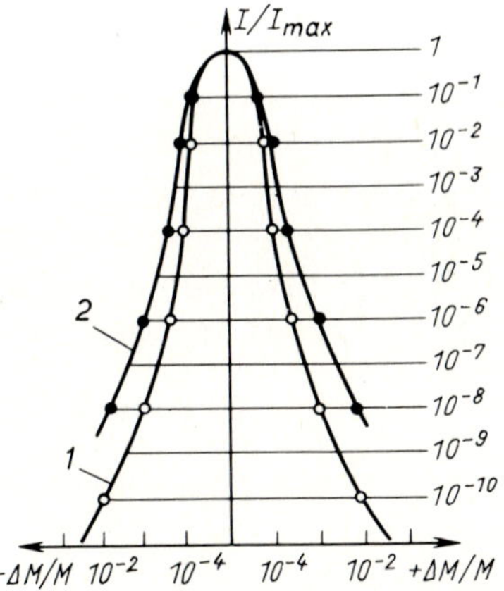

Fig. 5.5. Relationship between the shape of the MRMS mass-spectrum peak and the gas pressure in the analyzer (Khabarin, 1975).
1 = residual pressure of $3 \cdot 10^{-8}$ Torr; 2 = pressure of $7 \cdot 10^{-7}$ Torr.

from eq. 5.5. Assuming $D \approx 80$ mm per 1% of mass, $\Delta M/M \approx 2 \cdot 10^{-3}$, $P \approx 10^{-5}$ Torr, $Q \approx 5$ mm^2, and $l \approx 600$ mm, we obtain:

$$\frac{I_{\text{scat}}}{I_0} \, (HD + H_3) \approx 7 \cdot 10^{-4}$$

At a given D and $\Delta M/M$ the separation of authentic and scattered ions in the lead-out system is almost complete. The scattered ions, whose energies significantly differ from those of the authentic ions, shift from the axis of the lead-out system and hit the deflecting plates; only a few of the rescattered ions are reflected by the plates in such a way that they reach the input of the electron multiplier.

Due to their extremely low proportion, the HD and H_3 scattered ions accompanying the ^{3}He ions ($\ll 10^{-4}$) do not affect the measurement of the lowest ^{3}He/^{4}He ratios in terrestrial helium.

The operation mode of the MRMS belongs to the two-beam type (see Fig. 5.2). The ^{4}He ion current being much higher than that of ^{3}He, its value is measured by collector 3 (see Fig. 5.2) and a D.C. amplifier. If the residual pressure in the instrument is of the order of 10^{-8} Torr, the background multiplet peaks at mass 4 (primarily $^{12}C^{3+}$) are weak; this enables one to measure the ^{4}He ion current without any restrictions. The transmission and gain coefficients of the ^{3}He-detecting system sharply differ from those of the ^{4}He system, and therefore it is necessary to calibrate the instrument by standard helium. Normally, helium of the earth's lower atmosphere (see section 2.3) serves as such a standard.

The duration of the analysis under the two-beam operation mode depends mainly on auxiliary procedures (the leak-in and pumping of the sample) as the measurement itself lasts only a few minutes.

One may think that in the course of time the "memory" effect in the MRMS will increase owing to strong currents in the ion source. However, in practice it is shown that there is neither a considerable difference of the rate of helium pumping out nor a decrease of its residual ion current as compared to conventional static instruments.

Summarizing the analytical parameters of the MRMS discussed above we may conclude that such a system appears to be the best suited one for isotope analysis of terrrestrial helium.

5.3. A laboratory mass spectrometer

Mamyrin et al. (1969b) described a device for helium isotope study; more recently its analytical parameters were slightly improved (Mamyrin et al., 1972a, b).

The main analytical parameters of the device are the following:

(1) The resolution of the dynamic time-of-flight stage $R_{10\%}$ is $2 \cdot 10^3$.

(2) The ^{4}He scattered ion current which interferes with the ^{3}He ion current is less than 10^{-11} of its total value.

(3) The share of the HD and H_3 scattered ion current which interferes with the ^{3}He ion current is less than 10^{-4} of its total value.

(4) The sensitivity threshold of ^{3}He detection (signal equals noise) is about $5 \cdot 10^5$ atoms in the analyzer; this value being bounded mainly by the noise of the secondary electron multiplier.

(5) The sensitivity of ^{4}He measurements is bounded by the background of the gas extraction system and the analyzer; it amounts to 10^9 atoms.

(6) The minimum measured ^{3}He/^{4}He ratio is about $3 \cdot 10^{-10}$; further decrease of this value is bounded by the sensitivity of ^{3}He measurement.

(7) The linear relationship between the ^{3}He and the ^{4}He ion currents and the pressure is observed to be up to $\sim 10^{-7}$ Torr. A further increase in pressure (up to $2 \cdot 10^{-5}$ Torr) requires a calibration of the MRMS by standard mixtures similar in composition to the sample.

(8) A root mean square deviation of an analysis (1σ) amounts to about $\pm$ 2% for the high resolution stage, ^{3}He $\approx 7 \cdot 10^7$ atoms, and about $\pm$ 1% for the low resolution stage, ^{4}He $\approx 10^{10}$ atoms.

(9) The "memory" of the instrument does not interfere with the measurements when the pressure of any isotope varies over a range of three to four orders of magnitude.

(10) The time required for a routine analysis of a sample enclosed in a glass ampoule placed in a multi-position ampoule-breaker is about 10 min.

The scheme of the mass-spectrometer analyzer is illustrated in Fig. 5.2. Its dimensions are: ρ_1 = 47.5 mm; ρ_2 = 57 mm; ρ_3 = 71 mm; $(2\Delta\rho_1)_{\mathrm{max}}$ = 43

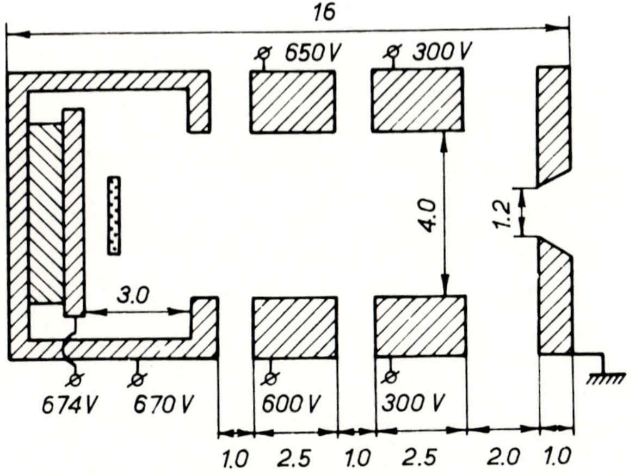

Fig. 5.6. Diagram of the MRMS ion source (Mamyrin et al., 1972a). Dimensions are given in mm, electrode potentials are given relative to ground potential.

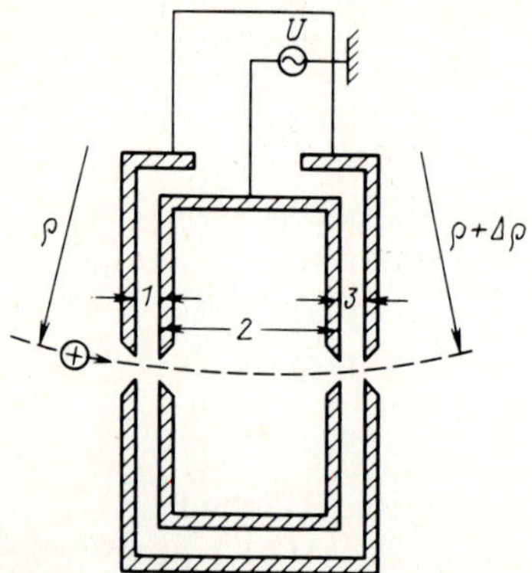

Fig. 5.7. Diagram of the MRMS modulator (Mamyrin et al., 1972a).

mm; Δl_2 = 21 mm; S_1 = 1 mm; S_2 = 1.5 mm; S_3 = 5 mm; S_4 = 7.5 mm; S_5 = 1.2 mm.

The scheme of the ion source is shown in Fig. 5.6. Fig. 5.7 illustrates the modulator. The size of gap *2* is chosen so that the flight time through the middle electrode should be about $T_g/2$ at an employed harmonic number, n, of about 30 and an ion trajectory radius in the drift slit of about 50 mm. The following ratio can be calculated:

$$\frac{\Delta l_2}{2\pi\rho} = \frac{T_c}{2T_g} \approx \frac{1}{2n} = \frac{1}{60} \tag{5.8}$$

Gaps *1* and *3* are approximately equal to S_2.

The amplitude of the modulator h.f. voltage is determined by eq. 5.9 (Mamyrin, 1966):

$$\frac{U_m}{U} \approx \frac{(\Delta\rho_I)_{max}}{\rho_I} \tag{5.9}$$

where U is the accelerating potential of the source, about 700 V.

A scheme of the complete mass spectrometer is shown in Fig. 5.8; details are given in the figure captions.

The device is appropriate for: (1) analysis of helium samples placed in special metal vessels supplied with a valve and a nipple for coupling with the mass spectrometer; (2) analysis of samples contained in flame-sealed glass ampoules placed in ampoule-breakers; (3) continuous analysis of helium released by heating and melting of solid samples in the furnace — the heating temperature can be changed and controlled automatically. Several samples

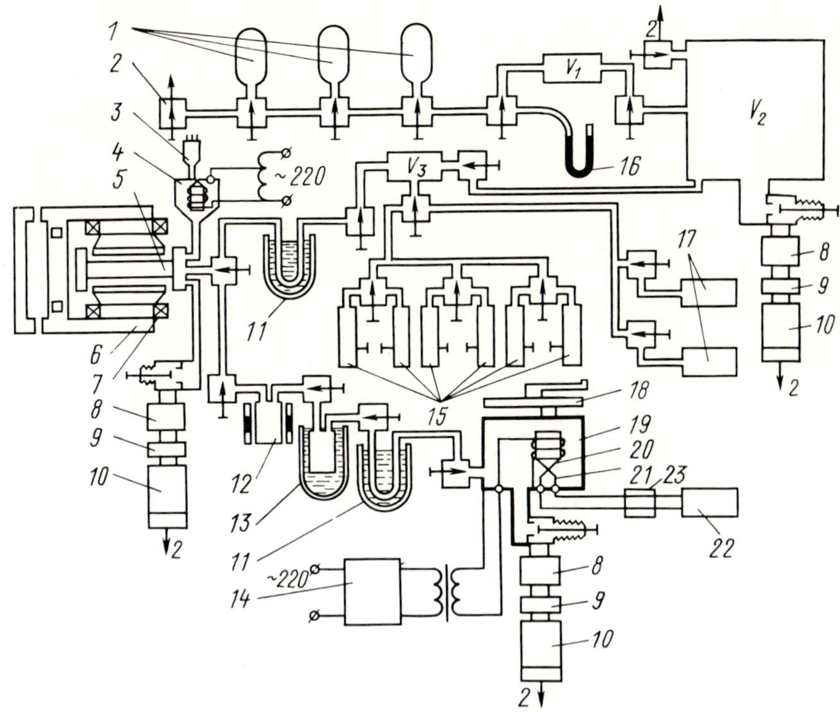

Fig. 5.8. Diagram of MRMS mass-spectrometer system for helium isotopic investigations (Mamyrin et al., 1972a, b).

1 = vessels for storage of standard mixtures; 2 = arrows indicate fore pump; 3 = ionization gauge; 4 = Ti getter coupled with the cold trap; 5 = analyzer chamber; 6 = magnet; 7 = auxiliary magnet coils; 8 = liquid nitrogen trap; 9 = water-cooled trap; 10 = diffusion pump; 11 = liquid nitrogen trap; 12 = heated titanium getter; 13 = trap with activated charcoal; 14 = device for automatic change of temperature rise; 15 = ampoule breakers; 16 = mercury gauge; 17 = portable metal ampoules; 18 = sample loading section; 19 = vacuum container; 20 = crucible and furnace; 21 = thermocouple; 22 = potentiometer; 23 = thermostat for thermocouple ends.

can be dropped in a crucible, one after another without disturbing the vacuum. A more detailed description of the system is given elsewhere (Mamyrin and Khabarin, 1977).

The device provides for the dosage and leak-in of helium standards and samples into the mass-spectrometer chamber.

In addition to dynamic pumping by a diffusion oil pump coupled with a trap filled with liquid nitrogen, a "quasistatic" pumping is provided. Most measurements of small amounts of helium are made under this pumping mode, when the diffusion pump is disconnected and pumping is accomplished by means of sorption on the Ti mirror cooled by liquid nitrogen. The mirror is renewed by titanium spraying when the Ti-coated molybdenum wire is

68

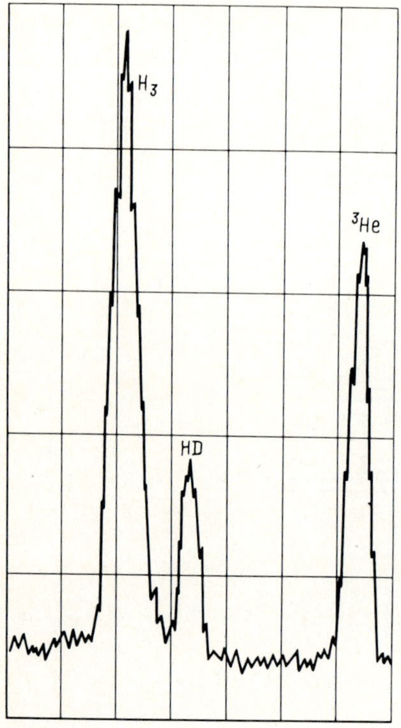

Fig. 5.9. Mass spectrogram of mass 3 multiplet. The residual pressure and the amount of ^{3}He in the analyzer are $2 \cdot 10^{-8}$ Torr and 10^7 atoms, respectively.

heated. The residual pressure under such a pumping mode is about $2 \cdot 10^{-8}$ Torr.

Fig. 5.9 shows the mass spectrum of the multiplet at mass 3 obtained by means of the described apparatus by Mamyrin et al. (1972a).

5.4. Industrial magnetic resonance mass spectrometer MI 9302

The Design Office of Analytical Instrument Engineering, USSR Academy of Sciences, together with the A.F. Ioffe Physico-Technical Institute, has manufactured a magnetic resonance mass spectrometer for two-beam analysis of gas mixtures within a mass number range of 2 to 170, and even to 340 amu at lower accelerating voltage (Anonymous, 1976; Alekseichuk et al., 1979).

A schematic drawing of the analyzer is shown in Fig. 5.10. It differs in some important features from that illustrated in Fig. 5.2. In the MI 9302 analyzer a constant (non-pulsed) ion beam that has passed through the slit of the switched off modulator can be led out to the multiplier input via deflec-

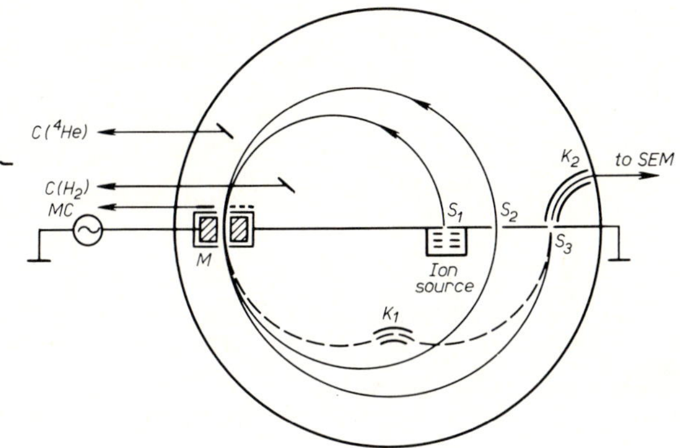

Fig. 5.10. Schematic diagram of mass-spectrometer MI 9302 analyzer (Alekseichuk et al., 1979).
S_1, S_2 and S_3 = ion source, drift and exit slits, respectively; M = three-chamber modulator; K_1 = deflection system; K_2 = lead-out deflection system; MC = movable collector; $C(H_2)$ and $C(^4He)$ = collectors to measure H_2 and ^4He ion currents, respectively.

tion systems K_1 and K_2. Hence a static mode of operation providing a high sensitivity becomes possible; the resolution of such a mode is sufficient for the separation of all adjacent peaks which differ by no less than 1 amu within the mass range of the instrument. The two-beam mode enables one to combine the measurement of: (1) an ion current of a given mass number by means of the high-resolution time-of-flight stage with an electron multiplier on its exit; (2) the ion currents of the ions slightly different in mass which can be led out to a movable collector, the latter being placed to the right or left of the modulator slit by means of a bellows drive. There are two additional fixed collectors which detect the ^4He or the H_2 ion currents when the ^3He or the $HD + H_3$ multiplet ions are measured in accordance with item (1).

The analyzer has accessory slits allowing a change of the resolution:sensitivity ratio.

When the source and exit slits are about 1—2 mm the analytical parameters are similar to those discussed in section 5.3; with narrow slits the best parameters are the following.

(1) The resolving power of the second dynamic high resolution stage is $R_{10\%} \approx 25\ 000$; $R_{10^{-2}\%} \approx 10\ 000$.

(2) The resolving power of the first static low resolution stage is $R_{10\%} \approx 250—350$.

(3) The peak shape is illustrated by Fig. 5.11 (taken at $R_{10\%} \approx 10\ 000$).

(4) The sensitivity threshold is $2\ 10^5$ Ar atoms in the analyzer ($7 \cdot 10^{-15}$ cm^3 of gas under normal conditions; $R_{10\%} \approx 10,000$).

(5) The sensitivity of the low resolution stage is $2 \cdot 10^{-3}$ A Torr^{-1}.

(6) The error of measurement of noble gas isotopic ratios at the static

mode of operation is about $\pm$ 0.2% (1σ); even if the amount of analyzed gas is extremely small ($\sim 10^{-7}$ cm^3 of air argon), the error is low enough ($\sim 1\%$).

To record mass multiplets on tape and/or to display them on the oscilloscope, several methods of mass-spectrum scanning are employed. The use of the oscilloscope makes the adjustment of the mass spectrometer to the required mode of measurement easier and allows fast changes in multiplet mass peak ratios.

The mass spectrometer can operate under three pumping modes — that is: (1) when the analyzed gas continuously passes through the analyzer, pumping is carried out by a diffusion oil pump with a special baffles trap providing a residual pressure of 10^{-8}—10^{-9} Torr; (2) when small amounts of non-sorbing gases (helium, neon) are admitted, the pumping is performed by means of a Ti mirror cooled by liquid nitrogen, the residual pressure being about $2 \cdot 10^{-10}$ Torr; (3) when small amounts of any gas are admitted, the analyzer is disconnected from all pumping means, and the vacuum-proof system provides a residual pressure of about $2 \cdot 10^{-8}$ Torr with the ion source working for one hour (Anufriev et al., 1979).

Electric heating of all the elements of the vacuum system is provided, as well as automatic control of temperature.

A universal leak-in system enables one to prepare precisely measured amounts of a gas, to admit them into the analyzer, to store pure gases and

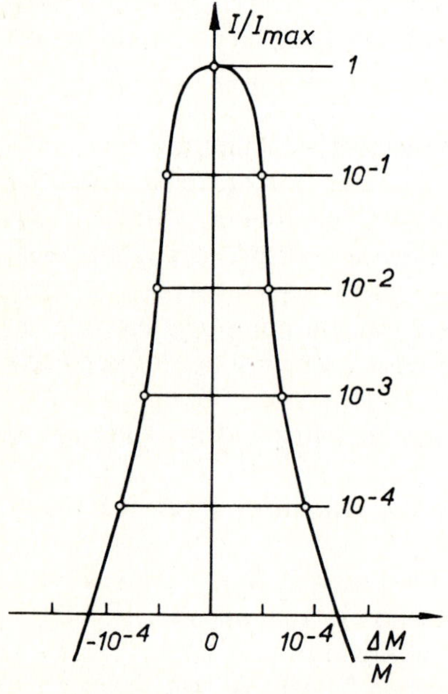

Fig. 5.11. Shape of the mass-spectrum peak in mass spectrometer MI 9302.

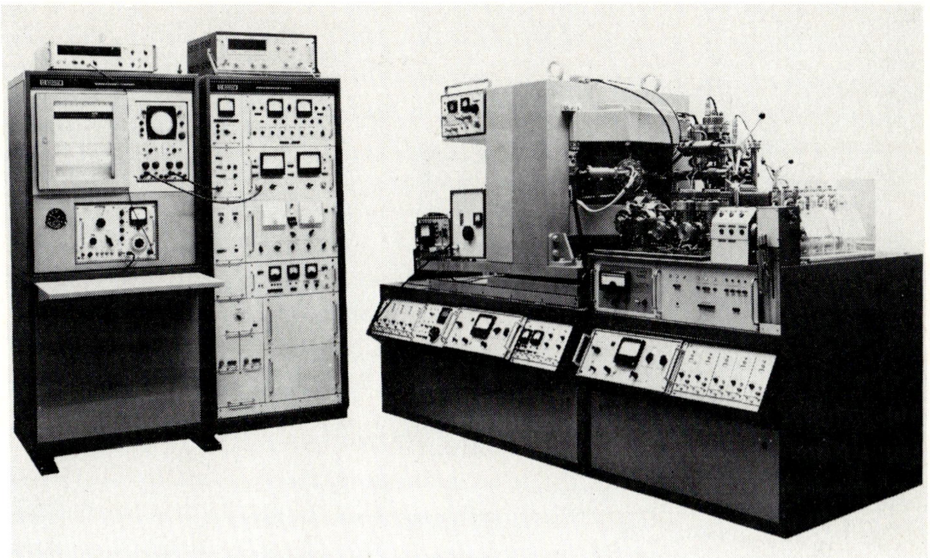

Fig. 5.12. Mass spectrometer MI 9302 (front view).

gas mixtures and to prepare standards with precisely measured component ratios (see Fig. 2.6).

Fig. 5.12 presents a general view of the instrument; a detailed description is given in Anonymous (1976) and in Alekseichuk et al. (1979).

5.5. Solution of the most difficult problems of helium isotope studies achieved with the MRMS

Several thousand isotope analyses of natural helium have been made by means of the MRMS; they are discussed in part II of this book. The following examples illustrate the analytical possiblities of the instrument.

(1) Isotopic analysis of extremely small amounts of helium has been carried out by Boltenkov et al. (1974), who investigated samples of lunar regolith ("Luna-16") on the mass spectrometer discussed in section 5.3. The high sensitivity of the MRMS allowed them to achieve the isotope analysis of separate regolith particles. For example, sample No. 5 is a microcrystal of about $3 \cdot 10^{-4}$ mm^3 (weighing 10^{-6} g). The ^{3}He/^{4}He ratio in the extracted helium was found to be $4.33 \cdot 10^{-4}$, the concentrations of ^{3}He and ^{4}He $15.9 \cdot 10^{-9}$ cm^3 g^{-1} and $36.8 \cdot 10^{-6}$ cm^3 g^{-1}, respectively.

The sensitivity was high enough to obtain outgassing curves for both helium isotopes released through continuous heating of $3 \cdot 10^{-4}$ g of the regolith fraction (Fig. 5.13). It is noteworthy that the incompletely resolved peaks of

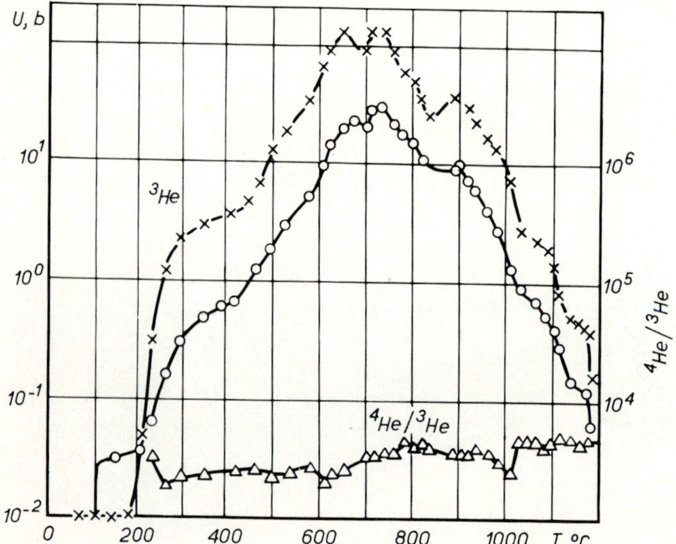

Fig. 5.13. Kinetics of helium release from the magnetic fraction of a lunar soil sample (weight 3 10^{-4} g) (Boltenkov et al., 1974).

the curves visible in Fig. 5.13 were not caused by instability in the operation of the instrument (this was confirmed by an additional checking experiment).

(2) When metal targets are irradiated (that is, in space or in nuclear physical experiments), it is necessary to know the background of the measured isotopes in the target material. There were 10^5—10^6 ^{3}He atoms and $\sim 10^9$ or even fewer ^{4}He atoms in the investigated samples of aluminium and nickel foils. These measurements were successful due to the high sensitivity of the instrument (Boltenkov, 1973; Mamyrin et al., 1972a, 1978).

The measurement of extremely small amounts of ^{4}He requires a separation of multiplet peaks at mass 4, ^{12}C^{3+} and ^{4}He$^+$. Fig. 5.14 shows the multiplet mass spectrum of 10^9 ^{4}He atoms in the analyzer at a residual pressure of about 10^{-8} Torr. Despite a relatively large amount of ^{4}He, it would be impossible to determine without separation of the multiplet peaks, as the error would exceed 100%.

(3) The reproducibility of helium isotope analyses is illustrated by the following example (Mamyrin et al., 1972a). Some samples[1] of natural gases were collected from the same thermal springs of Iceland in different years; they were subjected to gas purification, extraction of a neon/helium mixture and determination of ^{3}He/^{4}He ratios. Table 5.1 shows a good reproducibility of the results, especially if we take into account possible natural variations of

---

[1] Gas samples were collected by V.I. Kononov and B.G. Polyak.

TABLE 5.1

^{3}He/^{4}He ratios from some Icelandic thermal springs

Thermal springs	^{3}He/^{4}He ($\times$ 10^{-5})			
	1970	1972	1973	1976
Deildartunga	2.0	2.10	—	2.17
Reykir Fjoskadalur	—	0.91	0.92	—
Tjorsalaug	—	3.20	3.10	3.23
Raudomeljolkelda	—	—	1.50	1.54
Englandskver	—	—	2.60	2.66

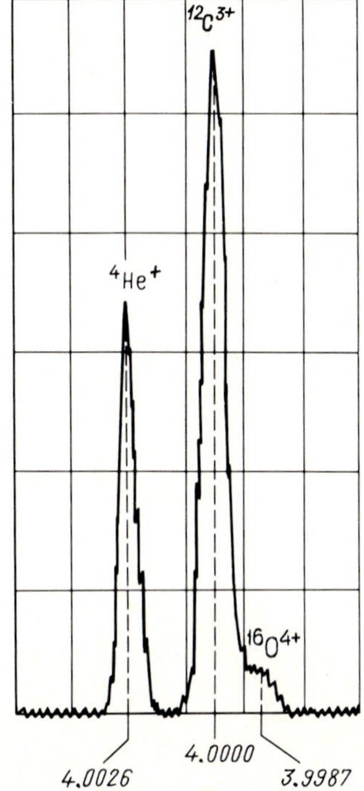

Fig. 5.14. Mass spectrogram of mass 4 multiplet. The amount of ^{4}He in the analyzer is about 10^9 atoms (Boltenkov, 1973).

the ratios, as well as errors caused by the sampling procedure and the preparation of noble gases for the analysis.

(4) A high accuracy of determination of the ^{3}He/^{4}He ratio in natural helium is shown in the works by Mamyrin et al. (1970a, b); these authors were the first to obtain the precise ^{3}He/^{4}He value of $(1.399 \pm 0.013) \cdot 10^{-6}$ for helium of the lower atmosphere. The preparation of mixtures with a known helium

74

isotope ratio presented the main difficulty in this investigation; errors in mass-spectrometric measurements themselves are much smaller than the above-mentioned final uncertainty.

If the ^{3}He/^{4}He ratio in a sample is similar to that in a standard, its value is about 10^{-6}, and in order to decrease the error of measurement to $\pm$ 0.2% it is necessary to take several measurements of one and the same sample.

(5) The lowest ^{3}He/^{4}He ratios were observed in the studies of uranium minerals (Kamensky, 1970) and uranium-rich rocks (Tolstikhin, 1975a).

Fig. 5.15 shows the ^{3}He/^{4}He ratio curve versus the uranium content in a sample. It must be borne in mind that while studying helium from uraninites, the sensitivity of the instrument should be higher than that of the MRMS (Mamyrin et al., 1972a). The possibility to reach such sensitivity has already been mentioned (see section 5.2).

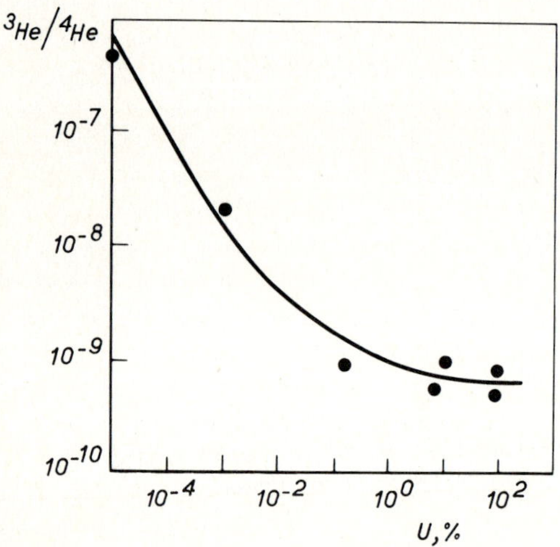

Fig. 5.15. Helium isotope composition in minerals as a function of uranium content (Kamensky, 1970).

Chapter 6

LIGHT NOBLE GAS ISOTOPES IN METEORITES

The three types of nuclear processes known to provide the noble gas isotope composition in space matter (first of all in meteorites) are: element synthesis, radioactive decay and high-energy nuclear reactions. A careful study of the observed isotopic abundance may lead to an estimate of the contribution of each of these processes and extend our knowledge about the origin and history of both meteorites and trapped gases. The importance of the information gained from noble gases stimulates further investigations in this field and accounts for the scores of recently published scientific papers devoted to this problem.

A comprehensive review of these works is beyond the scope of this chapter, it will be confined to a brief discussion of recent results obtained in isotope cosmochemistry of trapped noble gases as they are closely related to the problems considered in the following chapters. Sections 6.1 and 6.2 contain a brief description of meteorites (in the light of their possible contribution to earth matter) and primordial noble gases in the most primitive carbonaceous and enstatite chondrites. Meteorites appear to have supplied the earth with volatiles (section 6.4). Attention is paid to the close genetic relationship between primordial helium and neon resulting in a constant $(^4He/^{20}Ne)_{prim}$ ratio (sections 6.2 and 6.3). Examination of meteorites enables one to determine another ratio of primordial isotopes, namely $(^3He/^4He)_{prim}$, which also varies in a very narrow range; particular attention is given to the most probable value of $(^3He/^4He)_{prim}$ trapped by earth (section 6.3). The very important fact that the $(^3He/^4He)_{prim}$ and $(^4He/^{20}Ne)_{prim}$ ratios are constant, which proves that they are derived from the same source, is frequently used in chapters 7, 10 and 11.

6.1. Meteorites and the earth

The study of meteorites can yield extremely important information on the early solar system, and on proto-matter of the earth in particular. Of the four principle classes of meteorites (chondrites, achondrites, irons and stony irons) chondrites are the most primitive; hence, they contain a proportion of nonvolatile elements that is similar to that of the solar system (Mason, 1962; Anders, 1971). According to more comprehensive classifications (Urey and Craig, 1953; Anders, 1964; Keil and Fredriksson, 1964; Van Schmus and Wood, 1967), chondrites are subdivided into five major groups on the basis of some chemical criteria (such as Fe/Si, Mg/Si, Fe_{metal}/Fe_{total} ratios, etc): carbonaceous chondrite (Fe is practically completely oxidate), ordinary chondrite (further subdivided into three minor types LL, L and H according to the increasing proportion of metallic Fe), and enstatite chondrite ($Fe_{metal}/Fe_{total} \approx 0.8$). Achondrite is a highly differentiated and metamorphosed sort of matter, which appears to have undergone a planet stage of evolution (Mason, 1962).

A comparison of terrestrial and meteoritic matter includes chemical (section 6.1.1), isotope (6.1.2), chronological (6.1.3) and other aspects.

6.1.1. Chemical composition

The fact that major elements in chondritic meteorites and in terrestrial matter have almost the same concentration led to the assumption of the chondritic composition of the earth. This was widely accepted in the period from 1950 to 1965 (Vinogradov, 1959, 1961; Ringwood, 1966, and others). However, some geochemical data were inconsistent with the chondritic model of our planet; in particular, attention was drawn to a very low content of uranium in chondritic meteorites which was hardly enough to provide the abundance of uranium in the crust (Birch, 1958; Smyslov, 1969). A somewhat higher content of heavy radioactive elements in the earth (equal to their content in the Ca-rich achondrite) was proposed to bring the calculated and the observed heat flow into agreement (Wasserburg et al., 1964).

6.1.2. Evidence from isotopic systems

Early isotope data did not contradict the traditional ideas of the chondritic composition of the earth: the abundance of stable isotopes such as sulphur (Trofimov, 1949), carbon (Galimov, 1968) and oxygen ($\delta^{18}O$; Vinogradov et al., 1958), and radioactive isotopes such as potassium (Rik and Shukolyukov, 1954; Burnett et al., 1966) and uranium (Chen and Wasserburg, 1981) in chondritic meteorites and in terrestrial materials was similar. However, isotopic investigations have shown that silicate earth is certainly nonchondritic in composition.

The isotope composition of lead indicates that the value of U/Pb is obviously higher in earth matter than in primitive chondrite (Gast, 1976). The agreement between the calculated and the observed isotopic composition of lead (Fig. 6.1) required a twelve-fold increase of the $^{238}U/^{204}Pb$ ratio in the earth's proto-matter at 4.55 billion years ago, a further three-fold increase of the ratio at 4.45 b.y. ago and an additional small increase (11%) at 2.8 b.y.

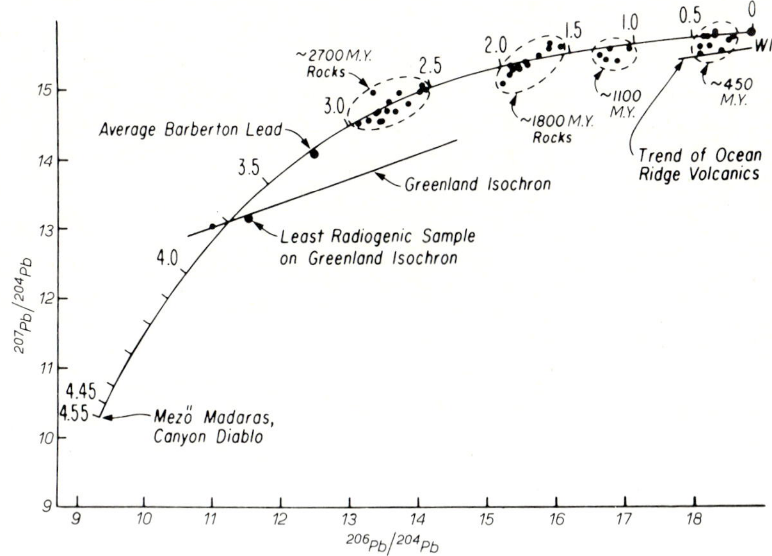

Fig. 6.1. Evolution of the isotopic composition of terrestrial lead (Wetherill, 1975). Measured data are fitted to a model in which the source of the lead undergoes a twelve-fold increase in the $^{238}U/^{204}Pb$ ratio over its primordial value of 0.25, about 4.55 b.y. ago. The ratio again increases to 8.96 about 4.45 b.y. ago and undergoes a further 11% increase about 2.800 b.y. ago. (Reproduced with permission from the *Annu. Rev. Nuclear Sci.*, 25; © 1975 by Annual Reviews Inc.)

(Wetherill, 1975). Similar results can be obtained from other episodic or continuous models, but in any case an early increase in the $^{238}U/^{204}Pb$ ratio is required. A much higher increase is necessary for explaining the lead isotope composition in lunar rocks. The U/K ratio in terrestrial rocks appears to be about 10^{-4}, which is in good agreement with the observed value of $(^4He/^{40}Ar)_{rad}$ in natural gases, whereas in chondritic meteorites the U/K ratio is considerably lower (Wasserburg et al., 1963, 1964).

The discrepancy between the Rb/Sr ratios in the earth's matter and in chondrites — recorded in the isotopic composition of Sr — is clearly shown in Fig. 6.2. The $^{87}Sr/^{86}Sr$ ratios in terrestrial samples are much lower than they were expected to be according to the chondritic model, but these ratios are close to those typical of achondrites (Faure and Powell, 1972).

Comparison of Sm-Nd and Rb-Sr systematics gives a terrestrial Rb/Sr ratio as ten times lower than the chondritic one (DePaolo and Wasserburg, 1976).

The U-He and K-Ar systems are also at variance with the chondritic abundance of potassium. Using the ^{40}Ar content in the atmosphere and the age of the earth, Hurley (1968) and Gast (1968) determined the potassium content in the solid earth to be 85 ppm, which is approximately ten times lower than the chondritic contents. Recent models of the earth outgassing and differen-

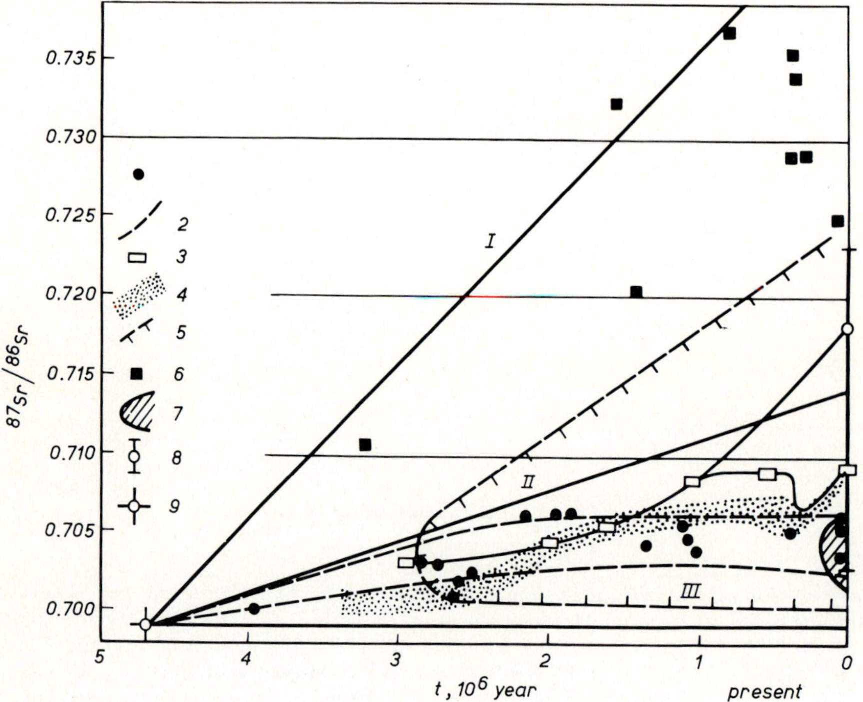

Fig. 6.2. Isotopic composition of terrestrial strontium confirms a low (achondritic) Rb/Sr ratio in the silicate earth (data are taken from Faure and Powell, 1972; Brass, 1976; Veizer and Compston, 1976).
I, *II* and *III* = evolution of isotopic Sr composition in chondrites, Ca-poor achondrites and Ca-rich achondrites, respectively; *1* = initial $^{87}Sr/^{86}Sr$ ratios in mafic rocks of probable mantle origin; *2* = assumed evolution of the ratio in the earth mantle; *3* = measured $^{87}Sr/^{86}Sr$ ratios in carbonates; *4* = ratios in ocean water; *5* = area of initial $^{87}Sr/^{86}Sr$ ratios in granitic rocks (about 120 values); *6* = several high ratios lying beyond boundary *5*; *7* = ratios typical of oceanic, circum-oceanic and island-arc basalts, andesites and dacites (more than 250 analyses); *8* = ratios in contemporary lakes and rivers (about 60 analyses); circle shows the ratio of $^{87}Sr/^{86}Sr$ = 0.718 assumed as an average value for the earth's crust; *9* = initial ratio of $^{87}Sr/^{86}Sr$ = 0.6989 (BABI).

tiation show that volatiles and radioactive elements (U, K, Rb) were mainly released by the mantle and accumulated in the crust and the atmohydrosphere. The K-Ar ages of crustal rocks are much lower than the age of the earth and, consequently, crustal radiogenic argon was also outgassed into the atmosphere. On the basis of these speculations the potassium abundance in the earth was estimated to be 150—200 ppm (Larimer, 1971; Tolstikhin, 1975b; Tolstikhin et al., 1975), which is significantly lower than in chondrites.

Investigations of the isotopic composition of oxygen (Clayton et al., 1976; Clayton and Mayeda, 1978) lent support to the achondritic model of the sili-

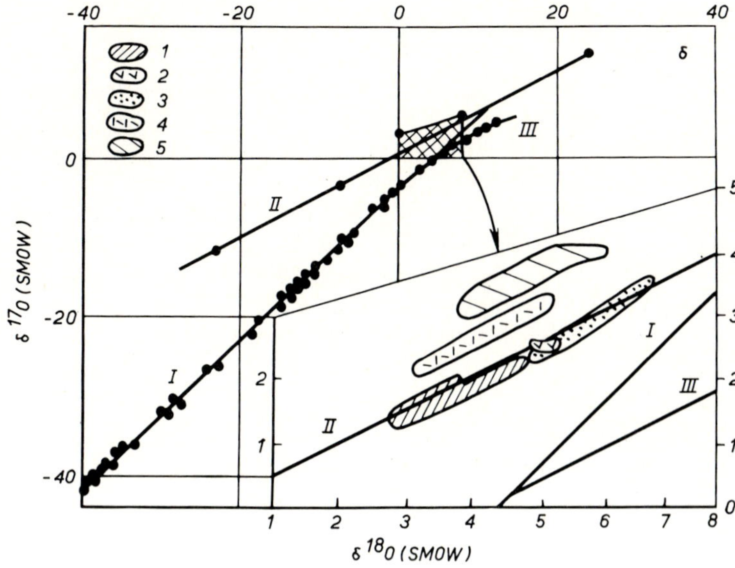

Fig. 6.3. Isotopic composition of terrestrial oxygen (line *II*) is similar to that of achondrites (*1*), enstatite chondrites (*2*) and some types of stony irons (*3*), but differs from carbonaceous (lines *I* = C1 and C2 and *III* = C3) and ordinary (*4* = H and *5* = L and LL) chondrites (data are taken from Clayton et al., 1976; and Clayton and Mayeda, 1978).

cate earth: only achondrites, enstatite chondrites, and stony-iron meteorites approach the line of mass fractionation of terrestrial materials (Fig. 6.3).

Thus, the achondritic composition of silicate shells of the earth appears to be more consistent with the isotopic systems studied than the chondritic composition. It should be noted that achondritic matter is impoverished in volatiles and does not contain planetary noble gases which are typical of the atmospheres of the terrestrial planets. Moreover, this matter is not rich in iron and the formation of the core required a contribution of iron and/or stony-iron meteorites. Correspondingly, matter of the chondrite-like type enriched in volatiles, as well as the achondritic and the stony-iron types, might have taken part in the accretion of terrestrial planets. A small amount of isotopically anomalous matter which formed before the condensation within the solar system might also have participated in this process (Clayton et al., 1976; Rees and Thode, 1977; Clayton, 1978; Podosek, 1978; McCulloch and Wasserburg, 1978).

6.1.3. Evidence from cosmo- and geochronology

Numerous age measurements of meteorites made by various methods suggest a coetaneous formation of meteorites 4.5—4.7 b.y. ago (Anders, 1964;

Wasserburg and Burnett, 1969; Faure and Powell, 1972; Wetherill, 1975; Kirsten, 1976; and others). This time interval is typical of all classes of meteorites, such as carbonaceous and ordinary chondrites, achondrites and stony irons (Wetherill, 1975).

Similar ages are reported (Fig. 6.4) for the most ancient lunar materials (Wasserburg et al., 1977). There is a very good correlation of ^{87}Sr/^{86}Sr and ^{87}Rb/^{86}Sr ratios (model age 4.6 b.y.) for highly varied samples from lunar soils of all lunar missions (Fig. 6.4, left). The model age of lunar rhyolite 12013 obtained for "total rock" chips is 4.52 b.y.; the internal isochron for a dunite clast indicated a very high age and a low initial ^{87}Sr/^{86}Sr ratio (Fig. 6.4, right). The Rb/Sr whole rock isochron for lunar anorthosites indicates that anorthositic crust was formed 4.6 b.y. ago (Kirsten, 1976). The Pb-U systematics also show that the early global differentiation of the lunar crust occurred between 4.3 and 4.5 b.y. ago (Wasserburg et al., 1977).

The oldest known terrestrial rocks were formed 3.8 b.y. ago (Black et al., 1971; Moorbath et al., 1972; Basu et al., 1981). The model age of the earth was reported to be 4.5 b.y. (Russell and Reynolds, 1965) or ranging from 4.35 to 4.5 b.y. (Wetherill, 1975).

The time interval of the accretion of terrestrial planets is estimated to be about 100 million years or less (Safronov, 1969; Weidenschilling, 1976; Safronov and Kozlovskaya, 1977; and others).

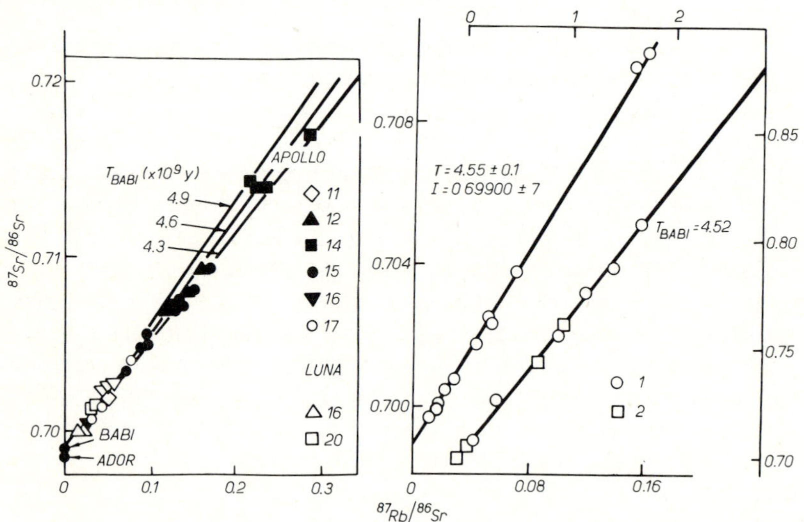

Fig. 6.4. Some highly differentiated lunar samples which show very old ages. Left: model Rb/Sr age of lunar soil sampled by all Apollo and Lunar missions approaches 4.6 b.y. Centre: Rb/Sr age for dunite clast 72417 is about 4.55 ± 0.1 b.y. Right: model age for rock 12013 (enriched in radioactive elements) is about 4.52 b.y. Legend: 1 = fragments; 2 = minerals. (From Wasserburg et al., 1977.)

These radiometric chronology data led to the conclusion that meteorites can be considered as a probable proto-matter of terrestrial planets.

6.2. Primordial noble gases in meteorites

Primordial (or trapped) noble gases were discovered by Gerling and Levsky (1956) in the achondrite Staroe Pesyanoe and later in many other meteorites. A large number of publications has been devoted to the problem of trapped noble gases in meteorites since. The most significant results have been described by Signer and Suess (1963), Pepin and Signer (1965), Mazor et al. (1970), Shukolyukov and Levsky (1972), Anders (1981) and others.

In discussing the origin of trapped noble gases, meteorites can be divided into three groups: (1) gas-rich meteorites (GRM); (2) carbonaceous chondrites (CC); and (3) ordinary chondrites (OC). In this section we review some regularities in the proportions of trapped light noble gases; the isotopic problems are discussed in section 6.3.

6.2.1. Noble gases in gas-rich meteorites

GRM are characterized by the following features:

(1) The relative abundance of rare gases in these meteorites is similar to that of lunar soil (Fig. 6.5) and to solar abundance. The gas content is usually high and approaches that of lunar soil (Fig. 6.6).

(2) GRM do not belong to meteorites of one particular type (Pepin and Signer, 1965); they include carbonaceous and ordinary chondrites as well as highly metamorphosed achondrites. The same is true for lunar soil: a high content of trapped gases is observed in particles differing in chemical composition.

(3) A distinct reverse correlation between the content of trapped gases and the grain sizes observed in some GRM suggests that trapped gases are concentrated on the surface of individual grains (Eberhardt et al., 1966). A similar distribution of gases is reported for lunar soil (Eberhardt et al., 1970; Eugster et al., 1973). Grains of GRM as well as grains of lunar soil have anisotropically distributed tracks suggesting irradiation on the surface of a parent body (Rajan, 1974).

(4) The ratio of $^4He/^{20}Ne$ in metal foils exposed on the moon surface (Geiss, 1976) and on space crafts (Gartmanov, 1974) is practically equal to that of GRM and lunar soil (see Fig. 6.6).

These data suggest that trapped noble gases were accumulated in GRM due to the implantation of solar wind or solar flare ions.

6.2.2. Noble gases in carbonaceous chondrites

CC as a rule show quite a different distribution of the trapped component.

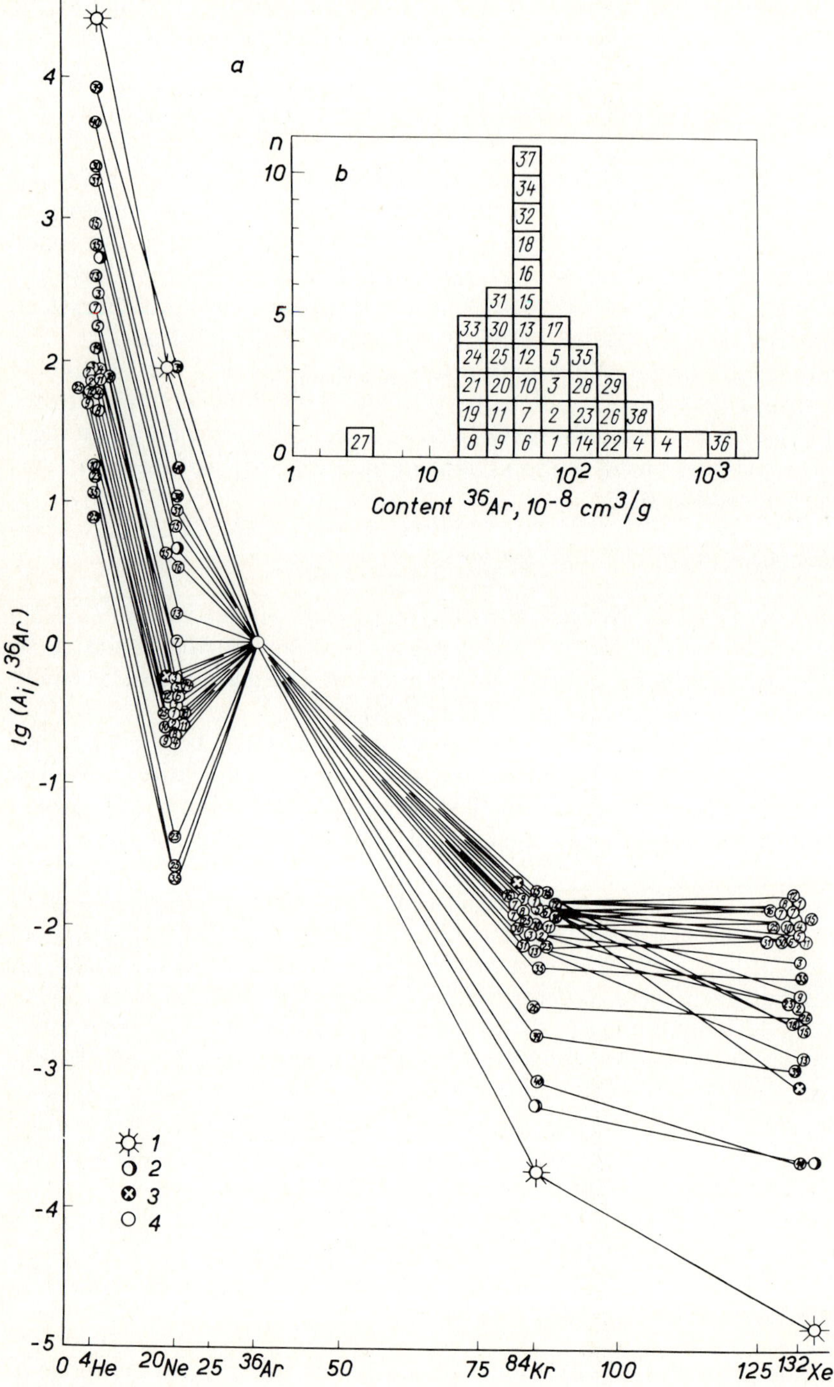
a
b
n
10
5
0
37
34
32
18
16
31 15
33 30 13 17
24 25 12 5 35
21 20 10 3 28 29
19 11 7 2 23 26 38
27
8 9 6 1 14 22 4 4 36
1 10 10^2 10^3
Content ^36Ar, 10^-8 cm^3/g
lg (A_i / ^36Ar)
4
3
2
1
0
-1
-2
-3
-4
-5
1
2
3
4
0 ^4He ^20Ne 25 ^36Ar 50 75 ^84Kr 100 125 ^132Xe

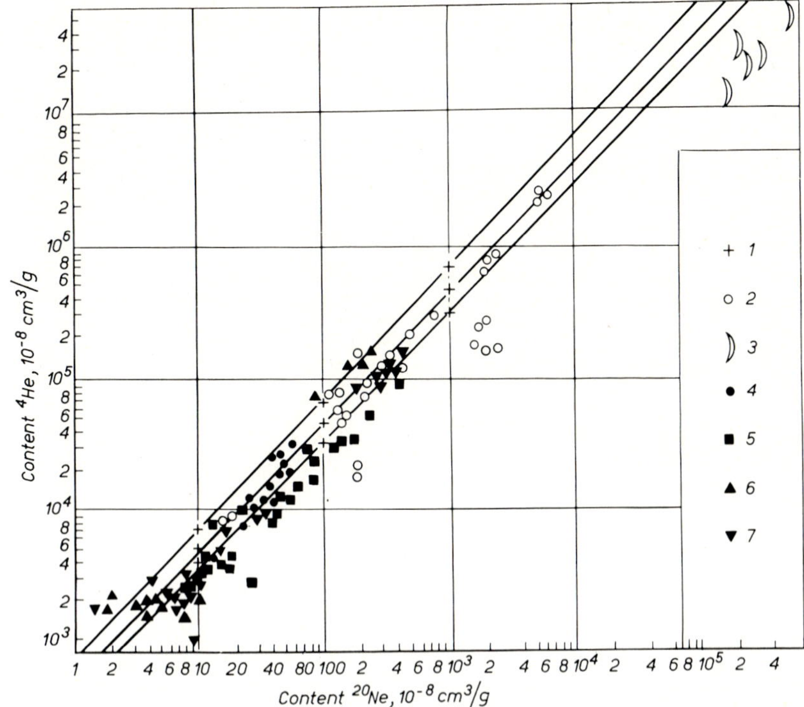

Fig. 6.6. Relationship between concentrations of the trapped light noble gases, He and Ne, in various cosmic matter.

1 = contemporary solar wind; 2 = meteorites rich in trapped noble gases; 3 = lunar soil. Carbonaceous chondrites: 4 = type 1; 5 = type 2; 6 = type 2 (Orgueil); 7 = type 3 (Virginia). (Data from Pepin and Signer, 1965; Eberhardt et al., 1970; Mazor et al., 1970; Geiss, 1976.)

Fig. 6.5. Relative abundance of trapped noble gases in meteorites (a) and concentrations of ^{36}Ar (b).

1 = cosmic abundance; 2 = lunar soil rich in noble solar gases; 3 = terrestrial atmosphere; 4 = meteorites (see the list below). Carbonaceous chondrites: 1 = Alais (C1); 2 = Ivuna (C1); 3 = Orgueil (C1, total); $3M$ = Orgueil (magnetite); 4 = Orgueil (silicate fraction); 5 = Tonk (C1); 6 = Al Rais (C2); 7 = Boriskino (C2); 8 = Cold Bokkeveld (C2); 9 = Erakot (C2); 10 = Essebi (C2); 11 = Haripura (C2); 12 = Mighei (C2); 13 = Murray (C2, silicate fraction); 14 = Murray (C2, total); 15 = Nawapali (C2); 16 = Nogoya (C2); 17 = Pollen (C2); 18 = Santa Cruz (C2); 19 = Allende (C4); 20 = Bali (C4); 21 = Coolidge (C4); 22 = Efremovka (C3); 23 = Felix (C3); 24 = Grosnaja (C3); 25 = Kaba (C2); 26 = Kainsaz (C3); 27 = Karoonda (C4); 28 = Lance (C3); 29 = Leoville (C3); 30 = Mokoia (C2); 31 = Pseudo St. Caprais (C2); 32 = Ornans (C3); 33 = Renazzo (C2); 34 = Vigarano (C3); 35 = Warrenton (C3); 36 = Dyalpur; 37 = Goalpara; 38 = Novo Urei. Meteorites rich in noble gases: 39 = Fayetteville; 40 = Staroe Pesyanoe (orbite). (Data from Pepin and Signer, 1965; Eberhardt et al., 1970; Mazor et al., 1970; Cameron, 1973.)

84

(1) The relative abundance of heavy noble gases (Xe and Kr) in CC is significantly higher than that in solar gases (see Fig. 6.5): $(^{84}Kr/^{36}Ar)_{CC} > (^{84}Kr/^{36}Ar)_{GRM}$; $(^{132}Xe/^{36}Ar)_{CC} \gg (^{132}Xe/^{36}Ar)_{GRM}$. The concentration of heavy gases in CC is also higher than in GRM. If we turn to light gases, the relative abundance of neon and helium would be much lower as compared with that of solar gases. Indeed, ^{20}Ne is less abundant in CC than ^{36}Ar, but the behaviour of helium in these meteorites has nothing in common with that of heavier gases: concentrations of trapped helium and neon (despite their wide variations, namely about five orders of magnitude) are perfectly correlated in meteorites of all types (CC, OC and GRM), the lunar soil and the solar wind (see Fig. 6.6). Somewhat lower $^{4}He/^{20}Ne$ ratios in lunar samples are due to the saturation of surface layer grains with solar helium.

(2) The described distribution of gases typical of chondritic meteorites is provided by a specific process of fractionation and trapping of light and heavy noble gases in the solar nebula as well as by the contribution of various host phases in a meteorite. Therefore, when Heymann and Mazor (1967) separated the CC Nogoya into two fractions, they observed that the dark fractions contained mostly light noble gases whereas the light fraction was enriched in heavy gases. Later Jeffry and Anders (1970) investigated the mineral fractions of the CC Orgueil. Heavy gases were concentrated in the silicate fraction and light gases (He, Ne and partially Ar) in magnetite; the $^{4}He/^{20}Ne$ ratio in magnetite was very close to the cosmic ratio and consistent with the relationship illustrated by Fig. 6.6.

Results reported for the same meteorite by Eberhardt (1974) show that in various silicate fractions the content of light gases ranges within one order of magnitude whereas the $^{4}He/^{20}Ne$ ratio varies by as little as 30%.

Further careful investigations performed by Prof. Anders and his co-workers in Chicago, Prof. Reynolds and his collaborators in Berkeley and Prof. Eberhardt and his co-workers in Bern (Herzog and Anders, 1974; Lewis et al., 1977; Srinivasan et al., 1977; Reynolds et al., 1978; Eberhardt et al., 1979; Alaerts et al., 1980; Anders, 1981) resulted in a very important discovery: almost all the primordial gases are concentrated in extremely rare minerals (host phases) whose total content in a meteorite is less than 0.1% by weight. An exact chemical identification of the host phases has not been obtained yet. One of the host phases, named Q, does not only contain the highest concentrations of all noble gases, but also serves as a major reservoir of heavy gases; another mineral (sulphide?) is a major carrier of He and Ne (Lewis et al., 1977; Anders, 1981). A more comprehensive discussion of the problem is beyond the scope of this book and we refer the reader to the original papers cited above and/or to the recent review by Anders (1981). Some results of investigations dealing with the distribution of helium and neon isotopes are presented in section 6.3.

For a further discussion we must focus our attention upon the $(^{4}He/^{20}Ne)_{trapped}$ ratio in the host phases of CC. There is no clear relationship between this ratio and the proportion of minerals-carriers of light and heavy gases indicated by the values of $^{20}Ne/^{132}Xe$ or $^{20}Ne/^{84}Kr$ (Fig. 6.7). High $^4He/^{20}Ne$ ratios do not go together with high ratios of $^{20}Ne/^{132}Xe$ ($^{20}Ne/^{84}Kr$). On the contrary, minerals characterized by a low proportion of light and heavy gases (similar to atmospheric) show high and slightly modified $(^4He/^{20}Ne)_{prim}$ ratios. It should be noted that, while the $^4He/^{84}Kr$ ratio varies widely (about two orders of magnitude), the $(^4He/^{20}Ne)_{prim}$ ratio is far less variable. The modal $(^4He/^{20}Ne)_{prim}$ value in host phases is approximately 300. However, bearing in mind (a) the very complicated history of meteorites (McSween, 1979), (b) the K-Ar ages of CC (Mazor et al., 1970) suggesting losses of radiogenic ^{40}Ar and other gases, and (c) the slightly preferential losses of ^4He as compared with ^{20}Ne, we can conclude that the initial $(^4He/^{20}Ne)_{prim}$ ratios might have been somewhat higher than the observed values, and we adopt a $(^4He/^{20}Ne)_{prim}$ value of about 500 for the following discussion. It is slightly lower than that inferred by Cameron (1973) from the cosmic abundance of elements.

(3) The temperature of ^{20}Ne release from CC determined under step-wise heating is more similar to that of ^4He than that of ^{36}Ar and heavier gases. This fact implies a similar position of trapped ^4He and ^{20}Ne in host phases. Fig. 6.8 illustrates the results of step-wise heating of the CC Nogoya (Black, 1972) and Orgueil (Herzog and Anders, 1974); similar results have been obtained by Eberhardt (1978) and others.

^{36}Ar appears to occupy an "intermediate" position and its losses are partially

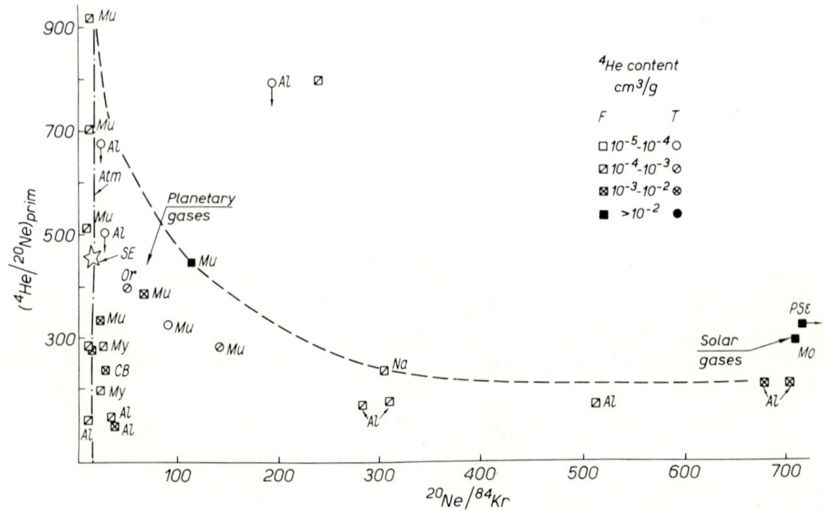

Fig. 6.7. Showing no clear relationships between the ratio of $(^4He/^{20}Ne)_{prim}$ and the ratio of light and heavy noble gases, $^{20}Ne/^{84}Kr$, in meteorites. Meteorites: *Al* = Allende; *CBo* = Cold Bokkeveld; *Mo* = Mokoia; *Mu* = Murchison; *My* = Murray; *Na* = Navapali; *Or* = Ornans; *PSC* = Pseudo St. Caprais. *F* = mineral fractions; *T* = total samples. (Data from Mazor et al., 1970; Bogard et al., 1971; Lewis et al., 1977; Srinivasan et al., 1977; Reynolds et al., 1978.)

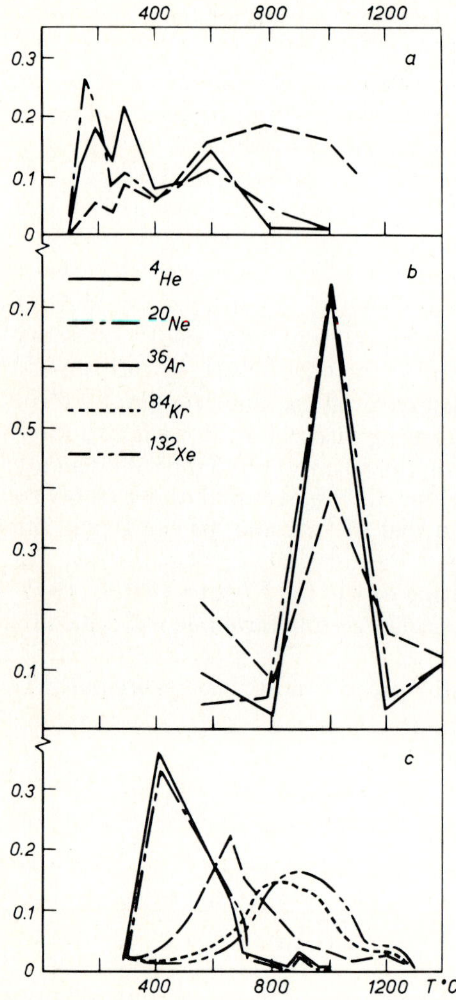

Fig. 6.8. Results of step-wise heating experiments with the meteorites Nogoya (a), a fraction of Orgueil (b) and lunar soil (c); vertical axis shows proportion of gas releases. (Data from Black, 1972; Vinogradov and Zadorozhny, 1972; Herzog and Anders, 1974.)

associated with light gases but mostly with heavier ones. It is noteworthy that the ratios of the square root of mass numbers are equal to:

$$\sqrt{M(^4\mathrm{He})} : \sqrt{M(^{20}\mathrm{Ne})} : \sqrt{M(^{36}\mathrm{Ar})} = 2 : 4.7 : 6$$

If the sites of gases were more or less similar and the losses were provided by diffusion, then $^{20}\mathrm{Ne}$ would have been released mostly together with $^{36}\mathrm{Ar}$ (but not with $^4\mathrm{He}$).

Finally, it should be emphasized that (a) the relative abundance of heavy noble gases (Xe and Kr) in CC is significantly higher than that in solar gases; (b) the content of ^{20}Ne is, as a rule, somewhat lower than that of ^{36}Ar; (c) the ratio of ^{4}He/^{20}Ne is practically constant and similar to that of GRM (see Fig. 6.6).

6.2.3. Noble gases in ordinary chondrites

The content of trapped noble gases in ordinary chondrites (including those which preserve gases well and are characterized by a K-Ar age approaching that of the solar system) is considerably lower than that in CC, and to supply the earth and other planets with noble gases ordinary chondrites must have been the major contributors to silicate planetary matter. This is, however, not in conformity with the data discussed in section 6.1.

In the following discussion we shall use the term "primordial" to denote gases trapped during the earlier stages of the evolution of the solar system (before the accumulation of the planets) and the term "planetary" to denote fractionated primordial gases trapped by chondritic meteorites.

6.3. Isotopes of light noble gases in meteorites

Now we turn to the problem of isotope composition of helium trapped by the earth. Because of helium escape from the terrestrial atmosphere the initial abundance of helium isotopes can be determined only on the grounds of the He-Ne correlation in meteorites as well as the amount of primordial isotopes of Ne in the atmospheres.

6.3.1. Neon isotopes in carbonaceous chondrites

Reynolds and Turner (1964) were the first to use a two-component diagram with coordinates of ^{20}Ne/^{22}Ne versus ^{21}Ne/^{22}Ne and distinguished planetary and cosmogenic components of neon. Later the neon of GRM was also plotted on the diagram and "the neon triangle" was suggested. The triangle allowed us to distinguish the solar (B, ^{20}Ne/^{22}Ne = 13, ^{21}Ne/^{22}Ne = 0.036), the planetary (A, ^{20}Ne/^{22}Ne = 8.2, ^{21}Ne/^{22}Ne = 0.024), and the spallogenic (C, ^{20}Ne/^{22}Ne = 0.9, ^{21}Ne/^{22}Ne = 0.9) components (Fig. 6.9) and it began to be widely used for classifying neon in meteorites (Pepin, 1967; Mazor et al., 1970; Shukolyukov and Levsky, 1972). However, later it was found that some experimental data were inconsistent with the three-component diagram.

(1) Some meteorites were shifted below line AC of the "neon triangle" and the isotopic ratios of type A neon determining this line were found to be dubious. Rather low ratios of ^{20}Ne/^{22}Ne were measured in neon released under step-wise heating of silicate fractions of CC Orqueil (Herzog and Anders, 1974).

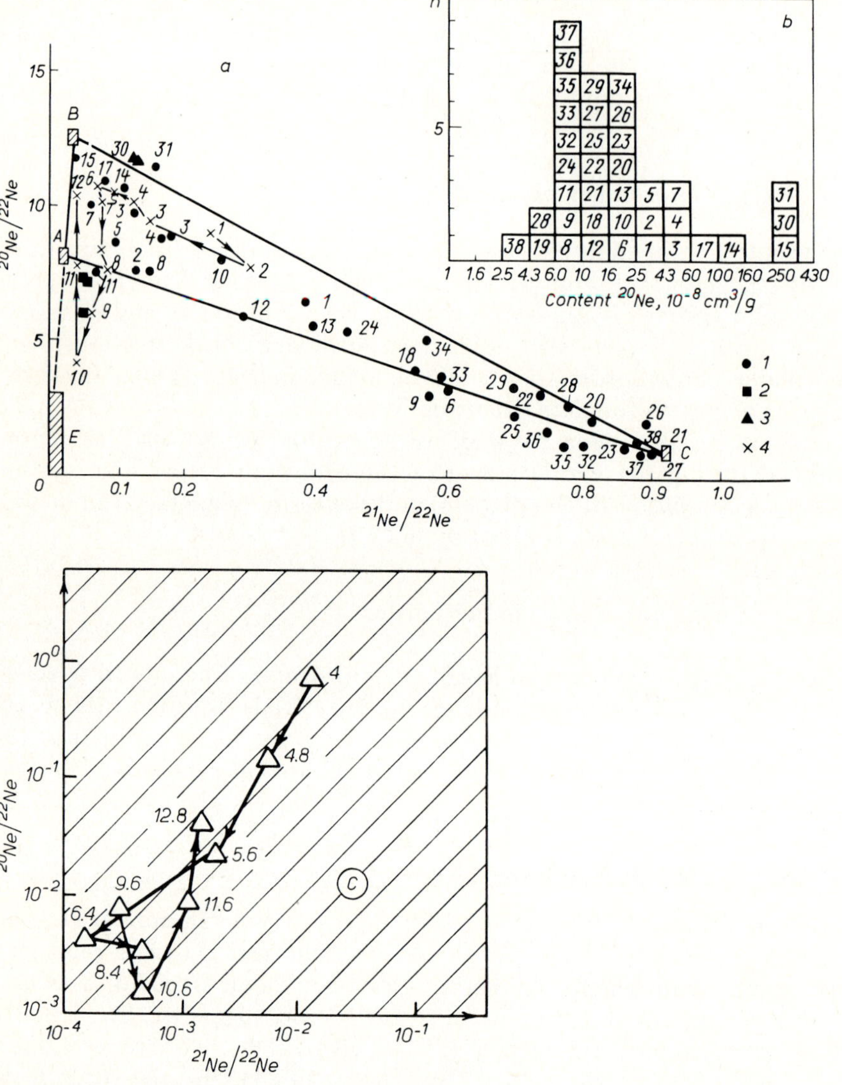

Fig. 6.9. Neon isotopes in carbonaceous chondrites.
a. The ratio of $^{20}Ne/^{22}Ne$ versus $^{21}Ne/^{22}Ne$ in meteorites: *1* = carbonaceous chondrites (numbers are meteorite names, see Fig. 6.5); *2* = silicate fractions of Orgueil; *3* = magnetite of Orgueil; *4* = results of step-wise heating of the meteorite (numbers near the symbols are proportional to the temperature, $\times$ 100°C); *B* = solar neon; *C* = spallogenic neon; *A* = planetary neon; *E* = E-type neon (see text).
b. Concentrations of ^{20}Ne in meteorites (numbers are meteorite names, see Fig. 6.5); *n* = number of cases.
c. Results of step-wise heating experiments with Ne-E rich fraction of Orgueil meteorite. (Data from Jeffery and Anders, 1970; Mazor et al., 1970; Black, 1972; Jungck and Eberhardt, 1979).

(2) Some meteorites contained heavy noble gases of the A type whereas their neon was attributed to the B type (according to its location in the diagram).

(3) Very important results were described by Black and Pepin (1969): isotopic ratios of $^{20}Ne/^{22}Ne \approx 3$, $^{21}Ne/^{22}Ne \approx 0.1$ were observed in high-temperature fractions of neon released from some carbonaceous chondrites. Such neon was denoted as neon-E. Later Black (1972) confirmed these results.

The isotope composition and origin of Ne-E were carefully investigated by Eberhardt (1974, 1978), Jungck and Eberhardt (1979), Lewis et al. (1977) and Meier et al. (1980). Two distinct host phases of Ne-E were found to be present both in Murchison (C2) and Orgueil (C1). One of these host phases (density < 2.3 g cm^{-3}) contained Ne-E$_L$ ($^{20}Ne/^{22}Ne < 0.01$; $^{21}Ne/^{22}Ne < 0.001$) released at a low temperature (500—700°C), another (density ~ 3.5 g cm^{-3}) contained Ne-E$_H$ ($^{20}Ne/^{22}Ne < 0.2$; $^{21}Ne/^{22}Ne < 0.003$) released at a high temperature (1100—1400°C). Ne-E is not associated with other noble gases and it has probably been produced by the β-decay of ^{22}Na.

The new neon triangle *BEC* includes all cosmic objects, and every known natural composition of neon isotopes may be interpreted as a mixture of the B, C and E types. Hence doubt arose as to the existence of Ne-A as a separate type of primordial neon and Black (1972) suggested that Ne-A might be a mixture of other types (Ne-B and Ne-E, for instance) formed directly in grains of CC. Later this suggestion was rejected on the grounds of numerous step-wise heating experiments because the isotope composition of neon in many temperature fractions was very similar to that adopted for neon-A. For example, more than 60% of neon released from Orgueil at 700—1000°C showed practically the same isotopic composition as in Ne-A (Eberhardt, 1978). Moreover, in some carbonaceous chondrites the total Ne is of a Ne-A like isotope composition. These data indicate a single source of Ne-A. The share of Ne-A (probably fractionated solar neon) in the solar nebula appears to have been sufficient to make it a gas reservoir from which primordial planetary gases have (partially) been derived. The mixing of Ne-B (Ne-A) and Ne-E results in the dependence between the ^{20}Ne content and the $^{20}Ne/^{22}Ne$ ratio (Fig. 6.10): a decrease in neon (A, B) content in a meteorite leads to a larger proportion of a high-temperature fraction of Ne-E, namely to lower ratios of $^{20}Ne/^{22}Ne$. A division of the ^{20}Ne content by a more or less constant value of the ^{36}Ar content (see Fig. 6.5) changes the scale on Fig. 6.10 alone. So, the relationship between the ratios of $^{20}Ne/^{36}Ar$ and $^{20}Ne/^{22}Ne$ discussed by Pepin and Signer (1965), Zähringer (1968), Mazor et al. (1970) and others seems farfetched; in fact, there is no connection between isotope abundance of neon and argon.

To conclude section 6.3.1, we can distinguish the following components of neon in carbonaceous chondrites: (1) presolar Ne-E, produced by ^{22}Na decay and preserved in ancient grains of presolar matter; (2) solar Ne-B, solar wind and solar flare ions implanted in solid matter; (3) planetary Ne-A trapped

90

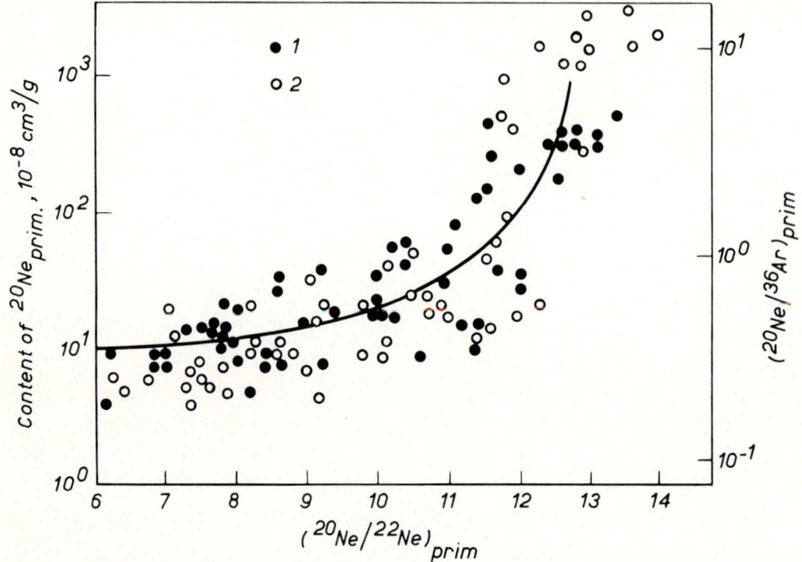

Fig. 6.10. Relationship between the ratio of $(^{20}Ne/^{22}Ne)_{prim}$ in carbonaceous chondrites and: 1 = concentration of primordial ^{20}Ne; 2 = the ratio of $(^{20}Ne/^{36}Ar)_{prim}$ (Tolstikhin and Khabarin, 1976).

by meteorites from a primordial gas reservoir; and (4) spallogenic Ne-C, produced in meteorites due to nuclear interaction between high-energy galactic nuclides and meteorite matter.

6.3.2. Helium isotopes in carbonaceous chondrites

Now we will turn to the isotope composition of primordial helium, bearing in mind the helium—neon correlation observed in all known objects of the solar system.

Early investigations of iron and stone meteorites led to the discovery of extremely high ratios of $^3He/^4He$, $\sim 10^{-1}$ (Paneth et al., 1952, 1953; Gerling and Levsky, 1956). It was established that helium enormously enriched in light isotopes was produced by nuclear reactions between high-energy galactic rays and meteoritic matter and was associated with spallogenic Ne-C. Moreover, the ratio of spallogenic isotopes of $(^3He/^{21}Ne)_{spall}$ in stony meteorites was found to be more or less constant and equal to about 5. A review of the origin, history and application of spallogenic Ne and He is beyond the scope of this chapter because atmospheres and probably solid matter of planets are completely free of the spallogenic component (see section 7.3).

Another unquestionable component is solar helium associated with solar Ne-B. Numerous analyses of solar gases carried out for GRM, lunar soil, as well as experiments with metallic foils exposed on the moon surface or on space crafts, show a constant $^3He/^4He$ ratio in solar helium (Table 6.1). Its average value can be assumed to be $3.5 \cdot 10^{-4}$. It should be noted that the samples listed in Table 6.1 are characterized by concentrations of $^{20}Ne > ^{36}Ar$ and the ratio of $^{20}Ne/^{22}Ne \approx 12-13$ — that is, by typical solar values.

The existence of a planetary component of helium, the He-A, is less certain than the existence of He-B and He-C. Anders et al. (1970) described a relation between $^3He/^4He$ and $^{20}Ne/^{22}Ne$ ratios: as for planetary Ne-A ($^{20}Ne/^{22}Ne = 8.2$), its $^3He/^4He$ ratio was found to be $1.4 \cdot 10^{-4}$, which was also suggested as a typical one of planetary helium. Due to the low content of primordial gases, the low values of K-Ar ages (suggesting significant losses of gases), and the considerable corrections for spallogenic isotopes, the relationship did not seem convincing enough and Tolstikhin and Khabarin (1976) preferred to use the solar value of $^3He/^4He = 3 \cdot 10^{-4}$ as a limiting factor in their model. Meanwhile, Anders et al. (1970) and Geiss (1976) discussed the possibility of an increase of the $^3He/^4He$ ratios in solar helium from the low

TABLE 6.1

Isotope composition of primordial helium in carbonaceous chondrites enriched in gases (Mazor et al., 1970)

Meteorite	^4He content $(10^{-6}$ cm^3 g$^{-1})$	$^3He/^4He$ ratio ($\times 10^{-4}$)	
		measured	corrected
Nawapali (C2)	285	3.2	3.2
	314	3.7	3.8
	483	3.3	3.3
Murray (C2)	312	4.2	3.6
Mokoia (C2)	1075	5.9	4.2
	990	6.0	4.0
Pseudo St. Caprais (C2)	1150	5.2	4.6
	1510	4.1	4.1
Nogoya, dark fraction (C2)	483	3.3	3.3
Lunar soil, Sea of Fertility	18 000	3.7	
	78 200	2.9	
Lunar soil, Sea of Tranquillity	64 000—124 000	3.8—4.3	
Grains of ilmenite, the same place	108 000—670 000	3.7—4.0	
Staroe Pesyanoe, achondrite	6300	2.5	
Kapoeta, achondrite	1400	2.7	
Fayetteville, chondrite	22 000	2.8	
Solar wind		4.25	

For comparison, similar data for lunar soil (Eberhardt et al., 1970; Vinogradov and Zadorozhny, 1972), gas-rich meteorites (Pepin and Signer, 1965; Shukolyukov and Levsky, 1972) and contemporary solar wind (Geiss, 1976). The same data corrected for spallogenic ^3He and radiogenic ^4He.

92

values of $\approx 1.5 \cdot 10^{-4}$ (typical of the young sun) to higher values of 3 or $4 \cdot 10^{-4}$ (observed in contemporary solar wind). Recent works (Reynolds et al., 1978) confirmed the early estimate obtained by Anders et al. (1970), namely mineral fractions enriched in primordial noble gases show the same relation (Fig. 6.11). Additional evidence in favour of planetary helium was reported by Eberhardt (1978) in the form of data obtained through stepwise heating of mineral fractions of Orqueil (Fig. 6.12). Cosmogenic ^{21}Ne together with ^{3}He are released at a temperature below 600°C; the contribution of cosmogenic isotopes in a high-temperature fraction does not exceed 1%. The uranium content ($U = 0.9 \cdot 10^{-9}$ g g^{-1}; Th/U = 3.6) and the K-Ar age of the meteorite (1.6 b.y.) yield a maximum concentration of radiogenic ^{4}He of $4 \cdot 10^{-6}$ cm^{3} g^{-1}, which is lower than 5% of the total ^{4}He content in the meteorite. Moreover, both radiogenic and spallogenic helium leave, when produced, more or less lengthy tracks stimulating similar losses of helium of these types. In our case it means that radiogenic helium was released at a temperature below 600°C as well and its contribution to high temperature fractions is negligible. Hence, Eberhardt (1978) concluded that: (1) helium

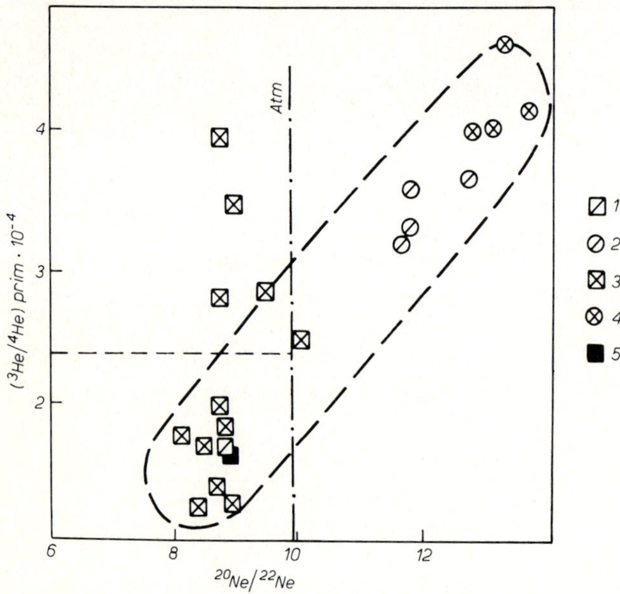

Fig. 6.11. Relationship between planetary neon and helium in carbonaceous chondrites. The ratio of ^{3}He/^{4}He = 2.4 $\cdot$ 10^{-4} corresponds to the atmospheric value of ^{20}Ne/^{22}Ne = 9.85. Circles and squares denote total meteorites and fractions rich in noble gases, respectively.
1 and *2* = concentrations of ^{4}He in 10^{-3} cm^{3} g^{-1}; *3* and *4* = 10^{-3} to 10^{-2} cm^{3} g^{-1}; *5* = over 10^{-2} cm^{3} g^{-1}. (Data from Mazor et al., 1970; Lewis et al., 1977; Srinivasan et al., 1977; Reynolds et al., 1978.)

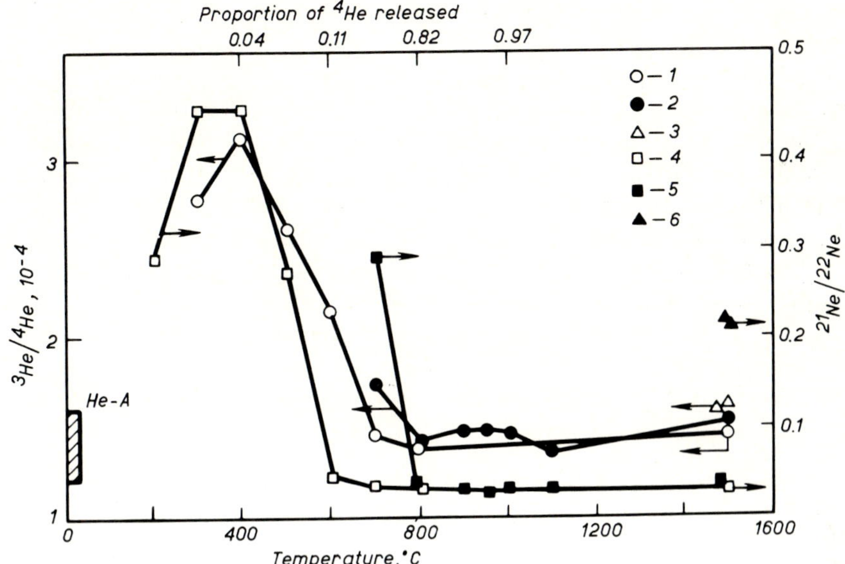

Fig. 6.12. Isotopic ratios of ^{3}He/^{4}He and ^{21}Ne/^{22}Ne in gases released under step-wise heating of Orgueil host phases (after Eberhardt, 1978).
1 and *4* = the ratios of ^{3}He/^{4}He and ^{21}Ne/^{22}Ne in temperature fractions, respectively; *2* and *5* = the same ratios in high-temperature fractions; *3* and *6* = the same ratios, gases released by melting of the sample.

characterized by the constant ^{3}He/^{4}He ratio of $1.5 \cdot 10^{-4}$ and released at a temperature of 600°C is very depleted in cosmogenic and radiogenic components; (2) it consists of pure trapped primordial helium; and (3) a homogeneous reservoir including planetary He-A is certain to have existed in the primitive solar nebula.

A correlation between ^{20}Ne/^{22}Ne and ^{3}He/^{4}He ratios (see Fig. 6.11) enables estimation of the isotope composition of terrestrial primordial helium. The atmospheric ratio of ^{20}Ne/^{22}Ne = 9.8 appears to be a mixture of solar and planetary neon. The ^{3}He/^{4}He ratio consistent with the isotope composition of atmospheric neon is approximately $2.4 \cdot 10^{-4}$. This ratio, intermediate between the solar and the planetary values, is used below.

Attention should be given to the absence of specific noble gases (including helium) which might be considered as accompanying Ne-E. Helium observed in Ne-E-rich fractions seems to be of the He-A and/or He-B type and there is nothing in favour of a He-E component.

Yet another consideration is that meteorites contain radiogenic helium produced by α-decay of heavy radioactive nuclides and accompanying nuclear reactions. An identification and utilization of this — normally minor in CC — component of helium characterized by a very low ^{3}He/^{4}He ratio of about 10^{-8} are beyond the scope of this work (see Shukolyukov and Levsky, 1972).

A detailed treatment of the origin of radiogenic helium in terrestrial rocks where it is most abundant is given in Chapter 7.

In summarizing section 6.3, it should be emphasized that primordial gases in CC are concentrated in micro-mineral fractions (host phases). Various fractions contain specific noble gases trapped (He, Ne-A, B) or produced *in situ* (Ne-E). The process of trapping gases of the A type as well as some other minor components is not clear yet. Gases could have been adsorbed on the surfaces of grains and buried by continuous accretion (Fanale and Cannon, 1972; Kerridge and Vedder, 1972). In some host phases they can be accumulated by a plasma implantation (Arrhenius and Alfvén, 1971) or due to solubility (Lancet and Anders, 1973). In any case, carbonaceous chondrites were formed in the early period of the evolution of the solar system (MacDougall and Kothari, 1976) and probably took part in the accretion of planetesimals.

6.4. Noble gases in the terrestrial atmosphere

The relative abundance of primordial noble gases in the atmospheres of the earth and planets is similar to that in CC (Fig. 6.12). Signer and Suess (1963) were the first to point out this important conformity; later it was verified by other authors (Wasson, 1969; Fanale, 1971; Shukolyukov and Levsky, 1972; Manuel, 1978). Though some results of meteorite research (Clayton et al., 1976; Kung and Clayton, 1978) point to the CC formation outside the zone of accretion of terrestrial planets (see Fig. 6.3), similar matter might have existed within this zone. This hypothetical matter might have been involved in the accretion of the earth and other planets. Based on this

TABLE 6.2

Concentrations ($\times$ 10^{-8} cm^3 g^{-1}) of primordial noble gases in carbonaceous chondrites and earth matter (Tolstikhin and Khabarin, 1976)

^{4}He	^{20}Ne	^{36}Ar	^{84}Kr	^{132}Xe	Remarks
Carbonaceous chondrites					
1000	0.1	2.5	0.018	0.011	minimum
10 000	8	50	0.6	0.4	mean
—	16	50	—	—	medium
150 000	460	128	6.9	0.48	maximum
Earth					
550[a]	1.1	2.1	0.044	0.00155	normalized to total earth
800[a]	1.6	3.1	0.064	0.00230	normalized to silicate shells of the earth

[a] Content of ^{4}He was assumed to be 500 times of ^{20}Ne (see text).

assumption the term "carbonaceous chondrites" (CC) will be used below in a broader sense to denote a very complex matter enriched in planetary noble gases which supplied the planets with volatiles.

The concentration of noble gases in CC is rather high (Table 6.2). To provide the atmospheres with planetary gases, a small contribution ($\sim$ 10% or lower) of carbonaceous chondrite-like matter might suffice and even a smaller contribution was enough to supply planets with other volatiles (C, N, S, etc). It should be noted that Table 6.2 illustrates measured concentrations of gases; however, the K-Ar ages of CC are as a rule well below 4.5 b.y. (Mazor et al., 1970) and, consequently, most of the radiogenic ^{40}Ar was released from meteorites. This proves that the initial contents of noble gases were much higher than the observed ones. Moreover, concentrations of noble gases in mineral fractions of CC imply that a matter extremely enriched in gases might have existed in the solar nebula if such a matter contained host phases in proportions larger than those observed in CC. That is why we believe that the atmospheres of terrestrial planets (including Venus whose atmosphere is enriched in primordial gases) might have been formed due to outgassing of solid planets only, though some authors (Izakov, 1979) search for an additional source of noble gases.

Thus, CC is the only class of meteorites that can provide the relative and the absolute contents of primordial noble gases in the atmospheres — without any restriction concerning the abundance of other elements.

Similarity in the isotope composition of noble gases in CC and in the atmospheres supports the idea that CC-like matter was the major carrier of volatiles for terrestrial planets (Table 6.3). Xenon is an exception, and there are several approaches to the enigma of atmospheric xenon (Manuel, 1978; Hamano and Ozima, 1980). In the light of the hypothesis of heterogeneous accretion (Orowan, 1969; Tolstikhin, 1980a, b), Manuel's observation of atmosphere-like xenon seems intriguing and stimulating.

To summarize, it may be said that carbonaceous chondrite-like matter was

TABLE 6.3

Isotope composition of primordial noble gases (Levsky, 1973)

Gas	Isotopic ratio	Type A	Type B	Terrestrial atmosphere
He	^{3}He/ ^{4}He	$1.4 \cdot 10^{-4}$	$(3-4) \cdot 10^{-4}$	$1.4 \cdot 10^{-6}$[a]
Ne	^{20}Ne/ ^{22}Ne	8.2	12—13	9.8
	^{21}Ne/ ^{22}Ne	0.020	0.036	0.0291
Ar	^{38}Ar/ ^{36}Ar	0.187—0.194	0.168—0.178	0.187
Kr	^{78}Kr/ ^{86}Kr	0.019	0.030	0.020
Xe	^{124}Xe/^{130}Xe	0.027	0.018	0.020
	^{126}Xe/^{130}Xe	0.025	0.018	0.022

a The helium influx and escape are responsible for this value (see Chapter 10).

the major carrier which supplied terrestrial planets with primordial rare gases and other volatile components. A considerable contribution of GRM, carriers of solar gases, is also imprinted in the isotope composition of atmospheric neon and argon. The constant $(^4He/^{20}Ne)_{prim}$ ratio in all available objects of the solar system stimulates utilization of ^{20}Ne as an equivalent of primordial helium in cases when it is difficult or even impossible to deal with helium itself, for instance, in the atmospheres of terrestrial planets (see Chapters 10 and 11).

In accordance with section 6.2.1 the ratio of $(^4He/^{20}Ne)_{prim} \approx 500$ will be used in this book. On the grounds of the isotope composition of He, Ne-A, He, Ne-B, atmospheric neon and the $^{20}Ne/^{22}Ne$—$^3He/^4He$ correlation, the value of $^3He/^4He \approx 2.4 \cdot 10^{-4}$ appears to be the best approximation of the isotope composition of terrestrial primordial helium.

Chapter 7

HELIUM ISOTOPES IN THE EARTH'S MANTLE

In the previous chapter it has been shown that terrestrial atmosphere (as well as Venus' and Mars's atmospheres) contains planetary primordial gases. These gases and other volatiles are generally believed to have been outgassed into the atmosphere from the solid earth. Thus, some fraction of the volatiles appears to be preserved in the earth's interior as it is difficult to imagine a planet totally degassed. However, numerous unsuccessful attempts to discover juvenile volatiles in the earth's interior have led to the conclusion of a very early catastrophic degassing event which resulted in a total initial outgassing of the earth (Fanale, 1971).

In 1968 Mamyrin et al. (1969a) observed an extremely high ^{3}He/^{4}He ratio of 10^{-5} in helium of thermal fluids of the south Kuril Islands. This ratio was 10 times greater than that of the atmosphere and about 1000 times higher than the radiogenic ratio. Somewhat later W.B. Clarke et al. (1969) reported a ^{3}He excess in oceanic water. They considered that these data bear evidence of primordial helium preservation in the most primitive deep earth's interior.

In this chapter the possibility of radiogenic origin of both helium isotopes will be described. In ordinary terrestrial rocks radiogenic ^{3}He/^{4}He ratios are about $2 \cdot 10^{-8}$ (section 7.1). The occurrence of very high ^{3}He/^{4}He ratios (equal to 10^{-5}) proves that there must be an additional source of terrestrial helium (section 7.2). The most probable cause of the high ^{3}He abundance is shown to be the trapping and preservation of primordial rare gases, a significant fraction of which has been retained to the present time (sections 7.3 and 7.4).

7.1. Abundance of ^{4}He and ^{3}He in radiogenic helium

The discovery of radioactivity at the turn of our century changed the notions of the structure of matter. The detection of helium atoms in the radioactive mineral cleveite by W. Ramsay was part of this great discovery. During many decades generations of investigators were convinced that terrestrial helium is produced by radioactive decay of heavy nuclei; in the 1960s this view was universally accepted.

98

Although one of the purposes of this work is to disprove this view, recent data on the problem of helium origin give ground to the conclusion that ^{4}He is generated mostly by α-decay, and it is important to describe the processes which yield radiogenic ^{3}He and to estimate the ^{3}He/^{4}He ratio in terrestrial radiogenic helium.

7.1.1. Fission

^{3}H and ^{3}He are characterized by an extremely low nuclear binding energy and so they are unlikely to have resulted from radioactive decay. Tritons and ^{3}He nuclei are known, however, to be emitted by fission of heavy nuclei (Fluss et al., 1972; Kugler and Clarke, 1972). Fission of uranium and transuranic isotopes (up to ^{252}Cf) yields ^{3}H(^{3}He) in proportion of $Y(^3$H, ^{3}He) = $2.2 \cdot 10^{-4}$ atoms per fission. Assuming that spontaneous fission of ^{238}U is characterized by the same yield of ^{3}H (^{3}He), it is possible to estimate the number of ^{3}He atoms produced by fission:

$$^3\text{He} = {}^{238}\text{U} \left[\exp \lambda_\alpha t - 1\right] \cdot (\lambda_f/\lambda_\alpha) \cdot Y(^3\text{H}, {}^3\text{He}) \tag{7.1}$$

where $\lambda_f = 8.6 \cdot 10^{-17}$ yr^{-1} and $\lambda_\alpha = 0.153 \cdot 10^{-9}$ yr^{-1} are constants of ^{238}U fission and α-decay, respectively. During the same time, t, the following number of ^{4}He atoms might be produced due to α-decay:

$$^4\text{He} = {}^{238}\text{U} \left[\exp \lambda_\alpha t - 1\right] \cdot N_\alpha \cdot N_\alpha' \tag{7.2}$$

where $N_\alpha = 8$ denotes the number of ^{4}He atoms formed due to uranium family decay; $N_\alpha' \approx 2$ is a coefficient which takes into account the contribution of the actinouranium and thorium families; the ratio of ^{238}U/^{232}Th is assumed to be equal to 3.

Now it is possible to calculate the fraction of (eq. 7.1)/(eq. 7.2):

$$^3\text{He}/^4\text{He} \approx \frac{(\lambda_f/\lambda_\alpha) \cdot Y(^3\text{H}, {}^3\text{He})}{N_\alpha \cdot N_\alpha'} \approx 8 \cdot 10^{-12}$$

Another approach to the problem is to compare fissiogenic ^{3}He$_f$ and Xe$_f$. The yield of ^{136}Xe$_f$ is known to be $Y(^{136}$Xe$) \approx (4.5-6.5) \cdot 10^{-2}$ (Srinivasan et al., 1969; Shukolyukov, 1970; Flynn et al., 1972). So, the fissiogenic ^{136}Xe$_f$/^{3}He$_f$ ratio might be about $2.5 \cdot 10^2$. This value is at variance with early results obtained by Shukolyukov and Tolstikhin (1965) and Funkhouser and Naughton (1968)[1] as well as with recent data for the richest ^{3}He

---

[1] The Xe isotope analyses obtained by these authors should be compared with the He isotope analyses from Table 7.5.

samples (Saito et al., 1978; Kaneoka and Takaoka, 1980), which show a very low ratio of $^{136}\text{Xe}_\text{f}/^3\text{He} \approx 10^{-2}$; this is $\sim 10^4$ times lower than the ratio yielded by fission. It is also possible to show that the abundance of $^3\text{He}_\text{f}$ which has been produced by ^{244}Pu fission is considerably smaller than that needed for the explanation of high $^3\text{He}/^4\text{He}$ ratios ($\sim 10^{-5}$) in terrestrial helium.

If we compare the results obtained here with those presented in the next section, we may conclude that fission produces a negligible part of radiogenic terrestrial ^3He.

Now let us consider the contribution of nuclear reactions which take place under natural conditions and produce the light helium isotope.

7.1.2. Neutron reactions

In 1955 Morrison and Pine showed that the isotope ratios of $^3\text{He}/^4\text{He}$ measured earlier by Aldrich and Nier (1948) in some minerals (beryl, spodumene) and in natural gases may be explained as follows: (1) The radioactive decay of uranium and thorium yields α-particles, the majority of which remain in the matter of a mineral. They instantly acquire two electrons, and turn into ^4He atoms. (2) A small proportion of α-particles reacts with nuclei in the minerals, mainly with nuclei of light elements, and some of these interactions are (α, n) reactions responsible for the appearance of a neutron flux in the rocks. Neutrons also appear from the spontaneous and neutron-induced fission of uranium isotopes. (3) A part of the neutron flux reaches epithermal energies, and the reaction of these neutrons with the nucleus of the light isotope of lithium gives ^3He:

$$^6\text{Li} + n \rightarrow \alpha + {}^3\text{H}; \quad {}^3\text{H} \xrightarrow{\beta^-} {}^3\text{He} \tag{7.3}$$

This explanation was later amplified by Gorshkov et al. (1966); in particular, they made a more precise estimate of the neutron yield from interaction of α-particles with light-element nuclei.

If the ^3He origin in a rock is due to reaction 7.3, then the $^3\text{He}/^4\text{He}$ ratio in the rock can be calculated from formula:

$$(^3\text{He}/^4\text{He})_\text{calc} = n_\text{f} \left\{ \frac{\sum\limits_k q_k S_k N_k}{\sum\limits_i S_i N_i} \right\} \cdot P_{t^0} \left[\frac{\sigma_{\text{Li}} N_{\text{Li}}}{\sum\limits_i \sigma_i N_i} \right] \tag{7.4}$$

Here n_f discounts neutrons produced by fission, $n_\text{f} \approx 1.15$; q denotes the neutron yield in (α, n) reactions; S is a relative brake capacity of elements; N is the number of atoms per gram of rock; σ is the cross-section of neutron capture, cm^2; P_{t^0} is the probability for a neutron to reach epithermal energy, $P_{t^0} \approx 0.8$ (Morrison and Pine, 1955). An example of such a calculation is given in Table 7.1.

TABLE 7.1

Calculation of the radiogenic ^{3}He/^{4}He ratio in Rapakiwi granite (see Fig. 8.1, Table 8.2)

Element	N (10^{20} atoms g^{-1})	S	q	σ	NS	qNS	σN
O	179	1.06	0.18	$2 \cdot 10^{-4}$	189	34	0.036
Si	71	1.61	0.38	0.13	114	43	9.2
Al	17	1.53	2.6	0.23	26	68	3.9
Fe	2.2	2.43	—	2.53	5.35	—	5.6
Mg	0.68	1.44	2.8	0.06	1.0	2.8	0.04
Li	0.032	0.47	3.6	71	0.015	0.054	2.27
K	6.5	1.98	—	2.0	12.8	—	13
Na	5.0	1.35	2.9	0.51	6.75	19.6	2.55
Ca	1.6	2.05	—	0.43	3.3	—	0.69
Ti	0.26	2.19	—	5.6	0.57	—	1.45
Be[a]	$4 \cdot 10^{-3}$	0.6	109	0.01	$2 \cdot 10^{-3}$	0.26	$5 \cdot 10^{-5}$
B[b]	0.011	0.73	24	755	0.008	0.19	8.3
F[c]	0.23	1.15	16	0.009	0.26	4.2	0.002
Gd[d]	$3 \cdot 10^{-4}$	—	—	$4.6 \cdot 10^4$	—	—	14

$$\sum_i S_i N_i = 359$$

$$\sum_k q_k S_k N_k = 172$$

$$\sum_i \sigma_i N_i = 61$$

$$(^3\text{He}/^4\text{He})_{\text{calc}} = 1.15 \cdot \frac{172}{359} \cdot 10^{-6} \cdot 0.8 \cdot \frac{2.27}{61} = 0.016 \cdot 10^{-6}$$

$$(^3\text{He}/^4\text{He})_{\text{meas}} = 0.016 \cdot 10^{-6}$$

[a] In this sample numbers of α-particles yielded due to α-decay of U and Th families are similar; so the average values of $q = (q_{\text{U}} + q_{\text{Th}})/2$ are used (Gorshkov et al., 1966).
[b] The cross-sections of thermal neutron capture are taken from Ebert (1968).
[c] The contents of these elements are taken from Gorshkov et al. (1966).
[d] The Gd content is taken from Haskin et al. (1966).

A comparison of the calculations from formula 7.4 and the measurements of the helium isotope composition from particular rocks was made for the first time by Gerling et al. (1971). The authors concluded that: (1) the measured ratios of ^{3}He/^{4}He in old granitic rocks are close to the calculated ratios and vary slightly from 10^{-8} to $3 \cdot 10^{-8}$; (2) the helium isotope composition of most terrestrial gases found within ancient plates varies over the same interval; (3) neutron reactions (Table 7.2) cannot produce helium with a high isotopic ratio in average terrestrial rocks. To prove the validity of this conclusion, an additional experimental test was carried out: some samples of ultrabasic rock were irradiated by a neutron flux, and after a year the tritium and ^{3}He contents were measured. The results obtained (Table 7.3) tes-

TABLE 7.2

Some neutron reactions yielding ^3H (^3He) (Kunz and Schintlmeister, 1965)

No.	Reactions	Reaction energy (MeV)	Interaction energy	Cross-section[a] (10^{-27} cm^2)	Calculated $^3He/^4He$ ratio
1	$^6Li + n \rightarrow {}^3H + {}^4He$	+ 4.8	thermal	$9 \cdot 10^5$	$\sim 10^{-8}$
2	$^2H + n \rightarrow \gamma + {}^3H$	+ 6.2	thermal	0.46	$< 10^{-11}$
3	$^7Li + n \rightarrow {}^3H + {}^5He$	− 3.4	4.5 MeV	30	$< 10^{-11}$
4	$^9Be + n \rightarrow {}^3H + {}^7Li$	−10.4	14 MeV	110	10^{-14}
5	$^{10}B + n \rightarrow {}^3H + {}^8Be$	+ 0.35	fission neutrons	30	10^{-12}
6	$^{11}B + n \rightarrow {}^3H + {}^9Be$	− 9.5	14.1 MeV	15	10^{-15}
7	$^{14}N + n \rightarrow {}^3H + {}^{12}C$	− 4.0	4.4 MeV	11	10^{-12}
8	$^7Li + n \rightarrow {}^3H + {}^4He^+$	− 2.5	4.4 MeV	0.1	10^{-14}

[a] The cross-section is shown for the neutron energy listed in the previous column.

TABLE 7.3

Tritium (^3H) yield by neutron irradiation of ultrabasic rocks (Tolstikhin et al., 1974a)

Sample	Content (atoms g^{-1})			Li/$^3H_{meas}$ (10^7)	^3He content (10^{-10} cm^3 g^{-1})	
	Li (10^{17})	^3H (10^{10})				
		calc[a]	meas		calc[b]	meas[c]
Peridotite	2.85	20	6	0.47	4.2	—
Bronzite	2.25	16	4.8	0.47	0.98	1.1
Peridotite	3.35	28	9	0.37	1.7	2.0

[a] Tritium content calculated from the formula:

$$^3H_{calc} = Li \cdot \sigma_{Li} \cdot \Phi_\Sigma$$

where Li is the lithium content in atoms g^{-1}; σ_{Li} is the cross-section of interaction between thermal neutrons and lithium, $\sigma_{Li} = 71 \cdot 10^{-24}$ cm^2; $\Phi_\Sigma = 10^{16}$ n cm^{-2} is the integral neutron flux. The systematic excess of calculated tritium concentrations over the measured concentrations is apparently accounted for by the presence of epithermal neutrons in the channel.

[b] The ^3He content that might have been formed in samples owing to $^3H_{meas}$ decay in time $t = 1$ year, is calculated from the formula:

$$^3He_{calc} = {}^3H_{meas} (e^{\lambda t} - 1)$$

where λ is the constant of tritium decay, $\lambda = 0.056$ yr^{-1}.

[c] The helium analyses were performed a year after the samples underwent irradiation. Prior to irradiation the ^3He content was low, about 10^{-12} cm^3 g^{-1}. It is essential that: (1) the $^3H_{meas}$ content is not higher than that produced by reaction 7.3; (2) the Li/$^3H_{meas}$ ratio is nearly constant; and (3) the $^3He_{meas}$ content was accumulated due to tritium decay alone.

102

tify to the absence of neutron reactions leading to higher ^{3}H (^{3}He) yields than reaction 7.3.

It could be assumed that at the earliest stages of the earth's evolution there was a neutron flux not related to (α, n) reactions, for example, a flux generated by the fission of a transuranium isotope. The fission of ^{244}Pu, whose traces were found in meteorites (Kuroda, 1969; Wasserburg et al., 1969) and in the earth's matter (D.C. Hoffman et al., 1971), might have been a possible source of the flux. It is important to note that $\lambda_f/\lambda_\alpha = 3 \cdot 10^{-3}$ for ^{244}Pu, which is 5300 times higher than that for ^{238}U. The contribution of ^{244}Pu to the production of ^{3}He may be estimated. For the calculation we take: (a) the abundance of U and Th in the silicate earth is equal to $3.3 \cdot 10^{-8}$ and $12 \cdot 10^{-8}$ g g^{-1}, respectively (Wasserburg et al., 1964); (b) the ratio of ^{244}Pu/^{238}U $= 0.03$ in the primary matter; (c) the age of the earth $t = 4.55 \cdot 10^9$ yr. The integral neutron flux $^{244}\Phi$ resulting from the fission of ^{244}Pu may be calculated from the formula:

$$^{244}\Phi = \frac{\bar{\nu} \cdot \lambda_f \cdot N_{Pu}}{(\lambda_f + \lambda_\alpha) \cdot \sum_i N_i \sigma_i} \tag{7.5}$$

where ν is the number of neutrons formed by the fission of ^{244}Pu, $\bar{\nu} = 2.5$; $\lambda_f/(\lambda_f + \lambda_\alpha)$ is the fraction of plutonium atoms which decay by fission, $\lambda_f = 0.28 \cdot 10^{-10}$ yr^{-1} and $\lambda_\alpha = 0.1 \cdot 10^{-7}$ yr^{-1}; N_{Pu} is the content of ^{244}Pu, $N_{Pu} = 0.03 \cdot U = 2.5 \cdot 10^{12}$ atoms g^{-1}; N_i is the number of nuclei per g; and σ is the neutron capture cross-section in a nucleus; $\sum_i N_i \sigma_i \gtrsim 4 \cdot 10^{-3}$ cm^2 g^{-1}.

Substituting numerical values in formula 7.5, we obtain $^{244}\Phi \lesssim 4.4 \cdot 10^{12}$ n cm^{-2}. Assuming that all the fission neutrons reached thermal energy, we conclude that the following amount of ^{3}He:

$$^3He = {}^3H = {}^{244}\Phi \cdot Li \cdot \sigma_{Li} = 3 \cdot 10^{-12} \text{ cm}^3 \text{ g}^{-1} \tag{7.6}$$

could be formed in matter of chondrite composition (Li $= 2.6 \cdot 10^{17}$ atoms g^{-1}) in the course of time equal to the age of the earth. During the same period the following amount of ^{4}He has been produced owing to the α-decay of uranium and thorium:

$$^4He = 68 \cdot 10^{-6} \text{ cm}^3 \text{ g}^{-1} \tag{7.7}$$

Combining eqs. 7.6 and 7.7, we obtain (provided the retention of helium was "ideal") the ratio of ^{3}He/^{4}He of $4.4 \cdot 10^{-8}$ (100—1000 times less than that measured in xenoliths).

7.1.3. α-particle reaction

Calculations show that a single possible reaction under natural conditions (Kunz and Schintlmeister, 1965):

$$^7\text{Li} + \alpha \rightarrow {}^8\text{Be} + {}^3\text{H} \tag{7.8}$$

produces helium with an isotope ratio of $^3\text{He}/^4\text{He} \sim 10^{-10}$ (Gerling et al., 1971).

To verify this estimation, three samples of ultrabasic rocks from the Monchegorsk pluton ($^3\text{He}/^4\text{He} = 2.5 \cdot 10^{-7}$) were irradiated in a cyclotron by He^+ with an isotope ratio of $^3\text{He}/^4\text{He} = 2 \cdot 10^{-8}$, energy $E_\alpha = 10$ MeV, and a flux value of $\Phi_\alpha \simeq 5 \cdot 10^{15}$ cm^{-2}. Then the tritium content and isotope composition of helium were determined in the irradiated samples. The following ratios were obtained:

$$^3\text{H}/\alpha < 7 \cdot 10^{-9}$$

$$(^3\text{He}/^4\text{He})_{\text{before irrad.}} > (^3\text{He}/^4\text{He})_{\text{after irrad.}} \tag{7.9}$$

Thus, the α-particle reactions in matter close in composition to the xenoliths do not produce a considerable share of ^3H (^3He). A quantitative estimate of the ^3H yield resulting from the α-particle reaction was not obtained as the flux was too low.

7.1.4. γ-quanta reaction

The following reaction (Kunz and Schintlmeister, 1965) may produce ^3H under natural conditions:

$$^7\text{Li} + \gamma \rightarrow {}^4\text{He} + {}^3\text{H} \tag{7.10}$$

This reaction has a threshold, E_n, of 2.5 MeV and quite a low cross-section of $\sigma_\gamma < 0.02$ mbar. It can take place only on γ-quanta emitted by the isotope ThC″ ($E_\gamma = 2.62$ MeV).

It is possible to show that, with the adopted abundances of uranium, thorium and lithium (section 7.1.2), this reaction occurring in matter similar in composition to ultrabasic rocks may lead to an accumulation of the following amount of ^3He during the time interval equal to the age of the earth:

$$^3\text{He} = {}^3\text{H} \simeq 10^{-17} \text{ cm}^3 \text{ g}^{-1} \tag{7.11}$$

Comparing the values in formulas 7.11 and 7.7, we obtain:

$$^3\text{He}/^4\text{He} \sim 10^{-13} \tag{7.12}$$

which is negligible compared to the measured ratio in xenoliths.

104

Thus, experimental data as well as theoretical considerations enable us to conclude that radioactive decay, fission and nuclear reactions (which are due to these processes) produce helium isotopes in proportion of $^3He/^4He \sim 10^{-8}$; the reactions between light lithium isotopes and thermal neutrons yield almost all terrestrial radiogenic ^3He. It is important to note that no nuclear process has been discovered so far which produces helium with a $^3He/^4He$ ratio $\gg 10^{-8}$ in terrestrial rocks.

The first comparison of measured and calculated $^3He/^4He$ ratios was carried out by Gerling et al. (1971, 1972) and later by Tolstikhin and Drubetskoy (1975, 1977). The conclusion of Morrison and Pine (1955) about the radiogenic origin of helium isotopes in ancient granitic rocks of the continental crust and natural gases associated with such rocks was confirmed: the observed and calculated ratios were similar and close to 10^{-8} (see Chapter 8 for a more detailed discussion). However the measured $^3He/^4He$ ratios exceeded the calculated ones by three orders of magnitude in young erupted rocks of possible mantle origin.

7.2. Helium isotopes in mantle materials

There is no single criterion so far which might testify to a mantle origin of terrestrial materials. Nevertheless, there are certain experimental data as well as some other arguments in favour of a mantle genesis of rocks, magmas or fluids. Among these are recent data from experimental petrology, temperature and pressure measurements in the source based on mineral associations, studies of microinclusions and the geochemistry of stable isotopes. To these also belong some results of investigations of geochronometric systems, such as the initial isotope composition of Sr and Nd in rocks and minerals, as well as considerations on tectonic structure of a region, heat flow, etc. All this produces some evidence that in many cases the mantle is a source of ultrabasic inclusions found in basalts and kimberlites, eclogitic rocks and diamonds, oceanic tholeiites, erupted ultrabasic rocks, some acid and alkaline intrusions, some components of thermal fluids, etc. In this section our aim is to check these more or less known data with some new data of the terrestrial helium isotope distribution and to estimate the $^3He/^4He$ ratio in the earth's mantle. With this end in view it is useful to keep in mind two peculiarities of helium isotope geochemistry: (1) Both helium isotopes (as well as argon and other volatiles) are released from mantle magma at the moment the latter intrudes into the earth's crust. (2) Accumulation of radiogenic helium in outgassed matter results in a $^3He/^4He$ ratio more or less rapidly approaching the radiogenic values. Thus, the samples which most likely have not been contaminated by radiogenic, trapped or produced *in situ* helium appear to be very young eruptive rocks as well as thermal fluids from regions where there is no old continental crust.

7.2.1. Helium isotopes in terrestrial fluids

The earth's fluids are complex mixtures of volatiles released from different sources, their chemical and isotopic compositions being averaged by migration processes. Correspondingly, investigations of fluids will lead to establishing general regularities in the isotope geochemistry of noble gases and other volatiles. They will also contribute to the study of the main properties of terrestrial sources of volatile elements.

The constant $^3He/^4He$ ratios in fluids within the same regional tectonic structure and extremely high $^3He/^4He$ ratios in areas which are characterized by high contemporary magmatic activity, seismicity and heat flow appear to be the most important tendencies in the distribution of terrestrial belium isotopes.

The highest $^3He/^4He$ ratios were found in subsurface fluids of Iceland, on the axis of the Mid-Atlantic Ridge (Mamyrin et al., 1972c; Kononov et al., 1974; Polyak et al., 1976). Helium isotope analyses were carried out for more than sixty hot springs in Iceland; the highest values of $^3He/^4He$ among all hitherto observed in fluids amount to $3.3 \cdot 10^{-5}$. The average value for the island, $^3He/^4He = 1.8 \cdot 10^{-5}$, is also considerably higher than that of other regions, except may be Hawaii. Ratios lower than 10^{-5} hardly occur in Iceland fluids (Fig. 7.1). High and more or less constant $^3He/^4He$ ratios throughout the territory of the island distinguish Iceland from most volcanically active zones where the ratios beyond the zones decrease to radiogenic values of 10^{-8}. Such a distribution of helium isotopes appears to be due to upwelling of deep mantle matter, including volatiles, on the one hand, and a specific crust structure of Iceland containing no "granitic layer" — that is, ancient rocks rich in radioactive elements — on the other (Kononov and Polyak, 1977).

Thus, it appears that the $^3He/^4He$ ratios in fluids of Iceland exceed those in radiogenic helium by over three orders of magnitude and that of the earth's atmosphere by about one order of magnitude. The helium content in many fluids is also high (up to 0.3%) and it rules out the possibility of ^3He excess due to technogenic ^3H β-decay (see section 8.6).

The spatial distribution of $^3He/^4He$ ratios in hot springs of Iceland shows no obvious relationship with the volcanic activity on the island, but it is consistent with the heat flow distribution whose maximum (2.2 UHF) is associated with the zone of highest isotopic ratios. Such relationships imply that both the helium isotope composition and the heat flow are affected by longer lived, internal factors than those controlling the recent volcanic activity.

High and very similar $^3He/^4He$ ratios are observed in gases of various chemical composition, namely carbon dioxide, nitrogen, hydrogen (see Table 8.13). This is due to various sources of macrocomponents and helium; the latter is derived from the deepest source and it is the only component of the fluids which avoids contamination by subsurface, air or oceanic volatiles.

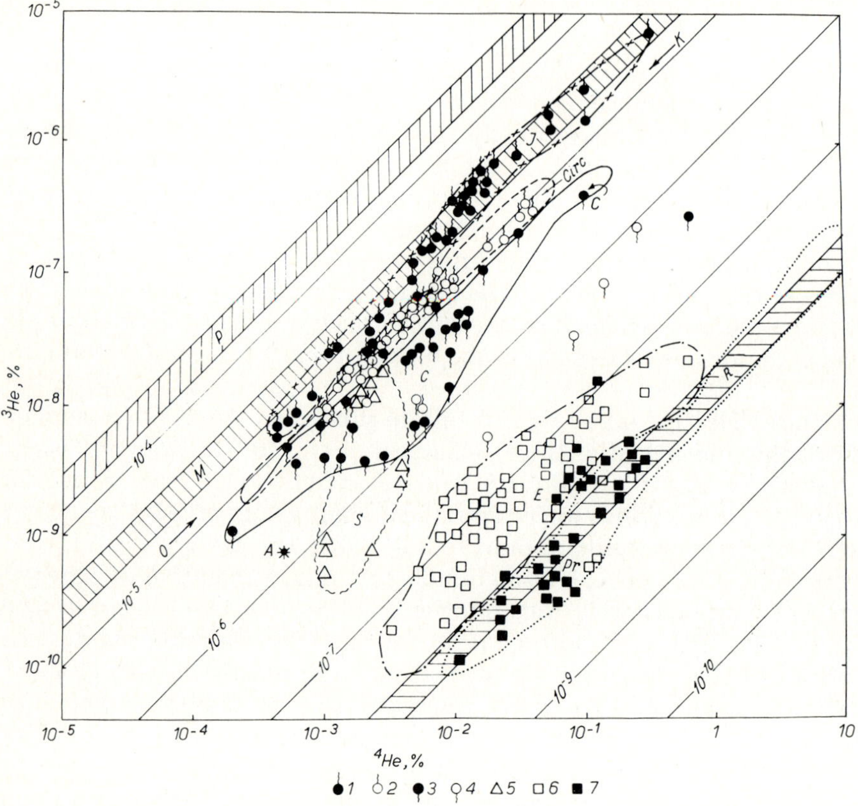

Fig. 7.1. Concentrations of helium isotopes and $^3He/^4He$ ratios in various terrestrial fluids. *1, 2, 3* and *4* = hot springs of Iceland (the area of concentrations is marked by *I*), the Circum-Pacific belt (*Circ*), the Caucasus (*C*) and the Baikal rift zone, respectively; *5* = springs, bed waters and gases of Sakhalin (*S*); *6* and *7* = bed waters and gases of epi-Hercynian (*E*) and Precambrian (*Pr*) plates, respectively. *P, M* and *R* = isotopic composition of primordial, mantle and crustal (radiogenic) helium, respectively; *O* and *K* = isotopic composition of helium injected in waters of the Pacific Ocean and the Red Sea, respectively, by sea-floor hydrothermal systems; *A* = the atmosphere (Tolstikhin, 1978).

The following example (see also sections 7.4 and 10.2) clarifies this point (Vinogradov et al., 1974): hot springs of inner areas of the island show high $^3He/^4He$ ratios and a juvenile isotope composition of oxide, reduced and native sulphur, $\delta\ ^{34}S \approx 0$; whereas coastal hot springs are characterized by the same ratios of helium isotopes though the isotope composition of sulphate, $\delta\ ^{34}S = +\ 10$ to $15^0/_{00}$, is related to that of oceanic waters, $\delta\ ^{34}S = +\ 20^0/_{00}$. The ocean appears to be a source of sulphates emanated by a hydrothermal system of coastal volcanoes. This example also shows that volatiles tend to be more readily outgassed in islands, circum-oceanic and mid-continental regions with a high tectonic and magmatic activity than through the oceanic crust.

A very high ratio of ^{3}He/^{4}He = 2.1 $\cdot$ 10^{-5} was reported by Craig and Lupton (1976) for a sample collected from the Sulphur Bank fumarole on the caldera edge of Kilauea Volcano, Hawaii. This value is very similar to the average ^{3}He/^{4}He ratio of the hot springs in Iceland. The authors showed that production of ^{3}He from technogenic ^{3}H is too small to affect the ^{3}He/^{4}He ratio in this sample; they used measurements of ^{3}H in the fumarole and estimated the maximum ratio of technogenic ^{3}He to ^{20}Ne equal to 6 $\cdot$ 10^{-6}, which was 6000 times lower than the observed ^{3}He/^{20}Ne ratio.

A large number of helium isotope analyses (about 200) have been carried out for the Circum-Pacific volcanic belt, namely Kamchatka (Tolstikhin et al., 1972a; Kamensky et al., 1976), the Kuril Islands (Mamyrin et al., 1969a; Tolstikhin et al., 1972a; Baskov et al., 1973), Japan (Craig et al., 1978a; Wakita et al., 1978; Nagao et al., 1979, 1981), the Mariana Islands (Craig et al., 1978a), Mexico (Prasolov et al., 1982), and Yellowstone and Lassen Park volcanic gases, U.S.A. (Craig et al., 1978b). Despite a sufficiently wide range of ^{3}He/^{4}He ratios, which is probably provided by the contribution of crustal, radiogenic helium, the average value is also high and amounts to about 1 $\cdot$ 10^{-5} (see Fig. 7.1). Similar results were obtained for the modern mid-continental rift and orogenic zones, such as the Baikal Rift (Lomonosov et al., 1976), the Caucasus (Matveeva et al., 1978), the Alpine-Apennine region (Polyak et al., 1979a), etc.[1] It is a noteworthy fact that hydrothermal systems of the oceanic crust supply sea water with helium of the same isotope composition (see Chapter 9).

From the above-mentioned data the following regularity emerges: helium of volcanic gases and thermal fluids from regions which are characterized by high contemporary tectonic and magmatic activity, deep faults, high seismicity and heat flow (that is, regions where mantle emanations can appear near the earth's surface), practically always shows the extremely high (as compared to radiogenic and atmospheric values) isotopic ratios of ^{3}He/^{4}He = (1—3) $\cdot$ 10^{-5}. As for helium of various natural gases and waters from stable regions[1], such as old continental plates, it shows ^{3}He/^{4}He values as low as (1—3) $\cdot$ 10^{-8}, similar to radiogenic values (see Fig. 7.1). Hence, the earth's mantle contains a peculiar variety of sources enriched in the light isotope ^{3}He. Investigations of rocks of possible mantle origin confirm this conclusion.

7.2.2. Helium isotopes in ultramafic nodules

The first results reported by Tolstikhin et al. (1972b) show that the ^{3}He/^{4}He ratios in helium of ultramafic inclusions of which the deep-seated origin

---

[1] Data for the Circum-Pacific volcanic belt and other regions are discussed in detail in Chapter 8 (section 8.5).

is widely accepted were about three times higher than the same ratio in atmospheric helium. A more thorough investigation (Tolstikhin et al., 1974a) was carried out on cognate inclusions of ultrabasic composition found in rocks of various type in tectonically different regions: in high-alumina basalt-andesites of Kamchatka (a region similar in tectonic structure to island arcs), in tholeiitic basalts of Iceland close in structure to mid-oceanic ridges, and in basalts of the Sverre Volcano (Spitsbergen) and Ross Island (Antarctica). Despite very different geological settings, the inclusions are homogeneous in composition and belong to a genetically common pyroxenite—peridotite—olivinite series.

Ultrabasic xenoliths are found mainly in large, essentially basaltic, or more rarely in basalt-andesitic, strato-volcanoes and occur more often in pyroclastic flows and scoria, while in lavas they are not so common. The size varies from several centimeters to one meter, averaging 3—8 cm; they are fragmentary and typically xenogenic in shape. The average size of the grains is 1—3 mm; sometimes the rock has a coarse-grained appearance and grains are 5—7 cm. The structure is massive, eutaxitic. Contacts with the country rock are usually abrupt, without signs of alteration or interaction. However, in some cases there is an amphibole rim 1—3 cm thick surrounding inclusions in rocks from the Kronotsky and Avacha Volcanoes (Kamchatka) in which rimmed and unrimmed inclusions occur together rather often.

Three mechanisms for the incorporation of ultrabasic inclusions are usually discussed: (1) accumulation of a crystalline phase at early stages of basaltic magma differentiation; (2) capture of upper crust xenogenic material; (3) capture of relicts of the original reservoir in which magma was generated.

The regular relationship of these inclusions to basalt-andesite assemblage rocks, the absence of such inclusions in acid volcanics, the similarity of the inclusion mineral assemblage, and the presence of chrome-diopside (which is not indicative of normal ultrabasic intrusions) are all arguments in favour of the third hypothesis.

All the samples (Table 7.4) of ultrabasic xenoliths studied so far are characterized by high values of $^3He/^4He$, $(3.2—14) \cdot 10^{-6}$. The high isotope ratios of $\gtrsim 10 \cdot 10^{-6}$ found both in xenoliths and in volcanic gases appear to characterize the isotopic composition of deap-seated helium. It is important to note that the helium with a high $^3He/^4He$ ratio is extracted from xenoliths at high temperatures; for example, the helium from samples *3*, *5* and *8* which yield a ratio of $^3He/^4He \gtrsim 10^{-5}$ was extracted at temperatures exceeding $1000°C$. We infer from this observation that this helium could not have entered into xenoliths due to low-temperature processes on the surface.

The results obtained by step-wise degassing of helium offer a possibility to subdivide the samples arbitrarily into two groups. The first group includes samples *1*, *3*, *5* and *6* characterized by a regular increase in the $^3He/^4He$ ratio as the temperature increases. For example, for sample *5* there is a sevenfold increase in the isotope ratio in a high-temperature fraction.

The variations observed in the isotope ratio cannot be accounted for by differences in the diffusion coefficients of ^{3}He and ^{4}He, as the observed effect is too large (Kalbitzer et al., 1969) and differs from that expected at diffusive fractionation — that is, ^{3}He should have been released preferentially in low-temperature fractions. These variations are more the result of differences in the sites of helium with a high ^{3}He/^{4}He isotope ratio and ^{4}He (or helium with low isotope ratio). In cases when a major loss of each "helium type" takes place at a different temperature, a well-defined change of the isotopic helium composition occurs in the temperature fractions. Unfortunately, the

TABLE 7.4

Contents and isotope compositions of helium in xenoliths (Tolstikhin et al., 1974a) [^{3}He/^{4}He × 10^{-6}; ^{4}He and ^{3}He in 10^{-8} cm^3 g^{-1} and 10^{-14} cm^3 g^{-1} respectively]

| No. | Sample description | In total sample | | | Heating temperature (°C) | | | | | |
| | | | | | 600 | | 900 | | 1200 | |
		$\frac{^3He}{^4He}$	^{4}He	^{3}He	$\frac{^3He}{^4He}$	^{4}He	$\frac{^3He}{^4He}$	^{4}He	$\frac{^3He}{^4He}$	^{4}He
1	Peridotite inclusion in andesites (Avacha, USSR)	3.0	6	18	1.6	3.5	3.5	1.5	7.6	1.0
2	Peridotite inclusion in basalts (Kronotsky, USSR)	3.9	—	—	4.1	—	—	—	3.8	—
3	Olivine inclusion in basalts (Avacha, USSR)	9.0	10	90	1.5	1.0	8.2	1.5	10	7.5
4	Olivine inclusion in basalts (Kamchatka, USSR)[a]	8.6	3.5	30	8.8	0.21	8.0	0.56	8.8	2.7
5	Dunite inclusion in basalts (Ross Island, Antarctica)[b]	14	30	420	6.4	20	20	5.5	45	4.0
6	Peridotite inclusion in basalts (Sverre, Spitsbergen)	3.2	30	96	2.6	—	3.2	—	7.2	—
7	Lherzolite inclusion in basalts (Spitsbergen)	4	10	40	4.5	3.0	3.8	6.0	4.0	1.0
7a	The same without pyroxene fraction[c]	2.8	10	28	2.5	6.0	2.9	3.2	6.3	0.8
8	Xenolith from Quaternary basalts (Iceland)[d]	9.5	1	9.5	1.8	0.25	12	0.50	12	0.25

[a] Collected by E.N. Erlikh.
[b] Collected by B.G. Lopatin.
[c] Collected by Yu.P. Burov.
[d] Collected by B.E. Jacobson.

large temperature intervals and consequent low resolution of the method does not at present allow us to determine the upper limit of the ^{3}He/^{4}He ratio in inclusions; the ratio for the high-temperature fraction of the dunite inclusion (5, Table 7.4) appears to be close to this limit (compare with samples 1—3, Table 7.5).

Samples 2, 4, 7 and 8, which are characterized by a constant isotope ratio in temperature fractions, may be assigned to a second group in which activation energies of both "helium types" are similar, as may result, for example, from a scattered homogeneous distribution of uranium and thorium in the samples.

Very interesting data were recently reported by Kaneoka and Takaoka (1978, 1980). They measured the content and isotope composition of all noble gases in some ultramafic samples and for the first time showed that the ^{3}He/^{4}He ratios in phenocrysts are considerably higher than those in nodules (Table 7.5) and the isotope composition of argon correlates with ^{3}He/^{4}He values (for a more detailed discussion see section 7.4). On the basis of these data Kaneoka and Takaoka arrived at the very important conclusion that noble gases trapped by nodules are similar to those observed in oceanic tholeiites and that they might have been derived from the same source, whereas noble gases of phenocrysts imply another source more enriched in juvenile components. These results are compatible with the hot spot hypothesis for

TABLE 7.5

Contents and isotope compositions of helium in Hawaiian phenocrysts and nodules (Kaneoka and Takaoka, 1978, 1980)

No.	Sample description	^{4}He (10^{-8} cm^3 g^{-1})	^{3}He/^{4}He (10^{-6})
Haleakala Volcano, Maui			
1	Augite phenocrysts	3.25	48.7 ± 6.1
2	Olivine phenocrysts	2.2	51.5 ± 9.7
3	Augite crystal	2.5	23.5 ± 4.8
Hualalai[a]			
4	Dunite	32.6	11.5 ± 1.0
5	Idem, another sample	17.4	14.8 ± 1.0
Oahu			
6	Spinel-lherzolite nodule, Salt Lake crater	32.4	10.5 ± 0.6
7	Idem, another sample	27.8	11.3 ± 0.7
8	Garnet-pyroxenite	132	11.0 ± 0.7

[a] Erupted in 1801.

the genesis of the Hawaiian chain as well as with the interpretation of Sr-Nd isotopic data (DePaolo and Wasserburg, 1976).

Other rocks which are believed to be derived from the earth's deep interior are ultramafic inclusions in kimberlites. Available results of helium isotope investigations of these samples are presented in Table 7.6. The isotopic $^3He/^4He$ ratios in samples of ultramafic inclusions differing in composition and location are markedly lower than those in phenocrysts and nodules in contemporary basalts but they are well above the typical radiogenic ratio of $^3He/^4He = (1—3) \cdot 10^{-8}$. On the other hand, the ^4He contents in the samples listed in Table 7.6 are one or two orders of magnitude higher than those presented in Tables 7.4 and 7.5.

For a correct interpretation of these data one needs to take into account the age and the uranium content of inclusions in kimberlites.

The age of the "Naked" pipe kimberlites (Siberia) determined by K-Ar and fission track methods is about 100—130 m.y. (Malkov and Gustomesov,

TABLE 7.6

Contents and isotope compositions of helium in inclusions of kimberlites and in diamonds

No.	Sample description	^4He $(10^{-6} cm^3 g^{-1})$	$^3He/^4He$ (10^{-6})
Inclusions in kimberlites, "Naked" pipe, east Siberia [a]			
1	Peridotite	1.2	0.87
2	Garnet-peridotite	1.4	0.45
3	Idem, another sample	2.3	0.43
4	Eclogite	2.9	0.44
5	Idem, another sample	2.0	1.3
6	Garnet-pyroxenite	1.7	0.67
7	Amphibole-garnet pyroxene	1.8	0.37
8	Micaceous peridotite	5.1	0.84
Inclusions in kimberlites, Kimberley, South Africa [b]			
9	Phlogopite nodule	27.5	< 0.7
10	Phlogopite-bearing peridotite	2.1	< 5.4
11	Clinopyroxene-ilmenite intergrowth	1.3	< 0.97
12	Olivine megacryst	0.5	< 6.2
Industrial diamonds, Kimberley, South Africa [c]			
13	Diamonds	3.48	8.2 ± 0.3
14	Idem, another sample	0.92	19.5 ± 0.7

[a] Unpublished data by L.V. Khabarin, I.N. Tolstikhin and A.V. Ukhanov.
[b] From Kaneoka et al. (1977, 1978).
[c] From Takaoka and Ozima (1978a, b).

112

1975; Komarov and Ilupin, 1978). A similar age is reported by Kaneoka et al. (1978) for South African kimberlites.

The uranium content in ultramafic inclusions varies widely due to their contamination by uranium from kimberlites. Heier (1963) estimated the average content of uranium in eclogitic inclusions to be $5 \cdot 10^{-8}$ g g^{-1}. Akimov et al. (1968) found about $3 \cdot 10^{-8}$ g g^{-1} uranium in inclusions carefully refined from crustal trapped uranium. Similar values are typical of ultramafic xenoliths in basalts (Komarov and Shitkov, 1973). In accordance with these estimations the following contents may be adopted: $U = 3 \cdot 10^{-8}$ g g^{-1} and $Th = 10 \cdot 10^{-8}$ g g^{-1}. Thus it is easy to calculate the content of radiogenic ^{4}He, $1 \cdot 10^{-6}$ cm^3 g^{-1}, which is approximately the same as that observed in inclusions (Table 7.6). It should be noted that practically all crustal processes reduce the ^{3}He/^{4}He ratio in rocks with time (see Chapter 8), and the initial ^{3}He/^{4}He ratios appear to be considerably higher than the contemporary ones. This conclusion is confirmed by the results obtained for diamonds (samples *13* and *14*, Table 7.6) as well as for the olivine megacryst (sample *12*), which imply a deep source of ultramafic inclusions and diamonds enriched in ^{3}He. In the source the ^{3}He/^{4}He ratio is probably similar to that in diamond (*14*) as it is the only sample among those listed in Table 7.6 that underwent no contamination by crustal and/or air volatiles (section 7.4).

7.2.3. Helium isotopes in oceanic basalts

According to recent petrological, tectonic and geochemical notions, oceanic basalts are believed to be derived from a "depleted" mantle reservoir. Nevertheless, much of these rocks are enriched in volatiles due to specific conditions of their cooling and crystallization: high water pressure and a rapid decrease of magma temperature after eruption (resulting in glassy rims and crusts of chilling) provide the preservation of various gases, including noble gases.

There, a search for ^{3}He in deep-sea basalts appears to be worthwhile, and Fisher (1970) was the first who attempted to detect ^{3}He in oceanic basalts but he only succeeded in estimating an upper limit of ^{3}He, 10^{-10} cm^3 g^{-1}.

The first measurements (Table 7.7) of ^{3}He/^{4}He ratios in oceanic rocks (sampled mainly near rift zones in the Atlantic, Indian and Pacific Oceans) were carried out by Krylov et al. (1974). The ^{3}He/^{4}He ratios in the basalts were found to be about 10^{-5} — in full agreement with the idea of a ^{3}He excess in the deep interior of the earth (Fig. 7.2). Some of the samples contained high contents of helium isotopes. The authors pointed out an interesting tendency in the helium isotope distribution in a spherical tholeiitic basalt (Table 7.7): both the helium content and the ^{3}He/^{4}He ratio decrease from the glassy edge towards the crystalline center. Such a distribution was also observed in step-wise heating experiments: the low-temperature fractions of helium (released from the glass parts of samples) are enriched in ^{3}He, whereas the high-temperature ones (released from the crystalline center of samples) are depleted in the light helium isotope. For instance, the following values were obtained for the crystalline part of basalt (Table 7.7, No. 5):

$$(^3\text{He}/^4\text{He})_{600°C} : (^3\text{He}/^4\text{He})_{900°C} : (^3\text{He}/^4\text{He})_{1150°C} =$$
$$= 7.3 \cdot 10^{-6} : 3.8 \cdot 10^{-6} : 2.0 \cdot 10^{-6}$$

TABLE 7.7

Contents and isotope compositions of helium in oceanic basalts[a]

No.	Location; coordinates; depth	Rock	^{4}He (10^{-6} cm^3 g^{-1})	^{3}He/^{4}He (10^{-6})
Atlantic Ocean				
1	Romansh Depression; 1°S, 18°W; 7300 m	gabbro-diabase	0.05	41
2	Same location	amphibole-plagioclase rock	0.54	8.0
3	Same location	hyperbasite	0.14	1.2
Indian Ocean				
4	5°22'S, 62°08'E; 1810 m	harzburgite	0.08	4.3
5	34°17'S, 77°57'E; 3080 m	spherical lava, basalt:		
		glass basalt, fraction 2—0.25 mm	6.5	11
		idem, fraction < 0.25 mm	1.1	11
		intermediate zone, fraction 2—0.25 mm	0.4	7.5
		idem, fraction < 0.25 mm	0.3	7.1
		crystalline central part	0.15	5.7
6	23°22'S, 62°36'E; 4400 m	lherzolite	0.02	3.5
Pacific Ocean				
7	20°42'N, 170°53'W; 1970 m	plagiobasalt	3.5	4.3
8	19°19'S, 173°09'W; 5700 m	andesite	0.34	11
9	24°04'N, 157°34'W; 4415 m	basalt	0.64	8.7
10	13°34'S, 112°21'W; 2830 m	tholeiite	0.64	14.4
11	21°30'N, 108°46'W	tholeiite 0—1 mm, glass	1.48	12.7
		0—1 cm, glass	0.95	13.2
		3 cm from rim, crystalline	0.01	—
12	15°10'S, 176°38'W; 1325 m	olivine tholeiite	0.17	17.2
13	17°21'S, 176°26'W; 2170 m	olivine tholeiite	0.54	13.0

[a] Samples 1—9 from Krylov et al. (1974); samples 10—13 from Lupton and Craig (1975).

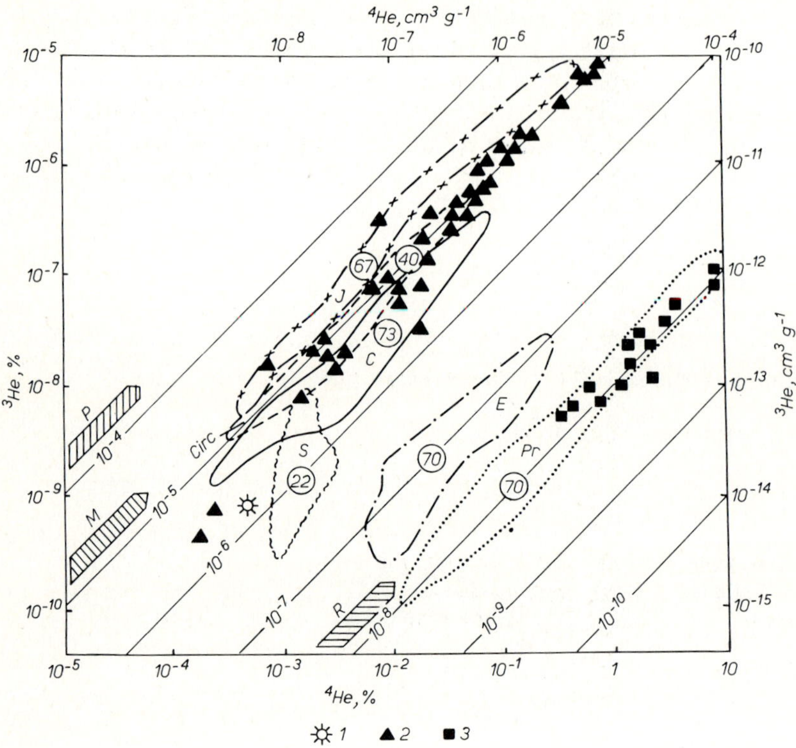

Fig. 7.2. Comparison of helium isotope distribution in rocks and terrestrial fluids.
1 = the atmosphere; *2* = oceanic basalts; *3* = ancient granitic rocks. The helium concentrations are given in cm³ g⁻¹ for rocks (upper and right axes) and in % by volume for gases (left and lower axes). The boundaries of the regions (marked by letters) and other symbols are the same as in Fig. 7.1; numbers of helium isotope analyses for each region are shown in circles (Tolstikhin, 1978).

In a pure glass fraction of the same sample the ^{3}He/^{4}He ratio was constant for all temperature intervals. Thus, the results obtained for temperature and mineral fractions are in good agreement.

The results of Krylov et al. (1974) were important experimental evidence for the origin of excess ^{3}He in seawater discovered by W.B. Clarke et al. (1969).

Later Lupton and Craig (1975) confirmed both main results of the previous work, namely a high value of ^{3}He/^{4}He, 10^{-5}, and small variations of the value in oceanic basalts. At that time it was known (Devirts et al., 1971; Tolstikhin et al., 1972a; Craig et al., 1975; see Chapter 9 for a more detailed discussion) that the isotope composition of helium injected in seawater is the same as in oceanic basalts. It was clear that oceanic basalt could be the source of excess helium isotopes in the ocean. Lupton and Craig (1975) compared the flux of helium isotopes through ocean water, their content in oceanic basalts and the areal rate of new crust formation. They arrived at the convenient conclusion that the ^{3}He and ^{4}He concentrations observed in basalts are residuals retained after outgassing of the rocks. In more recent works (Craig and Lupton, 1976; Anufriev et al., 1978; Kurz and Jenkins, 1981; and others) both the experimental data and the interpretation of the previous investigations were confirmed.

7.2.4. Helium isotopes in young erupted rocks

It seems likely that magmas and young erupted rocks of continental (and island) rifts and volcanic zones should be degassed more efficiently than those of the oceanic crust. Intense degassing occurs at the moment of magma eruption and during the comparatively slow solidification. Additional degassing of such rocks may take place due to active metamorphic processes caused by a high thermal activity of the regions (see section 8.5). Continental rocks are more likely to be contaminated by "crustal" radiogenic helium (trapped or accumulated *in situ*) because of the high concentration of radioactive elements in the crust. All these considerations indicate that young rocks erupted within the limits of continents must contain comparatively low helium contents and (in spite of their probable mantle origin) a wide range of ^{3}He/^{4}He ratios.

As shown in Fig. 7.3 and Table 7.8, continental rocks display a wide range of helium contents and isotopic compositions. In examining Fig. 7.3 attention should be paid to the fact that the ^{3}He/^{4}He ratios in recently erupted continental rocks vary over a wide range, from $5 \cdot 10^{-5}$ to $5 \cdot 10^{-8}$, and that the low ratios are found in rocks of probable mantle origin. For example, the ^{3}He/^{4}He ratios in Icelandic rocks vary from 10^{-7} to 10^{-5} (Mamyrin et al., 1974), whereas nearly constant ratios of ^{3}He/^{4}He $= (1—3) \cdot 10^{-5}$ are found in hot springs (Kononov et al., 1974; Polyak et al., 1976). On their turn, surface rocks cannot be considered as the source of helium in the hydrothermal fluids of Iceland. Magmas, very young intrusions and deep-mantle fluids probably serve as such a source.

It is important to note that some samples of alkaline and acid rocks show very high ^{3}He/^{4}He values (Table 7.8) and thus contain evidence of a deep-mantle source of their trapped volatiles. An extremely high ratio of ^{3}He/^{4}He $= 4 \cdot 10^{-5}$ (which reached $5.6 \cdot 10^{-5}$ under step-wise heating experiments) was observed in ugandite from Ruanda (Mamyrin et al., 1974). A similar ratio was obtained for amphibole from alkali basalts, Kakanui, New Zealand (Saito et al., 1978). The ratio of ^{3}He/^{4}He $= 2.2 \cdot 10^{-5}$ observed in a dacite (sampled from a modern lava flow of Karymskiy Volcano, Kamchatka) was twice as high as the maximum ratios observed in Kamchatka hot springs (Kamensky et al., 1976; Tolstikhin et al., 1976) and inclusions in basalts (Tolstikhin et al., 1974a).

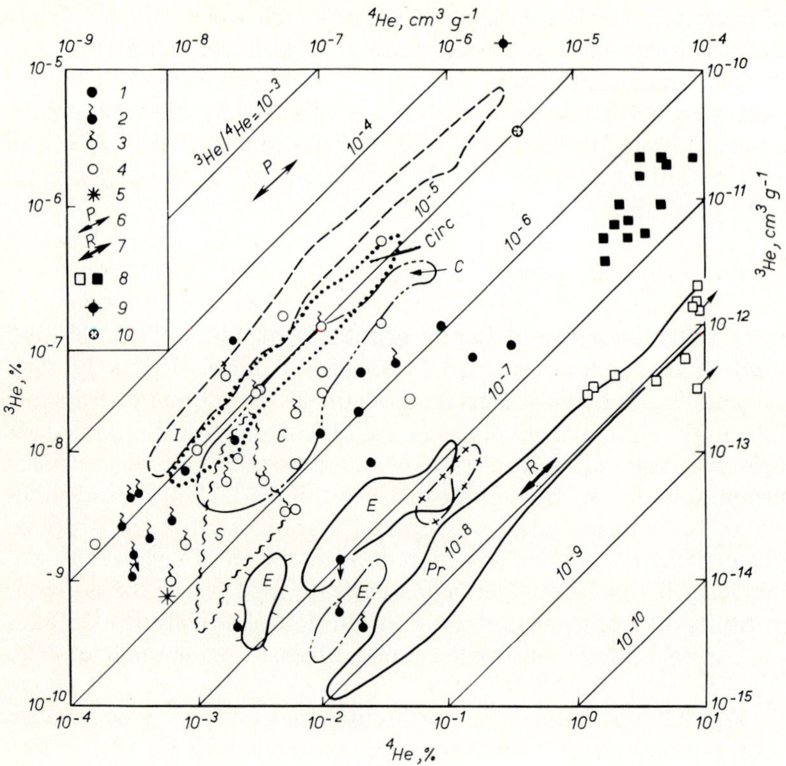

Fig. 7.3. Concentrations of helium isotopes in young continental and island rocks.
1 = Africa; *2* = Iceland; *3* = Kamchatka; *4* = some other regions; *5* = the atmosphere; *6* and *7* = isotope composition of primordial and radiogenic helium, respectively; *8* = ancient granitic rocks from the East European plate (open) and ancient ultrabasic rocks from the Kola Peninsula (solid), respectively; *9* = phenocryst of amphibole; *10* = diamond. For other symbols see Fig. 7.1 and 7.2. (Data from Gerling et al., 1971; Mamyrin et al., 1974; Tolstikhin et al., 1974a, 1976; Tolstikhin and Drubetskoy, 1975, 1977; Saito et al., 1978; Takaoka and Ozima, 1978a, b.)

These results disprove the idea that acid magmas generated from crustal matter. On the contrary, the sources of the volatile components of acid and alkaline magmas appear to be deep mantle layers (deeper than those represented by ultrabasic inclusions).

In addition, Fig. 7.3 shows the helium isotope distribution in some ancient ultrabasic rocks of Monche-tundra (Kola Peninsula). Gerling et al. (1963, 1968) reported a high content of radiogenic argon in these rocks and inferred unrealistic values of K-Ar ages up to 10 b.y. Later more precise measurements were carried out for these rocks (Shukolyukov and Tolstikhin, 1965; Kaneoka, 1974), and $^{36}Ar/^{40}Ar$ ratios of $(0.3-1.0) \cdot 10^{-4}$ were measured. Interpreting these data Kaneoka (1974) concluded that most of ^{36}Ar was already degassed from the earth's interior at least 2 or 3 b.y. ago.

The ^{3}He/^{4}He ratios in these rocks are about 1/100 of the modern mantle ratio and are only 1/500 that in the ancient mantle (Tolstikhin et al., 1975a). Assuming (1) that variations of ^{36}Ar/^{40}Ar as well as ^{3}He/^{4}He ratios in rocks

TABLE 7.8

Contents and isotope compositions of helium in some young erupted rocks[a]

No. Location	Rock	^{4}He $(10^{-6}\,cm^3\,g^{-1})$	^{3}He/^{4}He (10^{-6})
Africa, West Rift Zone			
1 Rwanda, Visoke Volcano	ugandite	0.023	31
2 Same location	leucitic basanite	0.008	8.3
3 Uganda, Katunga Lake	katungite	0.06	1.3
4 Uganda, Kaljango	carbonatite lava	0.02	5.2
Africa, East Rift Zone			
5 Tanzania, Khanang	melilitic picrite	0.2	2.4
6 Uganda, Moroto	olivine melane-phelinite	3.1	0.2
7 Kenya, Athi River	phonolite	1.4	0.35
Iceland, northern part of rift zone			
8 Viti Crater	granophyre	0.003	∿ 3.5
9 Little Krabla	rhyolite	0.003	< 5
Iceland, southern part of rift zone			
10 Hekla Volcano	andesite-basalt	0.004	∿ 15
11 Hove	dacite-rhyolite	0.003	∿ 15
East Iceland			
12 Reydharfjördhur	basalt	0.23	0.32
13 Same location	olivine basalt	0.40	1.2
14 Berufjödhurtindur	rhyolite	0.13	0.04
Kamchatka, volcanic zone			
15 Middle ridge	ignimbrite	0.05	0.63
16 Kaldera Uzon	rhyolite	0.017	3.17
17 Same location	rhyolite	0.021	4.1
18 Karymsky Volcano	dacite	0.015	22

[a] Samples 1—14 from Mamyrin et al. (1974); samples 15—18 from Tolstikhin et al. (1976).

occur due to the enrichment of these gases in radiogenic isotopes and (2) that $(^4He/^{40}Ar)_{rad}$ is approximately constant (see sections 8.1 and 10.2), one may use the above-mentioned coefficients to obtain a $^{40}Ar/^{36}Ar$ value for the ancient mantle, which is lower than that in the present atmosphere. Notably, this rough estimate is in agreement with calculations for the continuous degassing model (Tolstikhin et al., 1975), and it does not imply a catastrophic outgassing of the earth.

Some ancient acid rocks sampled from old platform regions (Ukraine, Kola Peninsula, etc.) are presented in Fig. 7.2. The agreement of the data obtained from rocks and gases of such regions as well as the low $^3He/^4He$ value of about 10^{-8} are remarkable and constitute valid evidence that in this case gas-well helium is really accumulated ". . . neither from radioactive minerals as such, nor from the atmosphere, nor from preplanetary materials, but from a large mass of ordinary granite rock . . ." (Morrison and Pine, 1955).

7.2.5. Conclusions

The analyses of the helium isotope distribution in thermal fluids, ultrabasic nodules in basalts, oceanic basalts, young continental erupted rocks (summarized in Figs. 7.1, 7.2 and 7.3) have led to the following conclusions.

(1) The earth's mantle is a helium source with an extremely high $^3He/^4He$ ratio, as compared to radiogenic values. This ratio varies only slightly in the mantle, and the distribution of helium isotopes is more or less homogeneous in the mantle. A ratio of $^3He/^4He = (3 \pm 2) \cdot 10^{-5}$ can be adopted as the average value[1] (see section 7.4 for a more detailed discussion). Within the limits of these variations the highest values were observed in some samples of phenocrysts in basalts, alkaline and acid rocks (up to $5 \cdot 10^{-5}$); the lowest values are typical of oceanic tholeiitic basalts ($\sim 1 \cdot 10^{-5}$).

(2) In spite of a wide range of ^3He and ^4He content in submarine basalts, the $^3He/^4He$ ratio is practically constant and very similar to that in thermal fluids of the Circum-Pacific volcanic belt. The "depleted" mantle is considered to be a source of oceanic basalts, and partial melting and degassing of these basalts may provide some volatiles, including helium, in thermal fluids of the Circum-Pacific belt.

(3) On the other hand, the $^3He/^4He$ ratios in hot springs of Iceland and Hawaii are about two or three times higher than the average ratio in basalts, and helium of Hawaiian phenocrysts is even five times more enriched in ^3He as compared to the helium of oceanic basalts. These data imply the presence of an "undepleted" mantle source of helium with the highest $^3He/^4He$ ratios.

---

[1] Compare with the first estimates by Kononov et al. (1974) equal to $(3 \pm 1) \cdot 10^{-5}$; this figure was assumed from Icelandic hot spring data.

(4) ^{3}He/^{4}He ratios in ultrabasic nodules in kimberlites are one or two orders of magnitude lower than those in recent mantle. This is caused by the contamination of the samples by radiogenic helium. Nevertheless, there is a set of data sufficient to confirm the high ^{3}He/^{4}He value in the subcontinental mantle (see Chapter 8).

7.3. On the nature of ^{3}He excess in the earth's mantle

In the earth sciences which deal with the reconstruction of events which occurred very long ago under conditions that have no parallel in human experience, it is hardly possible to speak of "exact and final" solutions of most questions or problems. The best thing we can do is to see that our theories should not contradict available experimental data. With this end in view we turn to the problem of ^{3}He excess in mantle helium.

It is known (see Introduction) that there are three main processes that produce isotopes in nature: nucleosynthetic processes, interaction of cosmic rays and matter, and radioactive decay and associated nuclear reactions.

The abundance of ^{3}He in radiogenic terrestrial helium is rather low (^{3}He/^{4}He $\approx 2 \cdot 10^{-8}$), and it is impossible to account for a considerably higher proportion of this isotope in mantle helium (^{3}He/^{4}He $\approx 3 \cdot 10^{-5}$) by known nuclear processes occurring inside the earth (section 7.1).

The ^{3}He/^{4}He ratio in cosmogenic helium is the highest in nature, amounting to $(1-3) \cdot 10^{-1}$. If we disregard the relationship between helium and other noble gases, it might seem probable that ^{3}He excess in solid earth is due to the contribution of cosmogenic helium. However, if we compare the data on helium isotopes and other noble gases we shall find a negligible amount of cosmogenic helium in terrestrial materials.

As was noted by many authors (Signer and Suess, 1963; Anders, 1964; Wasson, 1969; Shukolyukov and Levsky, 1972; and others), the isotope composition of noble gases in the earth's atmosphere (as well as in the atmospheres of terrestrial planets) is the result of mixing of primordial and radiogenic components (see Chapter 6). Despite the dissipation of helium from the atmosphere, a simple estimation confirms that this conclusion is in keeping with the origin of the helium isotopes. In Chapter 6 the following isotopic ratios were assumed: (^{4}He/^{20}Ne)$_{prim}$ = 500; (^{3}He/^{4}He)$_{prim}$ = 2.4 $\cdot$ 10^{-4}; the ratio of (^{3}He/^{21}Ne)$_{cosm}$ = 5 is also well defined in isotope cosmochemistry of rare gases (Herzog, 1973). Taking into account that all ^{20}Ne in the atmosphere is primordial (Shukolyukov et al., 1974) and that the ratio of (^{21}Ne/^{20}Ne)$_{atm}$ is equal to 0.00293, the value of ^{3}He$_{prim}$/^{3}He$_{cosm}$ in a "nondissipated" atmosphere can be calculated assuming all ^{21}Ne$_{atm}$ to be of cosmogenic origin (!):

$$\frac{^3\mathrm{He}_{\mathrm{prim}}}{^3\mathrm{He}_{\mathrm{cosm}}} = \frac{(^3\mathrm{He}/^4\mathrm{He})_{\mathrm{prim}} \; (^4\mathrm{He}/^{20}\mathrm{He})_{\mathrm{prim}} \; ^{20}\mathrm{Ne}_{\mathrm{atm}}}{(^3\mathrm{He}/^{21}\mathrm{Ne})_{\mathrm{cosm}} \; (^{21}\mathrm{Ne}/^{20}\mathrm{Ne})_{\mathrm{atm}} \; ^{20}\mathrm{Ne}_{\mathrm{atm}}} = 8.2 \qquad (7.13)$$

But actually the contribution of radiogenic $^{21}\mathrm{Ne}$ in the atmosphere is much lower than 10% (usually reported values are 2—4%, Shukolyukov and Levsky, 1972) and, consequently, the ratio (7.13) is higher that 100. Thus, only a small proportion of $^3\mathrm{He}$ in a "nondissipated" atmosphere can be due to the cosmogenic component, and if we share the widely accepted view that the atmosphere was formed due to outgassing of the earth, we may apply this conclusion to the solid earth as well. A similar estimate has been independently obtained by Bernatowicz and Podosek (1978) who arrived at the same conclusion.

In a broad sense nuclides produced by the interaction of cosmic rays with the earth's matter are also related to cosmogenic isotopes. Takagi et al. (1967) and Takagi (1970) attempted to explain isotope anomalies of noble gases by μ-mezon (muon) interaction with terrestrial materials; the cross-sections of the interaction was adopted to be equal to $2.4 \cdot 10^{-29}$ cm^2. As inferred from experimental data on muon absorption in the earth's matter and from theoretical results, the intensity of muon flux decreases substantially as depth increases (Bugaev et al., 1970): at depths of 3—5 km the muon flux is about eight orders of magnitude lower than that on the surface. This factor may be used as a basis for testing Takagi's assumption, for example, if $^3\mathrm{He}$ was produced by muon interactions with the earth's matter, its content in abyssal rocks should have been much less (by many orders) than in rocks of almost the same age occurring nearer to the earth's surface.

The data available in the field of isotope geochemistry of helium suggest, on the contrary, that $^3\mathrm{He}$ excess is in a first approximation proportional to the depth of sample formation. This thesis can be illustrated by the following example. Thorough investigations of quartz crystals from chambered pegmatites of Volyn (which were formed 1720 m.y. ago at a depth of about 3 km and at present, owing to erosion, are practically on the surface) have shown (Prasolov, 1972): (1) the recent $^3\mathrm{He}$ content in samples (free of microinclusions) is about 10^{-13} cm^3 g^{-1}; (2) the content of radioactive elements in crystals is extremely low, U $\approx 10^{-9}$ g g^{-1}; and (3) the diffusion cofficent of helium isotopes in crystalline quartz is extremely low, D $\approx 10^{-19}$ cm^2 sec^{-1} (Tolstikhin et al., 1974b). If we would follow Takagi (1970), then negligible, experimentally nonmeasurable contents of $^3\mathrm{He}$ would occur in abyssal rocks and nodules formed in the mantle at depths of more than 50 km and erupted on the earth surface during the last 10^2—10^5 years. Actually a much greater content of $^3\mathrm{He}$ is typical of rocks of mantle origin (see Tables 7.4—7.8).

Thus, cosmic ray interaction with the earth's matter cannot provide $^3\mathrm{He}$ excess in mantle helium, despite the fact that some products of such interaction are said to have been observed (see the recent studies of Hampel et al., 1975).

In the light of available experimental data and theoretical considerations the best explanation of the helium isotope composition in the earth mantle is a mixing of continuously accumulating radiogenic helium and primordial helium[1] trapped at the period of earth accretion. Isotopic ratios of $^3He/^4He$ in natural terrestrial materials lie between radiogenic and primordial values. The total mantle ratio of $^3He/^4He \approx 3 \cdot 10^{-5}$ is the result of mixing radiogenic ($^3He/^4He \approx 2 \cdot 10^{-8}$) and primordial ($^3He/^4He \approx 2.4 \cdot 10^{-4}$) helium in a proportion of about 1/7. In other words, mantle helium consists of radiogenic ^4He ($\sim 90\%$), primordial ^4He ($\sim 10\%$) and primordial ^3He ($\sim 100\%$).

This explanation of the helium isotope composition in deep earth's interior is in keeping with numerous geochemical observations, such as the content and isotope composition of planetary primordial gases in the earth's atmosphere, the contents and distribution of radioactive elements within the earth, the helium isotope balance in a "nondissipated" atmosphere, etc. It is important that recent models of earth degassing and differentiation show a quantitative consistency of data on helium isotopes and other terrestrial isotopic systems (see Chapter 11). Such explanation is widely accepted now.

At the end of this section we should like to emphasize that isotope composition of terrestrial helium is a unique key to the study of the earth's volatiles. Moreover, this ratio is and will always be the only clear tracer of juvenile fluids. Two well-established facts prove the above mentioned: (1) both helium isotopes escape from the earth's atmosphere into space and the helium content in the atmosphere is extremely low; the isotope ratios of $^3He/^4He$ in rocks and gases therefore are not distorted by contamination of atmospheric helium; (2) the isotopic ratio of $^3He/^4He$ in primordial helium is about 10 000 times greater than that in radiogenic helium; the ratios in natural helium therefore vary widely, making reliable interpretations possible.

In the following sections we shall, among other things, try to use this key to make clearer our notions about the neon and argon isotopic composition in the mantle.

7.4. $^3He/^4He$ ratio and ^{20}Ne excess in terrestrial samples

Search for other primordial gases in the earth's mantle was a logical continuation of helium investigations. We shall therefore, consider some recent results on the isotope geochemistry of neon (Fig. 7.4). The most impressive of these is the discovery of comparatively high ratios of $^{20}Ne/^{22}Ne$ equal to 10.6 — well above the atmospheric value of 9.85 (Verkhovsky and Shukolyukov, 1975). Later similar data were reported by Anufriev et al. (1976, 1978),

---

[1] Isotope fractionation of terrestrial rare gases (helium in particular) due to their migration does not exceed 10%, and there is no need to discuss these processes in connection with the origin of very large ^3He excess in the earth's interior.

122

and Craig and Lupton (1976). It is notable that excess ^{20}Ne was observed in the samples which varied in age, origin and location, but most samples showed high ratios of ^{3}He/^{4}He, and the authors considered the ^{20}Ne excess to bear evidence of the occurrence of primordial neon in solid earth.

However, such an interpretation is not convincing and some objections may be raised (Tolstikhin, 1978, 1980b). For instance, the isotope composition of atmospheric neon cannot be provided by mixing of neon from different terrestrial sources as there are no values in region X (Fig. 7.4). So we must assume that the mantle was heterogeneous, but there is very little likelihood that values in region X will ever be observed.

Other non-convincing assumptions have to be used for the explanation of a wide variation of the ratio of primordial isotopes ^{3}He and ^{20}Ne — in spite of the value of ^{20}Ne/^{22}Ne which is adopted for the mantle (Fig. 7.5). If we assume that the range is provided by the heterogeneity of the mantle, this assumption will be in disagreement with the constant ratio of ^{3}He/^{4}He which reflects the constancy in the ratio of ^{3}He/U in the mantle. If we as-

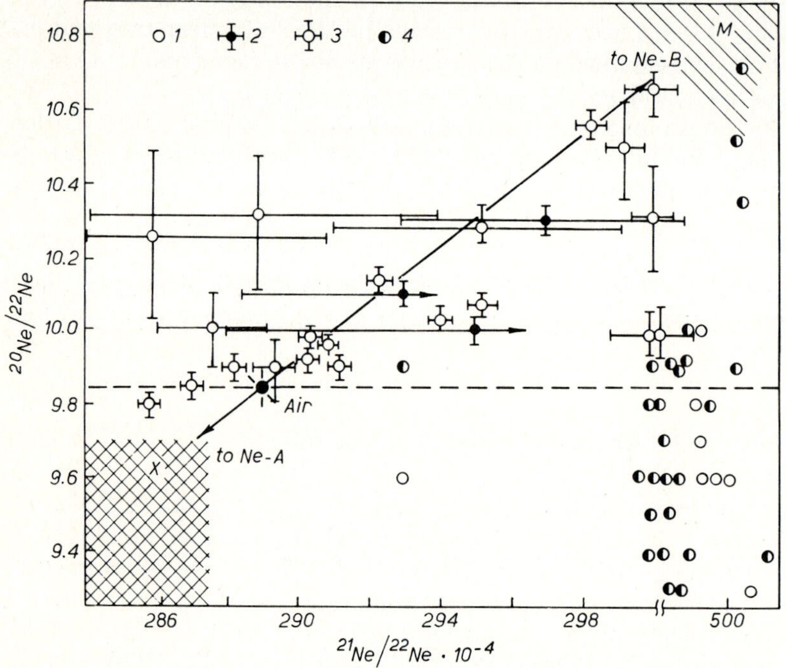

Fig. 7.4. Isotopic composition of neon in terrestrial samples. (Data from: *1*, Shukolyukov et al. (1974); *2*, Craig and Lupton (1976); *3*, Anufriev et al. (1976, 1978); *4*, Verkhovsky and Shukolyukov, 1975; and Verkhovsky et al., 1977.) If mantle neon is assumed to be located in the vicinity of *M*, atmospheric neon would not be produced by the mixing of known terrestrial sources of neon as points in the vicinity of *X* have not been observed.

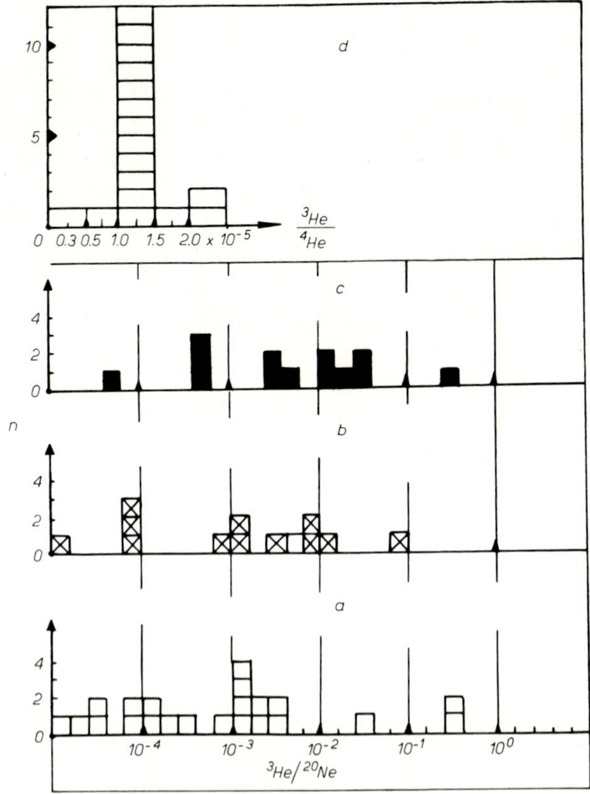

Fig. 7.5. Variations in terrestrial $^3He/^{20}Ne$ ratios (n = number of cases) (Tolstikhin, 1980b). Measured contents of ^3He are used with: (a) measured contents of ^{20}Ne; (b) calculated contents of "mantle" ^{20}Ne if the highest observed terrestrial $^{20}Ne/^{22}Ne$ ratio of 10.7 is taken as the "mantle" value; and (c) calculated contents of "mantle" ^{20}Ne if the mantle ratio is assumed to be similar to that in solar neon B, $^{20}Ne/^{22}Ne$ = 14. In the entire histogram the ratios of primordial isotopes vary widely, whereas the primordial $^3He/$ radiogenic ^4He ratios are more or less constant (d), indicating a more or less homogeneous distribution of primordial helium and uranium in the mantle.

sume further that this wide range is a result of isotope fractionation in migration processes, the assumption will be in disagreement (Fig. 7.6) with the constant ratio of radiogenic isotopes of He to Ne which reflects the constancy in their generation in rocks. Fig. 7.6 shows that noble gases of the same origin and dislocation undergo little fractionation during their migration.

Attention should be drawn to the fact that $^{20}Ne/^{22}Ne$ ratios in samples of the highest ^3He enrichment (see section 7.5) are similar to the atmospheric ratio within the limits of accuracy of measurements.

The difficulty in interpretating ^{20}Ne excess as originating from the mantle has led to the conclusion (Tolstikhin, 1978, 1980b; Nagao et al., 1979, 1981; Kaneoka, 1980) that somewhat higher ^{20}Ne/^{22}Ne ratios are provided by mass-fractionation processes. A satisfactory explanation for the cases when one and the same sample contains fractionated neon but unfractionated argon has recently been given by Nagao et al. (1981). These authors have shown that two natural processes, mass fractionation and mixing, account for the isotopic deviation of neon and argon, as well as elemental adundance patterns of noble gases in thermal fluids (Fig. 7.7). It should be noted that both end members in the model of Nagao et al. (1979, 1981) are ordinary natural samples derived mostly from the atmospheric reservoir: one of them

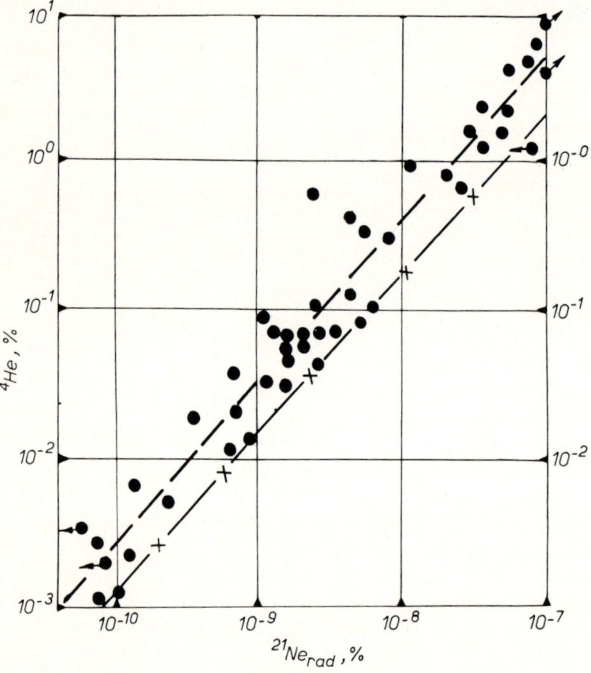

Fig. 7.6. Relationship between radiogenic isotopes of helium and neon in various terrestrial fluids (Verkhovsky et al., 1977).
Line with crosses represents the hypothetical ratio of $(^4He/^{21}Ne)_{rad}$ when ^{21}Ne is produced due to the reaction $^{18}O\,(\alpha, n)\,^{21}Ne$. This ratio is calculated by:

$$(^4He/^{21}Ne)_{rad} = \sum_i N_iS_i/q_oN_oS_o = 1.2 \cdot 10^7$$

where N and S denote the content of elements in rock and their brake capacity, respectively; q is the yield of the (α, n) reaction; subscripts o and i mark oxygen and other elements, respectively. Dashed line represents the observed ratio of $(^4He/^{21}Ne)_{rad} = (2.7 \pm 1.5) \cdot 10^7$ which is somewhat higher than the calculated one because of preferential losses of radiogenic helium.

is isotopically and elementally fractionated, another is enriched in heavy noble gases (Table 7.9). It is an important fact that heavy rare gases also show a significant mass fractionation. The mechanism of the mass-fractionation process has not been specified yet, but Nagao et al. (1979) noted that a single step of the Rayleigh process cannot produce mass-fractionation observed in samples 1 and 2 (Table 7.9) and that the actual process must ac-

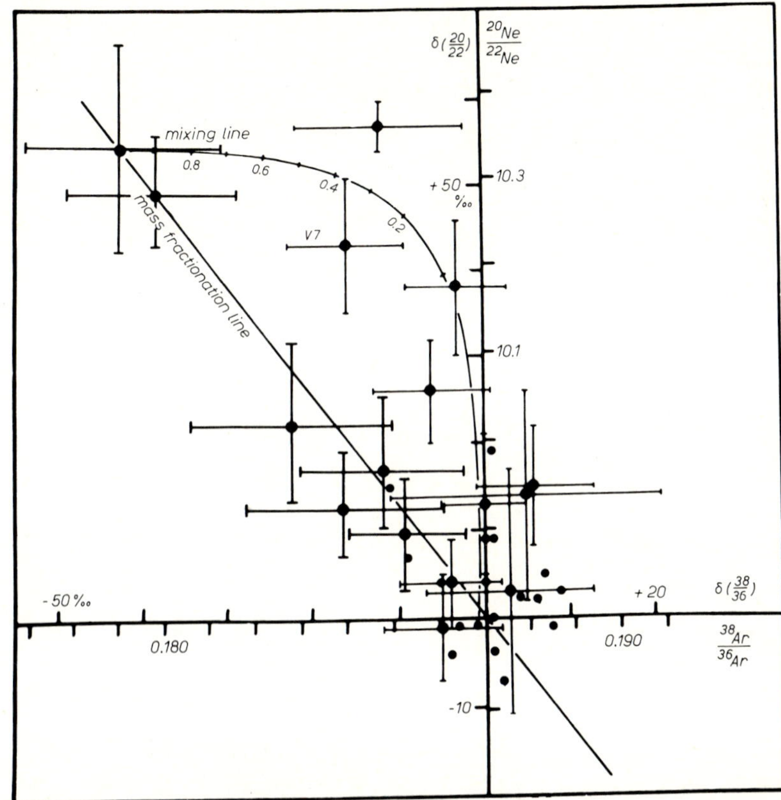

Fig. 7.7. Relationship between $^{20}Ne/^{22}Ne$ and $^{38}Ar/^{36}Ar$ isotopic ratios (Nagao et al., 1979, 1981). The slope of the mass-fractionation line is calculated as a ratio of $K(Ne)/K(Ar)$; K is the mass fractionation (%) per 1% mass difference. It is calculated from the following equation:

$$(R/R_{air} - 1) \cdot 10^3 = \delta\,(m/m_o) = K(N) \cdot [\,(m - m_o)/(m + m_o)/2\,]$$

where R and R_{air} denote measured isotopic ratio in a sample and in the atmosphere, respectively; N denotes noble gas, m and m_o are isotope masses. For instance, using the experimental data listed in Table 7.9 we obtain $K(Ne) = (54 : 21/2) \cdot 10^{-2} = 5.7$. Similar calculations for the abundance of argon isotopes in the same sample results in $K(Ar) = 7.9$ and the ratio of $K(Ar)/K(Ne) \approx 1.3$. The mixing line represents the mixture between two components: one is isotopically and elementally fractionated, and the other is enriched in heavy rare gases (samples 1 and 3 respectively, Table 7.9).

TABLE 7.9

Concentrations and isotopic ratios of noble gases in some volcanic and soil gases of Japan (Nagao et al., 1979, 1981)

No.	Sample, location	Concentration (ppm)			Isotopic ratios[a]			Fractionation coefficients	
		^{4}He	^{20}Ne	^{36}Ar	^{3}He/^{4}He (10^{-6})	^{20}Ne/^{22}Ne	^{38}Ar/^{36}Ar	K(Ne)	K(Ar)
1	Soil gas, Nigorikawa basin, Hokkaido	4.0	4.2	2.6	3.0	10.33/54	0.179/—43	5.7	—7.9
2	Same location, another sample	4.9	4.8	3.6	4.38	10.29/50	0.180/—37	5.3	—6.9
3	Free gas in hot spring, Koyohara, Shimane	5.3	0.034	0.53	9.15	9.79/—1	0.187/0.00	—0.1	0.00
4	Atmosphere	5.24	16.5	31.5	1.4	9.80	0.187	—	—

[a] The measured ratios and δ-values are given in the nominator and denominator, respectively; $\delta = (R_s - R_a/R_a)\,1000$, where R_s and R_a are the measured ratios in a sample and in the atmosphere, respectively.

count for the different fractionation of neon and argon. Further investigations in this field appear to be not only interesting for a new insight into fractionation processes but also useful for applied problems such as the origin and history of terrestrial fluids.

7.5. Helium/argon isotope relationship

The estimation of the $^{40}Ar/^{36}Ar$ ratio in the earth's mantle appears to be useful for K-Ar geochronology and isotope geochemistry. Young igneous rocks as a rule contain excess argon; a correction for "atmospheric" argon is calculated according to the ^{36}Ar content in the sample and the $^{40}Ar/^{36}Ar$ ratio in the atmosphere. In case ^{36}Ar is of juvenile origin (even partially), such a correction and calculated age would be wrong. So, the knowledge of the isotopic composition of argon in the upper mantle makes the geochemical base of K-Ar dating clearer.

A comparison of the argon isotopic composition in rock and in the upper mantle yields a more definite answer to the question of the origin of excess argon and other volatiles in this rock. For instance, if we can prove that ^{36}Ar in the rock has not been trapped from the atmosphere, then it follows that other gases might also be free from atmospheric contamination. A more likely interpretation of initial $^{40}Ar/^{36}Ar$ ratios (which can be obtained by K-Ar isochrone dating) becomes possible.

In the last decades the data of rare gas isotope geochemistry were used in models of the earth's evolution. Nearly all these models utilize the isotope composition of argon in the earth's interior. It is obvious that the absence of a valid $^{40}Ar/^{36}Ar$ value for the upper mantle decreases the reliability of the models and puts obstacles in the way of further investigations of the earth's degassing process. The abundance of primordial rare gases presently in the solid earth is connected with the temperature of its accretion, and models of high-temperature rapid accretion (Wetherill, 1975; S.P. Clarke et al., 1976) should be reshaped so that they will agree with the preservation of primordial gases in the earth's interior.

7.5.1. Review of the recent concepts

The definition of the $^{40}Ar/^{36}Ar$ ratio in the mantle is rather complicated due to strong contamination of samples by air and crustal argon. Argon and other rare gases (with the exception of helium) neither dissipate into space nor remain fixed within the crust. So all these gases, when released from solid earth, are accumulated in a most mobile reservoir, namely the atmosphere. The accumulation provides relatively high partial pressures of gases in the atmosphere and, correspondingly, a high probability of contamination of subsurface rocks and fluid in particular. Therefore investigated samples as a rule contain: (1) radiogenic ^{40}Ar trapped or accumulated *in situ*; (2) atmospheric ^{40}Ar, ^{38}Ar and ^{36}Ar isotopes; (3) mantle or juvenile argon. The ^{38}Ar/^{36}Ar ratio appears to be similar in all reservoirs — the upper mantle, the crust and the atmosphere — and may hardly be used as a criterion for the juvenile nature of argon. If the ^{40}Ar/^{36}Ar ratio in the mantle is higher than the atmospheric value, then a similar ratio can be obtained for a mixture of atmospheric and radiogenic crustal argon; in this case the ^{40}Ar/^{36}Ar ratio cannot be used for the identification of the argon source either. It has become clear that argon itself may be used for the identification of its own source only in the case of a $(^{40}\text{Ar}/^{36}\text{Ar})_{\text{mantle}} < (^{40}\text{Ar}/^{36}\text{Ar})_{\text{air}}$.

Ratios of ^{40}Ar/^{36}Ar smaller than 295.5 — that is, somewhat less than in the atmosphere — were firstly discovered in the hot springs of the Kuril—Kamchatka volcanic region by Cherdyntsev et al. (1967) and Cherdyntsev and Shitov (1976). These authors claimed that the source of such argon was (partly) the upper mantle. Later the experimental results of these authors were confirmed (Krummenacher, 1970; Baskov et al., 1973; Prasolov, 1976), but a different interpretation was put forward. Krummenacher carefully measured both ^{40}Ar/^{36}Ar and ^{38}Ar/^{36}Ar ratios in young surface igneous rocks and showed that these ratios approach the fractionation line within the limits of error and, correspondingly, the observed excess of ^{36}Ar (about 6%) can be due to fractionation of air argon isotopes during their migration and/or trapping. The same approximation of experimental points was later established by Smelov et al. (1975), Nagao et al. (1979), Kaneoka (1980) and others.

In some cases isochrone K-Ar dating results in low (as compared to the earth atmosphere) initial ^{40}Ar/^{36}Ar ratios (Brown et al., 1974). They were supposed to bear evidence of the low abundance of radiogenic ^{40}Ar isotope in the mantle. Brown et al. (1974) suggested that 10^9 years ago the ^{40}Ar/^{36}Ar ratio in the mantle was equal to 100 (!). This work was critically reviewed by Alexander and Schwartzman (1976) who showed that it was invalid as it was based on insufficient experimental data. Moreover, Brown et al. (1974) considered the mantle to be the main source of atmospheric argon (the contribution of the crust was assumed insignificant), but they apparently overlooked the fact that the suggested low ^{40}Ar/^{36}Ar ratio in the mantle was incompatible with the relatively high ratio in the atmosphere. In the light of the above-mentioned, the new measurements performed by Melton and Giardini (1980) must also be carefully checked.

Thus, it has not been proved so far that the mantle ^{40}Ar/^{36}Ar ratio is lower than that of the atmosphere. The presence of such argon in the mantle is hardly possible in view of the existing balance pattern of the mantle, crust and atmospheric reservoirs.

Ratios of ^{40}Ar/^{36}Ar $\geqslant$ 295.5 were often observed in young erupted rocks (Funkhouser and Naughton, 1968; Fisher, 1975 and others) and hot springs (Baskov et al., 1973; Kamensky et al., 1976; Matsubayashi et al., 1978; and others) — that is, in samples coming from regions of high tectonic and magmatic activity.

Investigating young erupted rocks appears to be the most promising way to estimate the argon isotope composition in the mantle, and many papers contain results obtained in this field. Some of these are devoted to oceanic basalts whose rapidly cooled portions are often rich in trapped rare gases (Krylov et al., 1974; Fisher, 1975 and others). A wide range of $^{40}Ar/^{36}Ar$ values, from atmospheric to 15 000, was observed in these rocks. Ultrabasic inclusions in basalts show a similar range (Funkhouser and Naughton, 1968; Kaneoka and Takaoka, 1978, 1980; and others). Usually the authors discuss two reservoir models which take for granted the possibility of mixing mantle and atmospheric argon (Dymond and Hogan, 1978). Because of this model they believe that the highest ratios of $^{40}Ar/^{36}Ar$ are typical of mantle argon while the intermediate ones may be seen as resulting from the mixing of mantle and atmospheric gases. Moreover, some authors (such as Kaneoka, 1974) measured $^{40}Ar/^{36}Ar$ ratios in old magmatic rocks and came to the conclusion that this ratio in the ancient mantle (about 3 b.y. ago) was as high as 10 000 (!).

To make a more or less happy choice of the $^{40}Ar/^{36}Ar$ ratio in the mantle, another isotopic ration, $(^{4}He/^{40}Ar)_{rad}$, was taken into consideration (Fisher, 1975; Dymond and Hogan, 1978; Hamano and Ozima, 1978), but an experimental approach to the $(^{4}He/^{40}Ar)_{rad}$ value in the mantle appears to be an even more complex problem than the estimation of the $(^{40}Ar/^{36}Ar)$ ratio.

Ozima (1975) summarized the situation best of all: ". . . our present knowledge of the $(^{40}Ar/^{36}Ar)$ ratio in the mantle is still far from conclusive. This is chiefly due to atmospheric Ar contamination (. . .) and to the difficulty in correction for radiogenic ^{40}Ar . . .".

7.5.2. Helium and argon isotope systematics

The above-mentioned considerations make it clear that an additional independent criterion is necessary for a reliable estimation of the $^{40}Ar/^{36}Ar$ ratio in the earth's mantle. It was shown that the $^{3}He/^{4}He$ ratio might serve as such a criterion (Tolstikhin, 1978, 1980b; Tolstikhin et al., 1978; Drubetskoy et al., 1979). In Fig. 7.8 the measured results of the isotopic ratios in young erupted rocks of probable mantle origin (corrected only for instrumental background) are plotted on a coordinate plane, $^{3}He/^{4}He$ versus $^{36}Ar/^{40}Ar$.

It should be noted the primordial isotopes ^{3}He and ^{36}Ar are in the nominator and the radiogenic ones, ^{4}He and ^{40}Ar, in the denominator. The range of the $^{3}He/^{4}He$ ratio in the mantle has been discussed above (see sections 7.2 and 7.3). Mantle samples without contamination by both atmospheric and crustal helium will occupy the mantle helium zone (for example, in the vicinity of point M). As radiogenic isotopes of ^{4}He and ^{40}Ar accumulate

130

in the samples owing to radioactive decay and/or trapping of crustal gases, the points in Fig. 7.8 shift from the mantle zone along the hypothetical line *MO* to the vicinity of *0*, where samples enriched in radiogenic gases must crowd together. Pure radiogenic ratios of $^3He/^4He \sim 10^{-8}$ and $^{36}Ar/^{40}Ar \sim 10^{-7}$ may be approximated by zero. Angle φ between line *MO* and the abscissa may be calculated from the formula:

$$\text{tg}\,\varphi = \frac{^3He/^4He}{^{36}Ar/^{40}Ar} = (^3He/^{36}Ar) \cdot (^4He/^{40}Ar)^{-1} \tag{7.14}$$

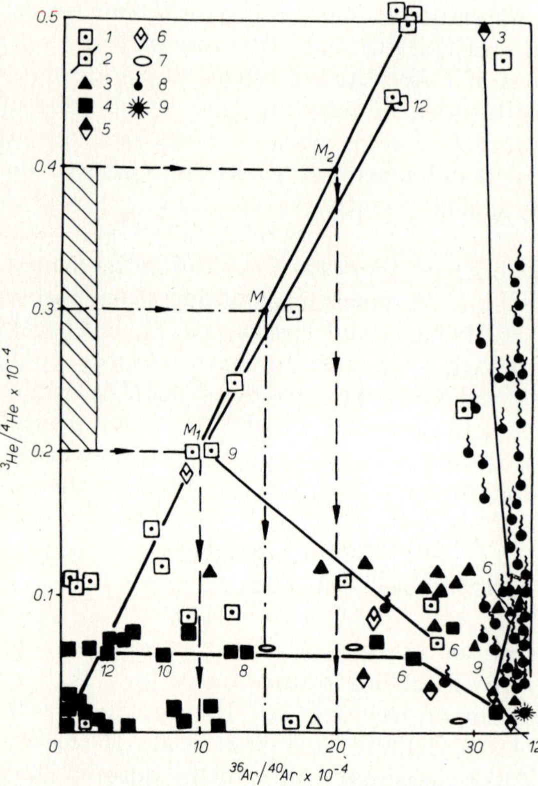

Fig. 7.8. Helium-argon isotopic systematics.
1 = ultrabasic nodules and phenocrysts; *2* = results obtained in step-wise heating experiments (figures near the symbols denote the temperature of heating, $\times\ 100°C$); *3* = oceanic basalts; *4* = crustal rocks with excess noble gases; *5* = young acid erupted rocks; *6* = diamonds; *7* = bed gases; *8* = thermal fluids; *9* = the atmosphere. The hatched rectangle shows the range of the $^3He/^4He$ ratio in the mantle assumed by Kononov et al. (1974). The figure shows that most samples of probable mantle origin are contaminated by atmospheric argon as well as by crustal gases. Samples without contamination approach the upper part of the boundary line *OM*. From these samples a $^{40}Ar/^{36}Ar$ value for the undepleted mantle of approximately 400 can be determined. For sources of data see the text.

The first ratio may be estimated from the ^{36}Ar content in normal atmosphere and the ^{3}He content in non-dissipated atmosphere (Tolstikhin et al., 1977b; see also section 10.2): $(^{3}\text{He}/^{36}\text{Ar})_{\text{prim}} = 0.8 \cdot 10^{19}$ cm^{3}/12.6 $\cdot 10^{19}$ cm^{3} = 0.063. The second ratio, ^{4}He/^{40}Ar, has been repeatedly discussed by many authors (Damon and Kulp, 1958b; Zartman et al., 1961; Wasserburg et al., 1963; Gerling et al., 1967a; Voronov et al., 1974; Craig and Lupton, 1976; see also sections 8.1 and 10.2). Summarizing in short the results of these works, it is possible to conclude that terrestrial gases show a $(^{4}\text{He}/^{40}\text{Ar})_{\text{rad}}$ ratio of about 10, which appears to be somewhat higher than the average one, and in the following discussion a total value of about 3 will be used.

The only assumption is that the ratios in released noble gases are similar to those in preserved ones. This assumption serves only for the tg φ estimation. Natural fractionation of rare gases will not change φ because the helium/argon ratios are both in the nominator and in the denominator of the fraction in eq. 7.14: for instance, if helium isotopes are released from a sample and argon is preserved, both ratios in the sample decrease but the value of fraction does not change.

Line $0M_2$ (Fig. 7.8) is drawn in accordance with the above-mentioned parameters; M_1 and M_2 are the points of intersection of this line and the boundaries of the zone of the mantle ^{3}He/^{4}He values, $2 \cdot 10^{-5} \leqslant {}^{3}\text{He}/^{4}\text{He} \leqslant 4 \cdot 10^{-5}$. If there was no air contamination of samples, the experimental points would have shifted in time along mixing line $M0$ from the mantle segment M_2M_1 towards the radiogenic segment in the vicinity of point 0 owing to the accumulation and/or trapping of radiogenic isotopes.

Systematic deflection of the points from line $M0$ is possible due to air contamination of samples. The low helium content in the atmosphere (because of its dissipation into space) makes the contamination of rocks and terrestrial fluids by air helium most improbable (see section 10.2). On the other hand, the effect of contamination by air argon is very important. If we neglect air helium contamination but take into account air argon contamination, we can rewrite formula 7.14 as follows:

$$\text{tg } \varphi_1 = \frac{^{3}\text{He}/(^{36}\text{Ar}_M + {}^{36}\text{Ar}_A)}{^{4}\text{He}/(^{40}\text{Ar}_M + {}^{40}\text{Ar}_A)} = \frac{^{3}\text{He}/^{36}\text{Ar}_M}{^{4}\text{He}/^{40}\text{Ar}_M} \cdot \left[\frac{1 + \overset{*}{^{40}\text{Ar}_A}/^{40}\text{Ar}_M}{1 + {}^{36}\text{Ar}_A/^{36}\text{Ar}_M} \right] \tag{7.15}$$

where indexes M and A denote mantle and atmospheric sources of isotopes, respectively. The ^{40}Ar/^{36}Ar ratio in the mantle appears to be higher than that of the atmosphere; in this case the inequality $^{36}\text{Ar}_A/^{36}\text{Ar}_M > {}^{40}\text{Ar}_A/^{40}\text{Ar}_M$ is correct, the second term ($*$) in formula 7.15 is less than 1 and $\varphi_1 < \varphi$. This means that air contamination deflects experimental points to the right from line $M0$, and this line should be the left boundary of the area containing the experimental points.

132

The experimental data in Fig. 7.8 are in good agreement with the fore-going consideration: despite the difference in types of samples (rocks, ultra-basic and alkaline inclusions in basalts, hot springs, bed gases), their location, genesis, etc., all samples which contain helium with the high ratio of $^3He/^4He > 1.5 \cdot 10^{-5}$ are placed to the right of line OM_2. The following samples approaching line MO are not contaminated and contain gases from the deep mantle:

(1) A megacryst inclusion of amphibole (kaersutite), picked up in an alkali basalt from Kakanui, New Zealand. The amphibole shows a very low $^{87}Sr/^{86}Sr$ ratio of 0.7029 and a high K/Rb ratio of 1142. These data suggest that the source of the sample is similar to that of the mid-oceanic ridge basalts (Saito et al., 1978).

(2) Industrial diamonds, believed to have come from the Kimberley Mines, South Africa (Takaoka and Ozima, 1978a, b).

(3) Olivine phenocrysts, separated from Kapuho lava erupted in 1960, Kilauea, Hawaii. The phenocrysts are greenish crystals (1—8 mm in size) including black impurities seen under a microscope (Kaneoka and Takaoka, 1978; Kaneoka et al., 1978).

(4) Olivine inclusion, sampled from basalts erupted in a glacier, Ross Island, Antarctica. The inclusion is about 5 cm in size, the olivine content being about 95%. High isotope ratios were observed in two stepwise heating experiments (Drubetskoy et al., 1979).

(5) Olivine phenocrysts from a lava flow, White Hill, Heleakala Volcano, Maui (Kaneoka and Takaoka, 1980).

Other samples appear to be contaminated by atmospheric and/or crustal radiogenic gases; some of these will be discussed below. Thermal fluids show a very high contamination by atmospheric argon and probably other air components.

Some important conclusions may be inferred from the results shown in Fig. 7.8.

The upper limit of the $^3He/^4He$ ratio in the mantle appears to be under-estimated in previous works (Kononov et al., 1974; Tolstikhin et al., 1975) because there is no doubt now that samples which show the very high $^3He/^4He$ ratios of $5 \cdot 10^{-5}$ represent the mantle reservoir. The source of the gases trapped by these samples may be an undepleted primitive mantle whose presence is indicated by comparison of the Sr-Nd isotopic systems (DePaolo and Wasserburg, 1976; Jacobsen and Wasserburg, 1979). A low $^{87}Sr/^{86}Sr$ ratio in the amphibole megacryst is not incompatible with such an interpretation because amphibole itself can be formed in the upper (depleted) mantle and at the same time it can trap volatiles from the deep mantle plume. As for the olivine and augite phenocrysts, some recent studies suggest that the Hawaii Archipelago is located on top of a hot mantle plume which may be considered as an upwelling of undepleted material derived from the primitive mantle. Evidence from the Sr-Nd isotope correlation (DePaolo and Wasserburg, 1976; O'Nions et al., 1977) also confirms that the volatiles in the samples can belong to a deep primitive reservoir. Kaneoka and Takaoka (1980) interpret their data in a similar way.

The ^{3}He/^{4}He ratio in helium of the upper (depleted) mantle appears to be within the range of $(1-1.5) \cdot 10^{-5}$, which is typical of oceanic tholeiites and hydrothermal systems of ocean rift (and rise) zones as well as of thermal fluids of the Circum-Pacific volcanic belt. It is an important fact that some samples containing such helium are enriched in radiogenic argon and deflected to the left of line MO. The enrichment may be explained in terms of the mixing model, as the result of co-occurrence of crustal rocks (pelagic sediments and/or rocks derived from continents in the subduction zone) and rocks of the depleted mantle and oceanic crust. It should be noted that crustal materials contain radiogenic ^{40}Ar in a high proportion to ^{4}He; the $(^4$He/^{40}Ar)$_{rad}$ ratio in a rock of the earth's crust is as a rule far lower than that inferred from the U/K ratio (Tolstikhin and Drubetskoy, 1975). So, the addition of such rocks to the upper mantle, which is the reservoir of oceanic crust, does not bring about a wide variation of helium isotope composition, whereas the ^{40}Ar/^{36}Ar ratio might greatly increase.

Intermediate ratios of ^{3}He/^{4}He, $(1.5-4.5) \cdot 10^{-5}$, may be due to the mixing of volatiles of depleted (^{3}He/^{4}He $\approx 1 \cdot 10^{-5}$) and enriched (^{3}He/^{4}He $\approx 5 \cdot 10^{-5}$) reservoirs or the continuous accumulation of radiogenic isotopes in outgassed matter of enriched reservoir. Occasionally in the following discussion we shall use this intermediate ^{3}He/^{4}He ratio of $3 \cdot 10^{-5}$ as typical of the total mantle. Fig. 7.8 enables estimating the ^{40}Ar/^{36}Ar ratio in the depleted and enriched mantle as 2000 and 400, respectively. Our earlier guess about this ratio in a homogeneous mantle (500—1000) is consistent with the former values (Tolstikhin et al., 1978; Drubetskoy et al., 1979). Similar ^{40}Ar/^{36}Ar values in the residual mantle were predicted by a model calculation of Tolstikhin et al. (1975) and O'Nions et al. (1979). These data led to the conclusion that the ^{40}Ar/^{36}Ar value in the mantle is higher than the atmospheric ratio, and it is probably within the limits of 400 to 2000 for all models of the mantle.

Attention should be drawn to the important fact that a very high ^{3}He/^{4}He ratio in an enriched mantle cannot be explained by the mixing of any terrestrial sources of helium, and the conclusion about the two types of reservoirs (depleted and enriched) in the earth's interior is inferred solely from the distribution of helium isotopes! The reason for this unique possibility is the existence of helium escape from the atmosphere: this global reservoir is practically devoid of helium, hence no mixing is possible. In fact, no other particular isotopic method can solve the problem of the earth's deep structure because the isotopic characteristics of the enriched reservoir could be obtained as the result of mixture of the crust and depleted mantle. The ideas of a three-reservoir solid earth are based mainly on the Sr-Nd correlation (DePaolo and Wasserburg, 1976), which is one of the most impressive results in recent isotope geochemistry. The helium isotope method (with respect to the above discussed problem) appears to be of similar importance.

If new experimental data will prove boundary line *M0* to be correct, it could be possible to discuss more carefully the values which determine tg φ: $(^{3}\mathrm{He}/^{4}\mathrm{He})_{\mathrm{prim}}$, $(^{4}\mathrm{He}/^{20}\mathrm{Ne})_{\mathrm{prim}}$ and $(^{4}\mathrm{He}/^{40}\mathrm{Ar})_{\mathrm{rad}}$.

The position of a sample in Fig. 7.8 appears to be the key to the origin of He and Ar as well as other volatile components: if the sample approaches boundary line *0M* and shows a high $^{3}\mathrm{He}/^{4}\mathrm{He}$ ratio, this means that the volatiles in this sample are not contaminated by both atmospheric and crustal gases. We believe that further investigations of such samples will lead to new interesting and important results.

Chapter 8

HELIUM ISOTOPES IN ROCKS, WATERS AND GASES OF THE EARTH'S CRUST

The crust is the only shell of the earth which is available for direct observation and investigation. Consequently, the most representative set of data has been obtained for the terrestrial crust, including data in the field of isotope geochemistry of helium. In this chapter the distribution of helium isotopes in various samples (rocks, minerals, terrestrial fluids, etc.) is interpreted from the genetic point of view, namely what sources and processes provide the abundance of helium isotopes observed in a sample?

The mixing of mantle, juvenile helium with pure radiogenic helium is the main process responsible for the helium isotope composition in any sample of the earth's crust (section 8.1), the share of each component (reflected in the ^{3}He/^{4}He ratio) depending on the history of the tectono-magmatic activity in the given region (8.5).

A specific chemical composition of a rock or mineral (8.3), peculiarities of losses or trapping (8.2) and a peculiar kind of distribution of radioactive elements (8.4) can lead to unusual isotopic ratios of ^{3}He/^{4}He in radiogenic helium.

Lastly, technogenic radioactive isotopes are widespread in nature; one of them, tritium (^{3}H), yields ^{3}He excess in terrestrial waters (8.6).

8.1. Distribution of ^{3}He and ^{4}He in rocks of the earth's crust

8.1.1. Measured and calculated ratios of ^{3}He/^{4}He

The most effective approach to the problem of the origin of helium isotopes in rocks — which are the main source of terrestrial helium — appears to be a comparison of measured ^{3}He/^{4}He ratios with calculated ones. Assuming a radiogenic origin for both helium isotopes, Morrison and Pine (1955) discussed the correlation of the chemical composition of a rock with the ^{3}He/^{4}He ratio. These authors compared the results of a calculation of the ^{3}He/^{4}He ratio in an ordinary granite with measurements of the same ratio in natural gases (Aldrich and Nier, 1948). Later, Gorshkov et al. (1966) improved the method of calculating the ^{3}He/^{4}He ratio in rocks and defined some of

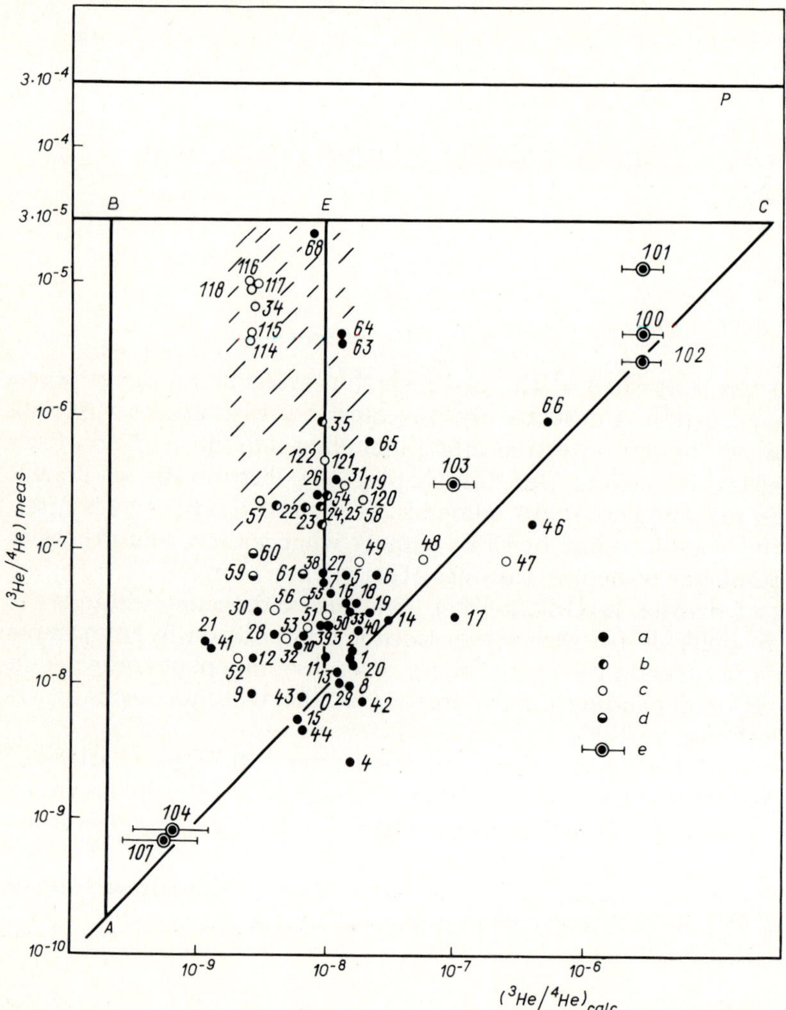

Fig. 8.1. Calculated and measured ratios of ^{3}He/^{4}He in different rocks and minerals (Tolstikhin and Drubetskoy, 1975a).

Isotope ratios of ^{3}He/^{4}He: P = in primordial helium; BEC = in mantle helium; AC = concordant line of (^{3}He/^{4}He)$_{calc}$ ratios equal to (^{3}He/^{4}He)$_{meas}$; OE = line of mixing of radiogenic (O) and mantle (E) helium in ordinary rocks. Types of rocks: a = acid; b = intermediate; c = basic; d = alkaline; e = uranium and lithium minerals. Age and location: $1—18$ = Precambrian, Ukraine; 19 = Paleozoic, near Lake Baikal; $20—25$ = Precambrian, Kola Peninsula and Karelia; $26—33$ = Neogene—Jurassic, Caucasus; $34—36$, $38—44$ = Precambrian—Paleozoic, Sayany; $46—53$ = Precambrian, Kola Peninsula; $54—61$ = Precambrian, Ukraine; $62—65$ = Quaternary, Kamchatka; 66, $100—103$ = lithium minerals (Adrich and Nier, 1948); 104, 107 = uranium minerals (Kamensky, 1970); $114—118$ = xenoliths in basalts (Tolstikhin et al., 1974a); $119—122$ = Precambrian, Kola Peninsula (Gerling et al., 1971). Some numbers on Fig. 8.1 and Fig. 8.2 are omitted due to the numbering adopted. The shaded region shows the possible area of samples of young magmatic rocks described by Krylov et al. (1974), Mamyrin et al. (1974) and Lupton and Craig (1975).

the attendant parameters more exactly. The first actual comparison of measured and calculated ratios was carried out by Gerling et al. (1971, 1972). The theory of the radiogenic origin of helium isotopes in rocks and a method of calculating radiogenic $^3He/^4He$ ratios are presented in section 7.1.

Fig. 8.1 and Table 8.1 (Tolstikhin and Drubetskoy, 1975, 1977) show measured versus calculated $^3He/^4He$ ratios of terrestrial samples. The points representing these ratios approach lines AC and OE. AC is a concordant line where $(^3He/^4He)_{meas} = (^3He/^4He)_{calc}$. Points corresponding to radioactive minerals, old acid rocks, minerals and rocks rich in lithium are scattered around AC.

The points on Fig. 8.1 which approach line OE represent ordinary rocks of different ages, locations and compositions. The low ratios of $(^3He/^4He)_{meas}$ are close to $2 \cdot 10^{-8}$ — that is, close to the pure radiogenic ratios (in the vicinity of point O). The high $^3He/^4He$ ratios of about $3 \cdot 10^{-5}$ (in the vicinity of point E) are typical of the "mantle mixture" of primordial and

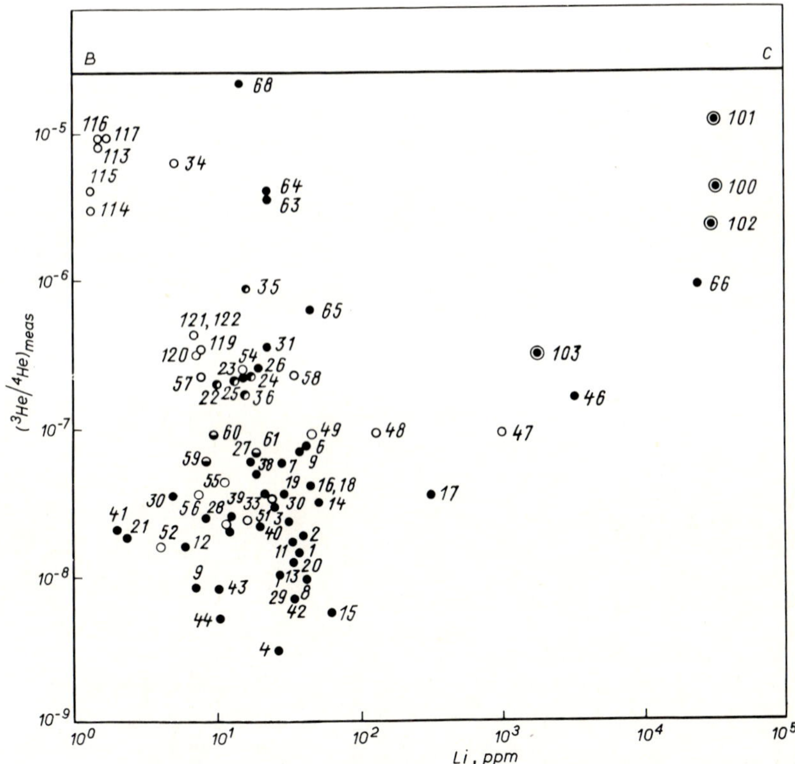

Fig. 8.2. Measured ratios of $^3He/^4He$ and lithium content in rocks (Tolstikhin and Drubetskoy, 1975). For legend see Fig. 8.1.

TABLE 8.1

Helium and argon isotopes, isotope ratios, and potassium, lithium, uranium and thorium contents in some USSR rocks (Tolstikhin and Drubetskoy, 1977)

No.	Sample	Measured concentrations						Measured ratio		Calculated ratio	Age[a]
		^{4}He (10^{-6} cm^3 STP g^{-1})	^{40}Ar$_{rad}$ (10^{-6} cm^3 STP g^{-1})	K (10^{-2} g g^{-1})	Li (10^{-6} g g^{-1})	U (10^{-6} g g^{-1})	Th (10^{-6} g g^{-1})	^{4}He/ ^{40}Ar$_{rad}$	^{3}He/^{4}He (10^{-6})	^{3}He/^{4}He (10^{-6})	(10^6 yr)
Ancient rocks of the Ukraine											
1	Granite	100	445	4.72	37	4.8	44	0.23	0.016	0.016	2000
2	Granite	84	470	4.87	38	9.5	42	0.18	0.019	0.017	2000
4	Granite	1135	331	4.04	26	11	85.4	3.43	0.003	0.014	2000
13	Albitite	2960	20	0.28	33	—	31.4	148	0.012	0.012	1750
15	Albitite	15700	14	0.23	61	—	105	1120	0.0055	0.006	1750
56	Labradorite	5.2	45	0.61	7.4	1.0	2.5	0.116	0.36	0.004	1750
59	Nepheline-syenite	130	300	4.59	8.3	2.5	16	0.43	0.06	0.003	1750
60	Alkaline syenite	18	310	4.37	9.3	1.0	2.6	0.06	0.09	0.003	1750
Ancient rocks of Kola Peninsula											
20	Rapakiwi granite	85	515	5.06	36	6.7	35	0.16	0.0145	0.016	1800
22	Diorite-gneiss	86	56	0.25	10	0.7	1.4	1.53	0.20	0.005	2700
25	Diorite-gneiss	145	350	1.6	15	1.3	6	0.41	0.22	0.009	2700
47	Amphibolite	2500	55	0.138	—	0.8	1.0	46	0.09	0.26	2700
48	Amphibolite	1300	71	0.19	125	0.9	1.1	18	0.09	0.05	2700

TABLE 8.1 (*continued*)

No. Sample	Measured concentrations						Measured ratio		Calculated ratio	Age[a]
	^{4}He (10^{-6} cm^3 STP g^{-1})	^{40}Ar$_{rad}$ (10^{-6} cm^3 STP g^{-1})	K (10^{-2} g g^{-1})	Li (10^{-6} g g^{-1})	U (10^{-6} g g^{-1})	Th (10^{-6} g g^{-1})	^{4}He/^{40}Ar$_{rad}$	^{3}He/^{4}He (10^{-6})	^{3}He/^{4}He (10^{-6})	(10^6 yr)
Rocks of Tuva										
34 Olivine-gabbronorite	7	6.0	0.17	5	0.4	1.0	1.17	6.7	0.003	—
35 Hypersthenic quartz-diorite	40	25	0.806	16	1.6	2.5	1.6	0.93	0.01	—
36 Hypersthenic granulite	18	5.2	0.304	14	1.3	3.5	3.5	0.17	0.008	—
43 Granite	42	85	4.4	10	1.6	10.0	0.5	0.008	0.007	—
Young acid rocks of the Caucasus										
26 Granodiorite	15	0.34	2.8	19	5.2	24	44	0.25	0.01	1—3
27 Granite	5	1.81	3.03	17	4.5	14[b]	2.8	0.06	0.009	30—50
31 Granodiorite	6	7.17	1.85	22	0.9	2.7[b]	0.84	0.34	0.012	120—140
Acid volcanic rocks of Kamchatka										
62 Dacite	0.017	0.263	1.31	14	1.1	5.4	0.06	22	0.008	10^{-5}
63 Rhyolite	0.017	0.106	2.12	22	1.6	5.4	0.16	3.17	0.013	10^{-2}
64 Dacite	0.021	0.327	2.85	22	3	6.2	0.06	4.1	0.013	10^{-1}
65 Ignimbrite	0.05	1.46	3.25	45	2	6.8	0.03	0.63	0.022	7—11

[a] Age adopted on the basis of geological data, on K/Ar (separate minerals) and/or Rb/Sr isochron and/or U/Pb methods.
[b] Th-content was estimated to be three times greater than that of U.

140

radiogenic helium. The problem of the origin of helium isotopes in the mantle was discussed earlier (Tolstikhin et al., 1974a; see section 7.2).

A distribution of points similar to that presented in Fig. 8.1 is observed when ratios of $(^3He/^4He)_{meas}$ are plotted versus the lithium content (Fig. 8.2). Such a presentation of data eliminates inaccuracies inherent in the calculation of $(^3He/^4He)_{calc}$, but in some specific cases high ratios of $(^3He/^4He)_{meas}$ may be produced by radioactive decay and nuclear reactions in minerals poor in lithium (see section 8.4). In the diagram with $(^3He/^4He)_{meas}$ versus Li coordinates, points corresponding to pure radiogenic helium may be found among those corresponding to a mixture of primordial and radiogenic helium.

It is possible to suggest the following sequence of processes in explaining the observed distribution of points on Fig. 8.1 (see also Table 8.1).

(1) At the moment of eruption the mantle magma releases most of its juvenile volatiles. Some traces of juvenile volatiles, however, are preserved in the solidifying magma and can be a significant component of the volatiles in very young rocks (this situation is well known from K-Ar dating). Data obtained from a ugandite (West Rift Zone of Africa) and a dacite (Kamchatka) illustrate this point. In these samples low contents of $^4He \sim 10^{-8}$ cm^3 g^{-1} samples are accompanied by high ratios of $^3He/^4He$, $> 2 \cdot 10^{-5}$ (Mamyrin et al., 1974; Tolstikhin et al., 1976).

(2) Known crustal processes (such as uranium and radiogenic helium accumulation or the metamorphism and outgassing of rocks) as a rule result in uncompensated losses of mantle ^3He and increases of the contribution of radiogenic helium (with a low $^3He/^4He$ ratio). In rocks initially high $^3He/^4He$

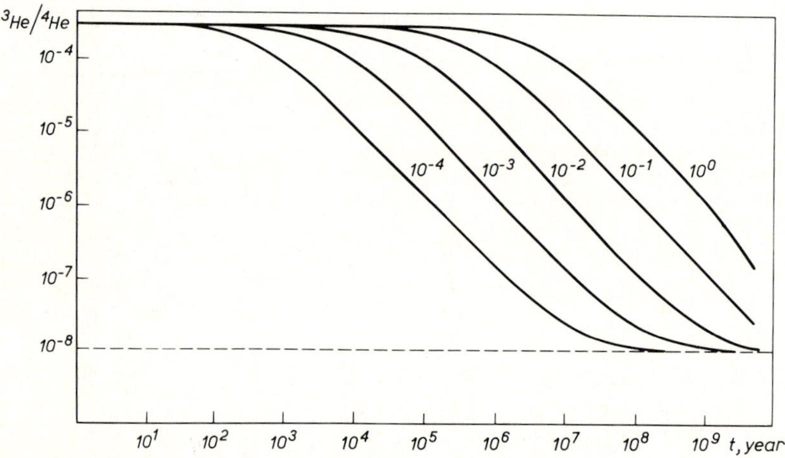

Fig. 8.3. Reduction of $^3He/^4He$ ratios in rocks with time (Tolstikhin and Drubetskoy, 1975). Numbers near the curves are the ratios of trapped primordial helium (cm^3 g^{-1}) and uranium contents (g g^{-1}, Th = 3U).

ratios decrease owing to these processes. Accordingly, the points on Fig. 8.1 will shift from the mantle region (point E) along EO and accumulate in the radiogenic region (in the vicinity of point O). Provided other things are equal, the rate of decrease depends upon the ratio of ^{3}He/(U + Th) in rocks (Fig. 8.3). For example, a ^{3}He/^{4}He ratio of $6.7 \cdot 10^{-6}$ is observed in the olivine gabbro-norite (Table 8.1, No. 34) characterized by the K-Ar whole-rock age of 750 m.y. and ^{3}He/(U + Th) = $3 \cdot 10^{-5}$ cm^3 g^{-1} (Tolstikhin et al., 1977a). Olivinite (Monche Tundra, Kola Peninsula) yields ratios of ^{3}He/^{4}He = $0.6 \cdot 10^{-6}$ and ^{3}He/(U + Th) = $6 \cdot 10^{-4}$ in spite of the very old age of the rock, 2800 m.y.

The comparison of measured and calculated ^{3}He/^{4}He ratios leads to the conclusion that the mixing of mantle (partly primordial) helium and radiogenic helium is the main process which produces the range in the helium isotope composition observed in rocks of the earth's crust.

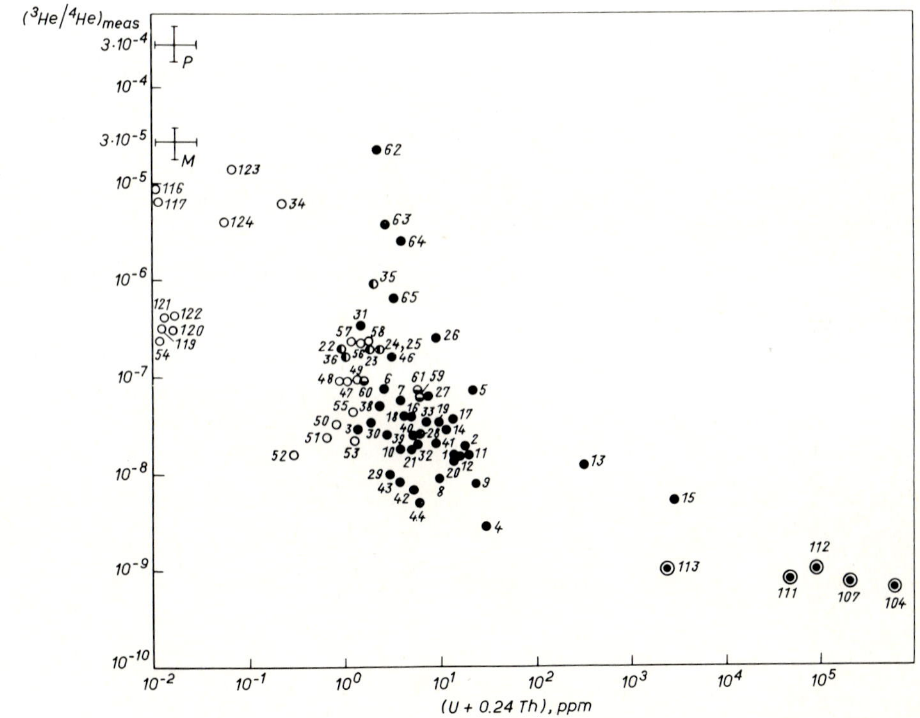

Fig. 8.4. Relationship between heavy-radioelement content and ^{3}He/^{4}He for various rocks and radioactive minerals (Tolstikhin and Drubetskoy, 1975).
P = chondrites rich in primordial helium; M = contemporary mantle of the earth. For legend see Fig. 8.1.

142

8.1.2. $^3He/^4He$, $(^4He/^{40}Ar)_{rad}$ ratios and the contents of radioelements

Both calculated and experimental data indicate a more or less constant ratio of $^3He/^4He \approx 2 \cdot 10^{-8}$ in radiogenic helium for all major types of rocks (see Table 8.1, Fig. 8.1). Nevertheless, in some cases there are variations in this ratio owing to peculiarities of the chemical composition of a rock, such as a high content of lithium (see Fig. 8.2) or radioelements. More clearly these variations are found in lithium or uranium minerals (see section 8.3). On the whole, there is a weak negative correlation (Fig. 8.4) between the radioelement contents and the isotope composition of helium: an increase in the contents of the parent elements by five orders of magnitude causes the $^3He/^4He$ ratio to fall by two orders. Least-squares fitting of the results (Hudson, 1964) gives the following regression equation:

$$\log (^3He/^4He) = -0.44 \cdot \log (U + 0.24\ Th) - 6.93 \qquad (8.1)$$

with an error of approximation $\sigma^2 = 0.5$.

These results may be compared with those on Fig. 8.5, where the ratio of the daughter isotopes, $(^4He/^{40}Ar)_{rad}$, is compared with that of the parent elements, $(U + 0.24\ Th)/K$. The measurements for rocks and minerals can be fitted to the regression equation:

$$\log (^4He/^{40}Ar)_{rad}^{rock} = 0.81 \cdot \log \left(\frac{U + 0.24\ Th}{K}\right) + 2.67 \qquad (8.2)$$

with an error of approximation $\sigma^2 = 0.27$.

Using the Rb-Sr, U-Pb or K-Ar (monomineral fractions) ages and the U, Th and K concentrations (see Table 8.1) Tolstikhin and Drubetskoy (1975) calculated the $(^4He/^{40}Ar)_{rad}$ ratio in: (1) rocks (assuming that there were no losses of ^4He and ^{40}Ar); (2) hypothetical gases derived from the rocks. The latter (Fig. 8.5) are approximated by the regression equation:

$$\log (^4He/^{40}Ar)_{rad}^{gas} = 0.93 \cdot \log \left(\frac{U + 0.24\ Th}{K}\right) + 4.60 \qquad (8.3)$$

with $\sigma^2 = 0.08$.

A comparison of the results presented in Fig. 8.4 and 8.5 and eqs. 8.1, 8.2 and 8.3 leads to the following conclusions:

(1) The $(^4He/^{40}Ar)_{rad}$ ratio in rocks is on the whole lower by two orders of magnitude than that in hypothetical gases "derived" from the rocks. This

reflects the well-known fact of a far better preservation of radiogenic argon as compared to helium (Fig. 8.6). Certain specimens possessing almost background levels of the parent elements (specimens *3, 7* and *65*, Fig. 8.5) show extremely low $(^4He/^{40}Ar)_{rad}$ ratios, close to those for potash salts. The low ratios in rocks (as well as the high ratios in gases) show that the gases have been dervied from rocks under conditions providing effective retention of argon but considerable losses of helium. Such conditions are typical of platform regions which are characterized by stable tectonics, low heat flow and the absence of magmatic activity. An example is the high ratios of $^4He/^{40}Ar$

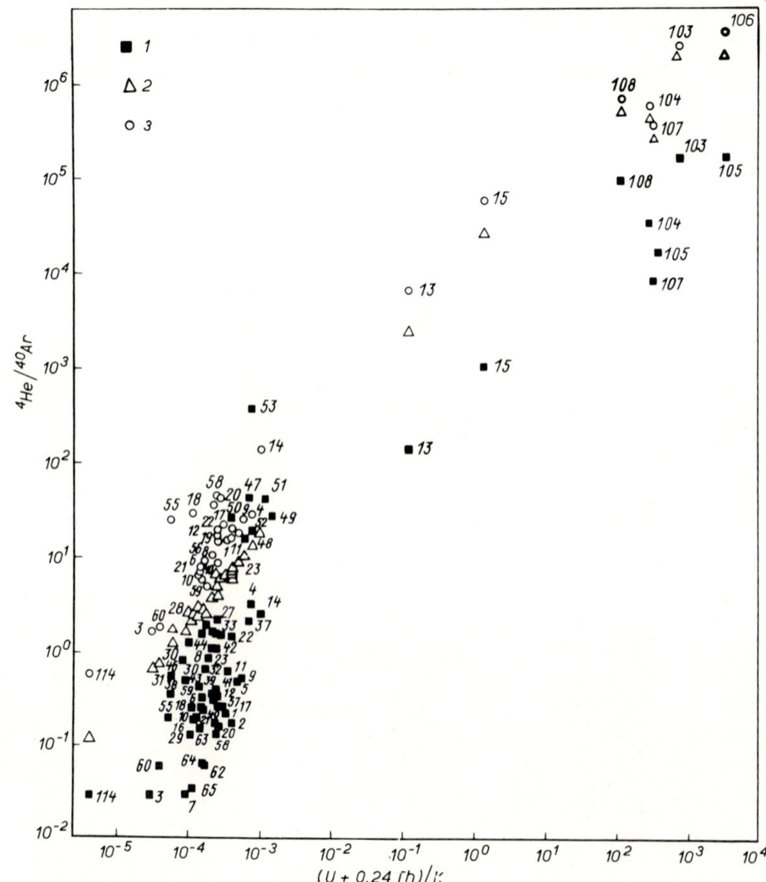

Fig. 8.5. Relationship between ratios of radioelements and radiogenic gases in crustal rocks and in hypothetical gas specimens associated with these rocks (Tolstikhin and Drubetskoy, 1975).

1 = measured $(^4He/^{40}Ar)_{rad}$ for rocks; *2* = calculated values derived from radioelement contents and age; *3* = calculated ratios for gases genetically related to rocks (ratios of released radiogenic helium and argon). For numbers see Fig. 8.1.

= 160—270, which have been found for hydrocarbon gases of the Khibiny intrusion, Kola Peninsula (Tolstikhin et al., 1967) in spite of the quite ordinary ratio of parent elements in the rock, $K/U = 1.2 \cdot 10^4$ (Kukharenko et al., 1968). Very high ratios of $({}^4He/{}^{40}Ar)_{rad}$ are observed in gas—liquid microinclusions in quartz from chambered pegmatite (see section 8.4); there is no evidence that these gases have at any time been derived from rocks enriched in U and Th. Besides, it is well known that the average ratio of $({}^4He/{}^{40}Ar)_{rad} \approx 20$ in natural gases from the sedimentary cover of ancient platforms is well above the theoretical value (4—7), which is consistent with the bulk ratio of parent elements in the crust (Zartman et al., 1961; Gerling et al., 1967a; Voronov et al., 1974).

(2) Conversely, if rocks which contained low proportions of ^{4}He and ^{40}Ar undergo a high-grade metamorphism, they release the gases in the same, low

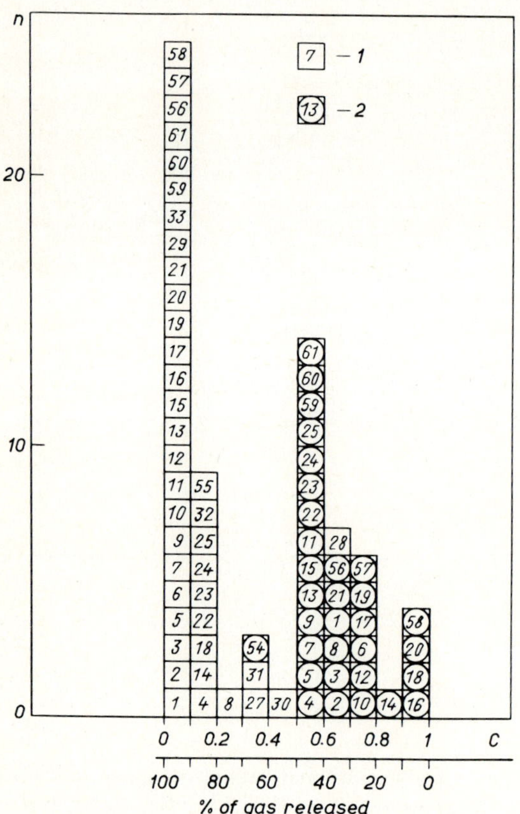

Fig. 8.6. Histograms for loss of radiogenic gases (Tolstikhin and Drubetskoy, 1975). *1* = He; *2* = Ar (for numbers see the specimen numbers in Fig. 8.1); *C* = conservation coefficient; *n* = number of specimens.

proportion. The situation of previously subsurface rocks (characterized by accumulated radiogenic ^{40}Ar and released ^{4}He) undergoing high-temperature heating and losing both radiogenic gases, is typical of regions of a high tectono-magmatic activity and heat flow. It should be noted that ratios of $(^4\text{He}/^{40}\text{Ar})_{\text{rad}} \approx 1$ observed in such regions (Iceland, the Kurils—Kamchatka volcanic zone, etc.) are lower as compared to the average value (Voronov et al., 1974).

(3) Thus, the $(^4\text{He}/^{40}\text{Ar})_{\text{rad}}$ ratio in natural fluids reflects the thermal history of rocks rather than the ratio of parent elements: Theoretically one and the same ordinary rock $(\text{U}/\text{K} \sim 10^{-4})$ can release radiogenic gases in a proportion varying between 0 and ∞; in practice the $(^4\text{He}/^{40}\text{Ar})_{\text{rad}}$ ratios in terrestrial fluids vary from $\sim 10^{-2}$ to $\sim 10^3$. That is why estimates of the U/K ratio in the earth's reservoirs (which serve as sources of natural fluids) based on the $(^4\text{He}/^{40}\text{Ar})_{\text{rad}}$ ratio do not deserve full credit (Dymond and Hogan, 1973; Fisher, 1975). However, study of this ratio may help penetrate into the geologic prehistory of the region.

8.1.3. ^{3}He/^{4}He versus ^{87}Sr/^{86}Sr relationships

In order to compare the origin of rocks and volatiles trapped by these rocks it is beneficial to combine the isotope analyses of helium with that of strontium. Strontium isotope data (as well as neodymium) are useful to clarify the origin of silicate matter: low ratios of $^{87}\text{Sr}/^{86}\text{Sr} \approx 0.702$—$0.704$ are attributed to the earth's mantle, while higher ratios of 0.709—0.718 are typical of the crust (Faure and Powell, 1972).

When comparing strontium and helium isotope data, one cannot but notice a striking difference in the behaviour of these elements in nature: (1) strontium is a typical refractory, rather immovable element and it represents the most "conservative" part of silicate matter whereas helium is the most mobile element in nature; (2) the isochron Rb-Sr method enables one to obtain the initial isotope composition of strontium, which preserve information on the origin of rocks during their entire history; whereas the isotopic ratios of ^{3}He/^{4}He decrease in the course of time (see section 8.5) and ancient rocks of mantle origin rarely show high ^{3}He/^{4}He ratios — no method of restoring the initial ^{3}He/^{4}He ratio in ancient rocks has been worked out yet, this being a rather arduous task.

Tolstikhin et al. (1976) interpreted the ^{3}He/^{4}He and $^{87}\text{Sr}/^{86}\text{Sr}$ ratios in contemporary erupted dacite from Kamchatka; both ratios were found to have typical mantle values. Polyak et al. (1980) reported a correlation between the $^{87}\text{Sr}/^{86}\text{Sr}$ ratios in young rocks from Italy and Iceland and the ^{3}He/^{4}He ratios in thermal fluids circulating in these rocks (Fig. 8.7). The correlation implies that silicate substance is the main carrier of juvenile helium and other volatiles, and thermal fluids owe their mantle helium to the outgassing of silicate melts and/or rocks.

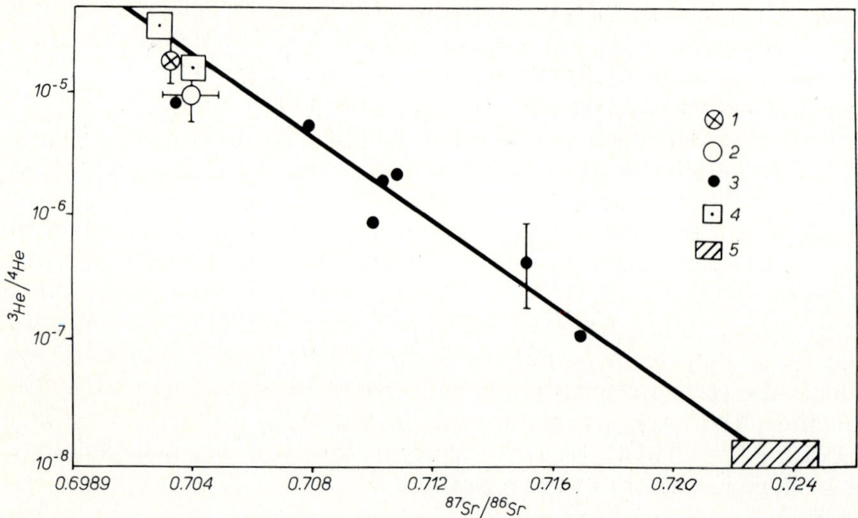

Fig. 8.7. Relationship between isotopic helium composition in fluids and isotopic strontium composition in rocks through which the fluids migrate.
1 = Iceland; *2* = Kamchatka; *3* = Appenines; *4* = two rock samples from Kamchatka and New Zealand; *5* = isotope composition of helium and strontium adopted for the earth's crust. (Data are taken from Tolstikhin et al., 1976; Saito et al., 1978; Polyak et al., 1979a.)

8.2. The distribution of ^{3}He and ^{4}He in mineral fractions

The observed ^{3}He/^{4}He ratios can differ from the calculated ones even if there is no mantle (primordial) helium. Such an effect may be due to differences in the sites of the helium isotopes in rock-forming and accessory minerals. Since the ^{3}He atoms localized in rock-forming minerals containing Li and ^{4}He atoms are produced within accessory minerals rich in U and Th, each of the two isotopes may have its own pattern of release. To study these one should analyze separated mineral fractions. The distribution of rare gas isotopes in the fractions should reflect the results of accumulation and migration of the isotopes into and out of the different minerals under the same P-T conditions. Such an investigation of three samples was carried out by Gerling et al. (1976).

The data obtained for a Rapakiwi granite — a case where $(^3$He/^{4}He$)_{\text{calc}}$ = $(^3$He/^{4}He$)_{\text{meas}}$ — are listed in Table 8.2. The distribution of elements (Li, K) and isotopes (^{40}Ar, ^{3}He) which are concentrated in rock-forming minerals enables obtaining a balance between their measured contents in the minerals and in the total rock. There is, however, no balance between the trace radioactive elements (U, Th) and the radiogenic ^{4}He, as all these species are con-

TABLE 8.2

Rare gas isotopes, radioactive elements and lithium in mineral fractions of Rapakiwi granite (from Gerling et al., 1976)

Fraction	Contents							Mineral contents in rock
	^{3}He $(10^{-12}$ cm^3 STP g$^{-1})$	^{4}He $(10^{-6}$ cm^3 STP g$^{-1})$	^{40}Ar $(10^{-6}$ cm^3 STP g$^{-1})$	K $(10^{-2}$ g g$^{-1})$	Li $(10^{-6}$ g g$^{-1})$	U $(10^{-6}$ g g$^{-1})$	Th $(10^{-6}$ g g$^{-1})$	(%)
Orthoclase	0.8	2	960	10.9	6.0	2.5	2.7	56
Biotite	2.3	18	622	6.4	620	7.2	4.8	3
Plagioclase	1.35	27	17.2	0.155	10	5.9	5.3	10
Quartz	0.3	2	11.4	0.186	12	5.1	3.0	30
Amphibole	16.6	368	135	1.31	37	3.5	12.7	1
Zircon	50	$25 \cdot 10^3$	6.4	—	—	990	150	0.003
Bulk sample calculated by mineral								
fractions	0.91	9.4	565	6.43	27.4	3.74	3.18	
measured	1.23	85	515	5.06	36	6.7	35	

148

centrated in accessory minerals, and can be lost in the process of mineral separation.

In spite of this discrepancy, the following important fact was established. The ratios of $^4\mathrm{He}/(\mathrm{U} + 0.24\ \mathrm{Th})$ in rock-forming minerals vary greatly while the ratios of $^{40}\mathrm{Ar}/^{40}\mathrm{K}$ are nearly constant. Biotite and amphibole are antipodes — both minerals contain approximately the same share of U and Th but differ greatly in the $^4\mathrm{He}$ content:

$$\frac{[^4\mathrm{He}/(\mathrm{U} + 0.24\ \mathrm{Th})]^{\mathrm{am}}}{[^4\mathrm{He}/(\mathrm{U} + 0.24\ \mathrm{Th})]^{\mathrm{bi}}} = 25.8$$

Moreover, the ratio of $^3\mathrm{He}/\mathrm{Li}$ in amphibole is about a hundred times greater than that in biotite:

$$\frac{(^3\mathrm{He}/\mathrm{Li})^{\mathrm{am}}}{(^3\mathrm{He}/\mathrm{Li})^{\mathrm{bi}}} = 120$$

Biotite contains almost all of the lithium in the rock but its $^3\mathrm{He}$ concentration is negligible. It is noteworthy that both minerals show a similar Ar/K ratio of 9.7 and 10.3 ($\times\ 10^{-3}\ \mathrm{cm}^3\ \mathrm{g}^{-1}$) for biotite and amphibole, respectively.

These results may be explained as follows: (1) $^3\mathrm{H}(^3\mathrm{He})$ is produced in the reaction $^6\mathrm{Li}\ (\mathrm{n},\ \alpha)^3\mathrm{H} + Q$; (2) the energy of the reaction ($Q = 4.8$ MeV) is split between the kinetic energies of tritium ($Q_{^3\mathrm{H}} = 2.75$ MeV) and the α-particle ($Q_\alpha = 2.05$ MeV; Zverev et al., 1972); $Q_{^3\mathrm{H}}$ implies a range of $^3\mathrm{H}$ tracks of $\sim 10^{-4}$ cm, which is much greater than the thickness of the monolayer of the biotite structure, and the atoms of $^3\mathrm{H}(^3\mathrm{He})$ can be released from the mineral along the track and then push forward between the layers of the crystalline structure.

A similar explanation of the $^4\mathrm{He}$ distribution appears to be reasonable assuming that: (1) a share of the atoms of U and Th are dispersed in the mica and/or are located in small accessory minerals with a diameter of less than the α-range; (2) other atoms are concentrated in comparatively large accessory minerals. In case (1) some atoms of $^4\mathrm{He}$ should have been released by biotite but preserved in amphibole (which is a complete analogy with $^3\mathrm{He}$), whereas in case (2) other atoms (located in accessory minerals) should have been released by biotite and amphibole in the same proportion. Such an interpretation is compatible with the above-mentioned ratios of daughter and parent atoms.

Analysis of monomineralic fractions of another ancient rock, a diorite-gneiss, shows a similar distribution of helium isotopes (Table 8.3). Biotite contains a small fraction of $^3\mathrm{He}$ (about 10%) and nearly all of Li (as much as 80%). Almost all of the helium (both isotopes) is concentrated in the am-

TABLE 8.3

Rare gas isotopes, radioactive elements and lithium in mineral fractions of diorite-gneiss (from Gerling et al., 1976)

Fraction	Contents							Mineral contents in rock
	^{3}He (10^{-12} cm^3 STP g^{-1})	^{4}He (10^{-6} cm^3 STP g^{-1})	^{40}Ar (10^{-6} cm^3 STP g^{-1})	K (10^{-2} g g^{-1})	Li (10^{-6} g g^{-1})	U (10^{-6} g g^{-1})	Th (10^{-6} g g^{-1})	(%)
Biotite	18	83	1100	7.4	64	0.8	4.0	20
Plagioclase	4.6	19	33	0.132	0.6	1.0	3.0	60
Quartz	5.3	29	64	0.14	0.1	0.8	2.7	10
Amphibole	250	1250	111	0.6	17	1.0	9.0	9
Chlorite	9.4	39	25	0.2	64	0.6	5.6	1
Bulk sample calculated by mineral								
fractions	29.6	144	257	1.64	15	0.93	3.7	
measured	32	145	350	1.6	15	1.3	6.0	

150

phibole, in spite of the comparatively low contribution to the Li, U and Th
content of the rock by this mineral.

It is noteworthy that the measured ^{3}He/^{4}He ratio in the rock ($0.20 \cdot 10^{-6}$)
is about 40 times greater than the calculated one ($0.005 \cdot 10^{-6}$). If we assume a radiogenic *in situ* origin of ^{3}He and ^{4}He, it is difficult to explain the
high content of the isotopes. The (U + Th)/^{4}He age of the amphibole is
somewhat greater than that obtained by the K/Ar method. The observed ^{3}He
content in this mineral is 30 times greater than the calculated one. These
results indicate that juvenile volatiles (excess ^{3}He among them) took part in
the metamorphism of the rock. The trapped ^{3}He and ^{4}He appear to be well
preserved by the host minerals of diorite-gneiss and olivine gabbronorite,
whereas the radiogenic isotopes are released more or less completely. Such a
conclusion is based on the constant ^{3}He/^{4}He ratio in the temperature fractions (Fig. 8.8) which reflect a homogeneous distribution of both helium isotopes in the rocks and a negligible contribution of radiogenic (^{3}He/^{4}He $\approx$
$2 \cdot 10^{-8}$) helium. The latter had probably been released earlier.

The olivine-gabbronorite (Table 8.4) is an excellent example of a comparatively old (300—700 m.y) rock rich in mantle helium. Again, nearly all of the
helium is concentrated in the dark-coloured fraction rich in amphibole. The
separation of pure dark-coloured minerals from this fraction was not carried

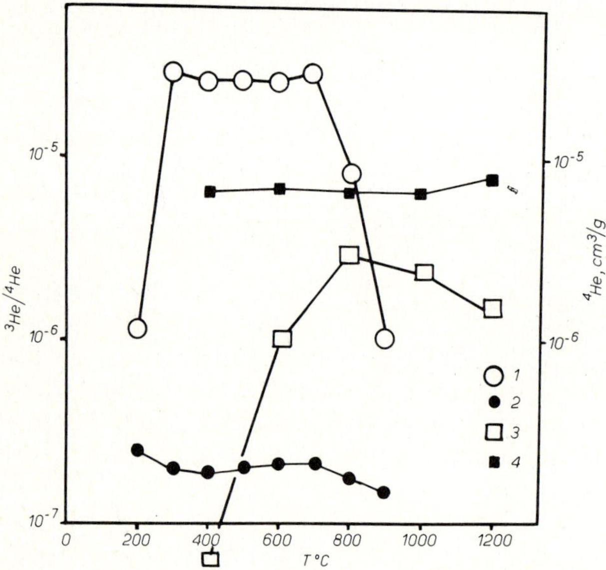

Fig. 8.8. Concentrations and isotope composition of helium released by step-wise heating
experiments from diorite-gneiss and olivine gabbronorite (shown by circles and squares,
respectively).
1, 3 = concentrations of helium released from diorite-gneiss and olivine gabbronorite,
respectively; 2, 4 = ^{3}He/^{4}He ratios (Tolstikhin and Drubetskoy, 1975).

TABLE 8.4

Rare gas isotopes, lithium and potassium in mineral fractions of olivine gabbronorite (from Gerling et al., 1976)

Fraction	Contents					Mineral content in rock (%)
	^{3}He (10^{-12} cm^3 STP g^{-1})	^{4}He (10^{-6} cm^3 STP g^{-1})	^{40}Ar (10^{-6} cm^3 STP g^{-1})	K (10^{-2} g g^{-1})	Li (10^{-6} g g^{-1})	
Plagioclase	1.39	0.19	7.3	0.26	3	65
Olivine	10.4	1.4	0.84	0.071	5.5	15
Clinopyroxene-hypersthene	107	16	0.86	0.061	4.5	15
Hypersthene + amphibole + clinopyroxene	280	42	9.1	0.43	—	5
Bulk sample calculated by mineral fractions	32	4.8	5.5	0.21	3.5	
measured	47	7	6.0	0.17	5	

out because of its complex structure. The rock is characterized by a high ^{3}He/^{4}He ratio equal to $6.7 \cdot 10^{-6}$ (compare with the calculated ratio of $0.003 \cdot 10^{-6}$) and a high content of ^{4}He, $7 \cdot 10^{-6}$ cm^3 g^{-1}. The following model of the genesis of the gabbronorite was put forward on the basis of these results (Tolstikhin et al., 1977a). The matter of the upper mantle (or the lower crust) had been squeezed up into the upper crust rather slowly and a decrease in pressure was accompanied by a proportional decrease in temperature. The conditions for degassing were unfavourable and the major part of the volatiles was preserved.

However, there is a possible alternative explanation. The rock was sampled near the outcrop of a deep fractured zone and it may have trapped volatiles when this zone was active. It is noteworthy that another (typical metamorphic) rock collected in the same region is very rich in helium (^{4}He = $1.5 \cdot 10^{-3}$; ^{3}He = $2.7 \cdot 10^{-9}$ cm^3 g^{-1}) with a comparatively high ratio of ^{3}He/^{4}He = $1.8 \cdot 10^{-6}$.

Thus, some difference between the calculated and the measured ratios of ^{3}He/^{4}He in radiogenic helium may be due to variation in the mineral composition, the peculiarities of Li, U and Th distribution in a rock, the intensity of helium migration, etc. But these discrepancies by no means disprove the theory of Morrison and Pine (1955) which, on the whole, correctly describes the origin of radiogenic helium isotopes. Amphibole (and most likely, some other dark-coloured minerals) may be considered as the best suitable mineral for the study of the origin and the sources of volatiles which are well trapped and preserved by it. The results obtained by Saito et al. (1978, see Fig. 7.8) appears to confirm this. Further investigations in this field, together with geochronological data will probably make it possible to reconstruct the mode and time history of the earth's degassing.

8.3. Helium isotopes in some minerals

As shown in sections 7.1 and 8.1, radiogenic ratios of ^{3}He/^{4}He vary in rocks over a very small range, $(1—3) \cdot 10^{-8}$ and may be considered constant for all practical purposes. Nevertheless, the ratios differ essentially from the average value $(2 \cdot 10^{-8})$ in some minerals, especially those enriched in radioactive elements (8.3.1) or lithium (8.3.2) as well as those containing excess rare gases. The peculiarities of the ^{3}He and ^{4}He origin in these minerals are briefly discussed below.

8.3.1. ^{3}He and ^{4}He in uranium minerals

Aldrich and Nier (1948) estimated the upper limit of the ^{3}He/^{4}He ratio in uranium minerals as equal to $0.2 \cdot 10^{-7}$. This value was compared with a calculated one by Morrison and Beard (1949). At that time Khlopin and Gerling (1948) published the result of a ^{3}He/^{4}He measurement in uraninite, ^{3}He/^{4}He $= 3.5 \cdot 10^{-10}$; a cyclotron was used for the isotopic analysis. To check these results Kamensky (1970) determined the isotope composition of helium in some radioactive minerals by a magnetic resonance mass spectrometer and found a value of about $5 \cdot 10^{-10}$ or somewhat less. Table 8.5 illustrates that the ^{3}He/^{4}He ratios in helium of uranium minerals are about two orders of magnitude lower than those found in typical radiogenic he-

TABLE 8.5

Isotopic helium composition released from uranium minerals (after Kamensky, 1970; Shukolyukov, 1960)

Mineral	Age (m.y.)	U (%)	^{4}He (cm^3 g^{-1})	^{40}Ar$_{rad}$ (10^{-6} cm^3 g^{-1})	^{4}He/^{40}Ar$_{rad}$ (10^4)	^{3}He/^{4}He (10^{-10})
Uraninite	1950	59	2.61	6.8	4	$\leqslant 8$
Pitchblende	230	69	0.350	3.7	9	$\leqslant 8$
Cleveite	2000	72.8	2.45			$\leqslant 7$
Samarskite	330	5.48	0.750			$\leqslant 7$
Betafite	293	9.39	0.307			9
Britholite	1800	0.264	0.158			9

lium. Morrison and Pine (1955) accounted for the decrease in the ratios referring to the radiogenic origin of ^{3}He and ^{4}He, namely that low abundance of the light helium isotope in uranium minerals is predetermined by their specific chemical composition.

Low concentrations of light elements (Si, Al) lead to a decrease in the neutron flux produced by (α, n) reactions. Gorshkov et al. (1966) pointed out that the neutrons in uraninite and pitchblende are yielded mostly by spontaneous and neutron-induced fission of uranium isotopes (in rocks this source of neutrons contributes only 0.1—0.2 of the total flux) and, consequently, the ratio of n/α in these minerals is from 5 to 10 times lower than in ordinary rocks. The resonance capture of neutrons by uranium isotopes offers the neutron little opportunity to decrease its energy to the thermal level. Thus, ^{238}U is characterized by a very high resonance capture in the region of 10 eV. According to Gorshkov et al. (1966), this probability is reduced to the values of 0.2—0.4, whereas in ordinary rocks it is approximately 0.8. The low lithium content in uranium minerals and a high content of rare-earth elements, which are strong absorbers of neutrons, also provide a tenfold decrease in ^{3}He/^{4}He. Taking into account all the above coefficients, a total decrease of the helium isotope ratio in radioactive minerals might be described by a factor of 100—400 — that is, in perfect agreement with experimental results (Table 8.5).

8.3.2. ^{3}He and ^{4}He in lithium minerals

Very high ^{3}He/^{4}He ratios (up to $\sim 10^{-5}$) in radiogenic helium are observed in lithium minerals, which as a rule also contain some amount of trapped gases. Because of a high lithium concentration the proportion of neutrons that react with ^{6}Li and yield ^{3}H(^{3}He) approaches the maximum value, 1, in these minerals. A high concentration of light elements in lithium-bearing pegmatite stimulates a high neutron production by the (α, n) reaction. The ^{3}He/^{4}He ratio in radiogenic helium, in its turn, increases by factors of 30—50 as compared to ordinary rocks and reaches the value of about 10^{-6}. A further rise in the ^{3}He/^{4}He ratios (up to 10^{-5}) is provided by irradiation of minerals by the neutron flux originating in country rock (see also section 8.4).

TABLE 8.6

Isotopic composition of helium released by step-wise heating of spodumene (Mamyrin and Tolstikhin, 1981)

Temperature (°C)	200	300	540	640	720	840	970
^{3}He/^{4}He (10^{-6})	0.8	2.5	10.3	6.1	2.9	2.0	1.2

154

Step-wise heating experiments show that the isotope composition of helium released in temperature fractions varies over a wide range (Table 8.6). Radiogenic ^{3}He (produced *in situ* by the ^{6}Li + n $\rightarrow$ α + ^{3}H reaction) is released mostly at the temperature of about $500°$C; the high and the low temperature fractions are characterized by essentially lower ^{3}He/^{4}He ratios.

8.3.3. Helium isotopes in trapped helium

As a rule, beryl, cordierite, spodumen, chromite and some other minerals contain excess rare gases. A detailed discussion of the nature of excess gases is beyond the scope of this work, but the generally accepted view that the gases were trapped and preserved by the minerals appears to be well grounded and consistent with observations.

An investigation of helium isotopes released from beryls shows that the ^{3}He/^{4}He ratio varies over a wide range, from $5 \cdot 10^{-8}$ to $1 \cdot 10^{-6}$, which is similar to the range typical of terrestrial fluids. An unambiguous interpretation of the ^{3}He/^{4}He ratios in beryl is complicated by the presence of two types of helium: the trapped excess helium and the radiogenic helium produced *in situ*. The initial ^{3}He/^{4}He ratio is as a rule unknown, as well as the proportion of the remaining radiogenic helium which is liable to considerable losses. Concerning the helium origin in beryl, qualitative considerations lead to the following supposition: if trapped excess helium was produced within the ancient crust, then its ^{3}He/^{4}He ratios ($\sim 10^{-8}$) had to be essentially lower than those typical of *in situ* produced helium ($\sim 10^{-7}$—10^{-6}). This is predetermined by two factors: (1) beryl is characterized by the maximum yield of neutrons in the (α, n) reaction (Gorshkov et al., 1966); (2) the li-

TABLE 8.7

Concentration and isotope compositions of helium in beryls (Aldrich and Nier, 1948; Prasolov, 1972)

No.	Location	^{4}He (10^{-6} cm^3 g^{-1})	^{3}He/^{4}He (10^{-6})
1	Finland, Eräjärvi	18 000	0.06
2	Finland, Kimito	11 000	0.05
3	USA, South Dakota	22 000	0.12
4	Sweden, Lappland	23 000	0.18
5	USA, New Hampshire	4 000	1.2
6	USSR, Siberia	7 000	0.20
7	USSR, Siberia	39 000	0.023
8	USSR, Volyn	1 700	1.1
9	USSR, Kola Peninsula	7 400	0.30
10	USSR, Kola Peninsula	1 600	0.11
11	USSR, Karelia	25 000	0.29

thium concentration in beryl is high and a significant amount of ^{3}He is bound to be produced by the reaction of ^{6}Li (n, α) ^{3}H. Table 8.7 illustrates the isotope composition of helium in various beryl samples. It should be noted that minerals rich in helium show a lower ^{3}He/^{4}He ratio: evidence for radiogenic, crustal origin of trapped helium. For instance, a Siberian beryl sample with a maximum concentration of helium contained radiogenic, crustal helium with a ^{3}He/^{4}He ratio of $2.3 \cdot 10^{-8}$. A ^{3}He/^{4}He ratio as low as $0.12 \cdot 10^{-6}$ with a very high ^{4}He concentration of $2.2 \cdot 10^{-2}$ cm^3 g^{-1} were found in an ancient cordierite from Dakota, U.S.A. This differs, however, from minerals formed in a region of high tectono-magmatic activity (see section 8.5). In this case the minerals might contain trapped isotopically light helium. For instance, a high ^{3}He/^{4}He ratio was observed in a magnetic fraction of cordierite from the Sangelen Massif, Sayan Mountains, Tuva (Drubetskoy et al., 1977).

Some additonal data on the genesis of helium isotopes in beryl and other minerals can be obtained by helium measurements in step-wise heating experiments. Table 8.8 illustrates that trapped helium isotopes and helium isotopes produced *in situ* are distributed through minerals more or less homogeneously: the isotope composition of helium in temperature fractions is approximately constant, not counting small amounts of helium released at high temperature.

TABLE 8.8

^{3}He/^{4}He ratios (10^{-6}) and concentrations of helium $(10^{-6}$ cm^3 g$^{-1})$ released by step-wise heating of beryls (from Tolstikhin and Drubetskoy, 1977)

No.	Age (m.y.)	Values	Temperature interval (°C)				
			200—400	400—600	600—800	800—1000	1000—1200
Siberia							
1	350	He	1220	3630	750	10	1
		^{3}He/^{4}He	0.18	0.19	0.21	0.25	0.23
2	390	He	5930	20200	5110	2.2	3.7
		^{3}He/^{4}He	—	0.027	0.025	0.32	0.11
Ukraine							
3	1800	He	1310	5830	265	9.7	3.1
		^{3}He/^{4}He	0.32	0.3	0.3	0.10	0.05
4	1800	He	57	540	560	95.5	0.4
		^{3}He/^{4}He	0.89	1.2	1.05	0.89	1.18

8.4. Distribution and origin of helium and argon isotopes in gas—liquid microinclusions in minerals

Helium, argon and other volatiles released by minerals occur as components of terrestrial gases; they may fall into two main categories: (1) gases in closed inclusions (porosity) of minerals and rock (more than 90% of gases); and (2) free and oil—water-dissolved gases (about 10%) (Sokolov, 1966).

In this section we discuss the first category of natural gases which has been the subject of many investigations (Lippolt and Genther, 1963; Leutwein and Kaplan, 1963; Rama et al., 1965; Lippolt, 1966; Naughton et al., 1966; Funkhouser and Naughton, 1968; Prasolov and Tolstikhin, 1969, 1970, 1972; Krummenacher, 1970; Harper and Schamel, 1971; Tolstikhin and Prasolov, 1971; Naidenov et al., 1972; Prasolov, 1976; and others). The ^{4}He contents in microinclusions vary over a wide range, from 10^{-8} to 10^{-3} cm^3 g^{-1} of mineral. The average value is $\sim 10^{-5}$ cm^3 g^{-1}. The corresponding values for the ^{40}Ar contents are 10^{-8} to 10^{-4} and $\sim 10^{-6}$ cm^3 g^{-1}. The isotopic composition of helium and argon indicates that they are mainly of radiogenic and/or atmospheric origin. These data confirm the prevailing notions about the main sources of volatiles in the earth's crust, being (1) rocks (releasing volatiles under metamorphism), and (2) atmospheric (oceanic) water and gases. The occurrence of radiogenic isotopes in microinclusions may be a major misleading factor in the dating of young rocks and minerals as well as minerals poor in potassium. The study of rare gas isotopes in microinclusions is helpful in resolving dating problems; it also increases our knowledge of the fluid regimes of the earth's crust.

A collection of minerals from a chambered pegmatite was carefully investigated. Some nontrivial results were obtained from this study. In the large chambers the core diameters amount to several meters; the chambers contain crystals of quartz, topaz, fluorine, etc., enriched in gas and gas—liquid microinclusions; the age of the pegmatites determined by the Rb-Sr method is 1715 ± 12 m.y. (Gorokhov, 1964).

The results of the measurements of the helium and argon concentrations and isotope ratios of ^{3}He/^{4}He, ^{4}He/^{40}Ar and ^{40}Ar/^{36}Ar are shown on Fig. 8.9 and in Table 8.9. The samples (Fig. 8.9) are arranged in accordance with the sequence of their formation: gabbrolabradorite — granite — the external zone of pegmatitie — the quartz core — honeycomb quartz — the "latter" quartz and topaz.

The ^{4}He concentration in pure quartz crystals varies within a wide range from $1 \cdot 10^{-6}$ to $1.6 \cdot 10^{-3}$ cm^3 g^{-1}. It should be noted that in most quartz samples (which are extremely poor in uranium and thorium), the helium concentration is higher than in country granites (Table 8.9). Almost all samples show ^{3}He/^{4}He ratios varying from $1 \cdot 10^{-7}$ to $6 \cdot 10^{-7}$ and about half of them show a practically constant ratio of $(2.0 \pm 0.5) \cdot 10^{-7}$ in spite of a wide range of ^{4}He concentrations. Only in a few samples of honeycomb quartz and topaz high ratios of ^{3}He/^{4}He were found: $(1-7) \cdot 10^{-6}$ (Table 8.10; Fig. 8.10). Comparison of calculated and measured concentrations (Table 8.9) showed that ^{4}He was trapped by the minerals; but the question of the origin of ^{3}He excess remained unresolved.

The pegmatites are located among ancient granites which contain radiogenic helium (Table 8.11). The distance between the pegmatites and the nearest region of recent volcanic and tectonic activity is more than 300 km. The ^{40}Ar/^{36}Ar ratios in trapped argon are about 10 000 and even higher (compare with similar ratios in volcanic rocks and gases as well as in mantle argon; section 7.5). Very high ratios of ^{3}He/^{4}He are found only in some samples of honeycomb quartz and topaz. All these facts are in disagreement with the hypothesis that the mantle is the source of excess ^{3}He in the minerals.

In order to account for this high ^{3}He/^{4}He ratio, the following model was proposed (Prasolov and Tolstikhin, 1972), based on the heterogeneous distribution of radioactive elements between the crystals (U $\sim 10^{-8}$ g g^{-1}) and the country rocks (U $\sim 10^{-6}$, Th $\sim 10^{-5}$ g g^{-1}).

(a) Since the range of α-particles produced by radioactive decay in the country rocks is about 10^{-2} cm, the α-particles do not penetrate deep into crystals whose sizes are approximately 10 cm. The particles, in their turn,

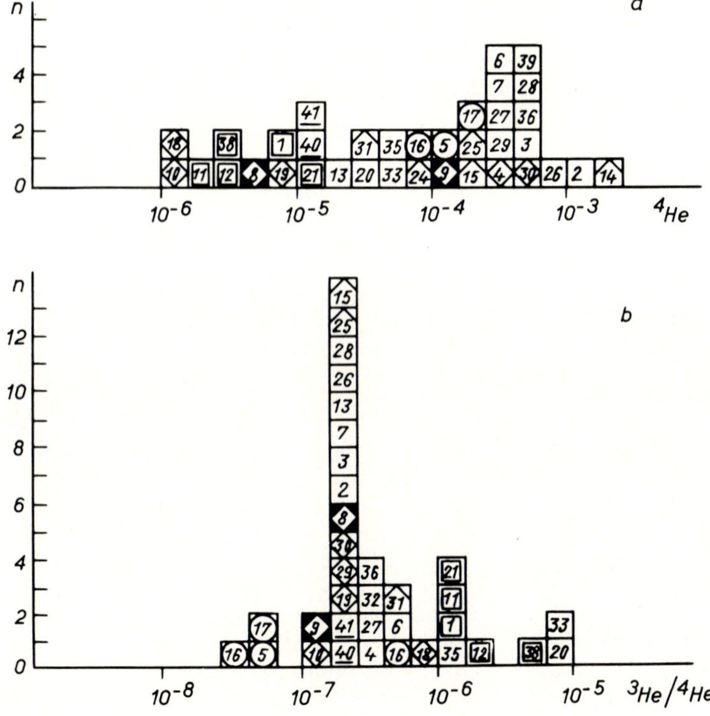

158

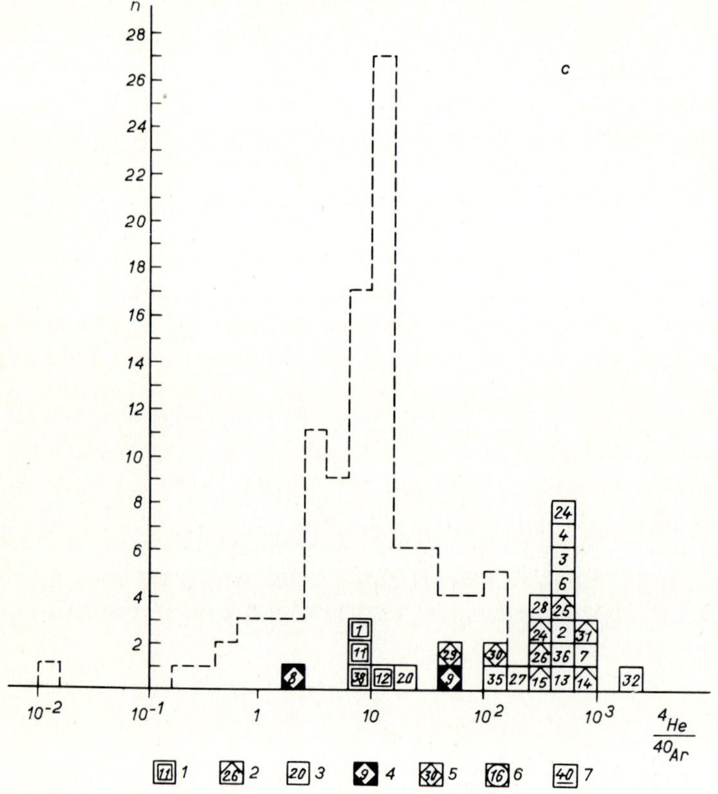

Fig. 8.9. Concentrations of helium, its isotope composition and ^{4}He/^{40}Ar$_{rad}$ ratios in minerals of chambered pegmatites (Prasolov, 1972). a. Concentrations of ^{4}He (cm^3 g^{-1}). b. ^{3}He/^{4}He ratios. c. ^{4}He/^{40}Ar$_{rad}$ ratios; dotted line shows the distribution of these values in terrestrial gases; n = number of cases.
Samples: *1* = topaz; *2* = "latter" quartz; *3* = honeycomb quartz; *4* = quartz core; *5* = minerals from external zone of pegmatite; *6* = granite; *7* = gabbrolabradorite. Numbers in squares show samples according to the following list: *1* = topaz, pegmatite body A; *2, 3* = honeycomb quartz, pegmatite body B; *4* = honeycomb quartz, pegmatite body C; *5* = granite, country pegmatite body D; *6—12* = samples from pegmatite body E; *6, 7* = honeycomb quartz; *8, 9* = quartz from the core; *10* = quartz from the external zone; *11, 12* = topaz; *13—19* = samples from pegmatite body F; *13* = honeycomb quartz; *14* = morion; *15* = "latter" quartz; *16* = metamorphic granite; *17* = country granite; *18 and 19* = feldspar and quartz from the external zone; *20—23* = samples from pegmatite body G; *20* = honeycomb quartz; *21* = topaz; *22* = lithium biotite; *23* = beryl; *24—32* = samples from pegmatite body H; *24, 25* = morion from the edge of a large crystal; *26* = honeycomb quartz, the same crystal, 4—5 cm from the edge; *27* = the same, 8—10 cm from the edge; *28* = wall of the crystal; *29* = morion from feldspar zone; *30, 31* = quartz and morion from the external zone of a crystal; *32* = honeycomb quartz from the same crystal; *33—39* = samples from quarry dump; *33—37* = honeycomb quartz; *38* = topaz; *39* = colourless quartz with large inclusions; *40, 41* = gabbrolabradorite.

TABLE 8.9

Concentrations of helium, argon (10^{-6} cm^3 g^{-1}) and radioactive elements (ppm) in minerals and country rocks of chambered pegmatites (after Prasolov, 1972)

Sample (see Fig. 8.9)	Concentration		Concentrations of helium and argon						Concentrations of excess ^{4}He, ^{40}Ar and ^{4}He/^{40}Ar in trapped gases		
	U	K	calculated		measured[a]						
			^{4}He	^{40}Ar	^{4}He		^{40}Ar		^{4}He	^{40}Ar	^{4}He/^{40}Ar
					1	2	1	2			
Pegmatite minerals											
13	0.020	23	8.5	0.26	47	1.4	0.26	0.1	40	0.1	400
20	0.040	23	17	0.26	46	1.0	0.85	1.5	30	2.1	11.5
33	0.0016	—	0.68	—	75	4	—	—	78	—	—
36	0.020	7	8.5	0.08	1100	10	1.3	1.5	1090	2.7	410
1	0.020	—	8.5	—	6.5	8.5	—	—	6.5	—	—
2	*	84	10	0.95	1090	—	3.0	—	1080	2	540
14	*	490	10	5.5	1640	—	5.4	1.0	1630	1.9	860
15	*	27	10	0.30	210	—	0.61	0.3	200	0.6	330
25	*	20	10	0.23	210	—	0.54	0.2	200	0.5	400
Country rocks											
16	7.2	47 000	3400	480	110		410		—	—	—
5	3.3	43 700	1150	450	128		505		—	—	—

[a] 1 = gas released by crushing; 2 = gas released by melting of fine fraction of crushed minerals.
* Concentration of U was assumed to be equal to the average value for the five samples listed above; concentation of Th $\approx$ 3U was also assumed.

160

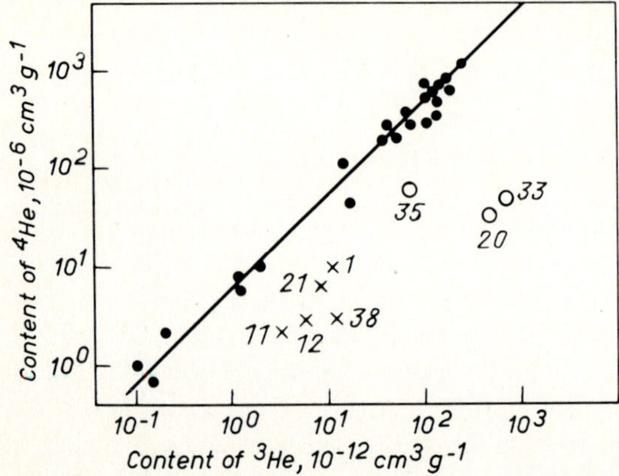

Fig. 8.10. Concentrations of helium isotopes in gas—liquid microinclusions in quartz and topaz sampled in chambered pegmatites. Numbers near points refer to short description of the samples given in Fig. 8.9 (Prasolov, 1972).

TABLE 8.10

Concentrations of ^{4}He, Li and ^{3}He/^{4}He and Li/^{4}He ratios in quartz and topaz from chambered pegmatites (after Tolstikhin, 1975a)

Mineral (see Fig. 8.9)	^{4}He (10^{-6} cm^3 g^{-1})	Li (ppm)	^{3}He/^{4}He (10^{-6})	Li/^{4}He (g cm^{-3})
Quartz samples				
15	210	1.5	0.19	0.0071
2	1090	40	0.25	0.037
36	1100	60.5	0.26	0.039
28[a]	640	43	0.23	0.067
27[a]	300	64	0.40	0.21
33	75	50	7.0	0.67
20	46	60.5	6.5	1.31
Topaz samples				
1	15	24	1.5	1.6
11	2.2	—	1.5	—
12	3.2	—	1.8	—
21	10	—	1.1	—
38	5.7	—	4.3	—

[a] Samples from the same crystal (see Fig. 8.9).

are not generated in sufficient quantities within the crystals because of a very low content of radioactive elements.

(b) The neutron flux produced by the (α, n) reactions and fission is characterized by a comparatively long range ($\gtrsim$ 10 cm) and penetrates deeper into the minerals. The ^{6}Li + n $\rightarrow$ α + ^{3}H reaction and ^{3}H $\xrightarrow{\beta^-}$ ^{3}He decay bring about a continuous accumulation of ^{3}He, which is proportional to the Li content provided other things are equal. The ^{4}He content is determined by an amount of trapped helium which does not increase in the course of time.

(c) This model seems to be consistent with the data listed in Table 8.10. The high ^{3}He/^{4}He ratios are observed in minerals with high values of Li/^{4}He. It is noteworthy that in crystals free from trapped helium, extremely high ratios of ^{3}He/^{4}He may occur (up to 10^{-3} and higher).

It was found that ^{3}He/^{4}He ratios in microinclusions and crystal lattice in quartz crystals are slightly different as well as the ratios in two samples selected from one large crystal (see Table 8.10). On the basis of this difference the helium diffusion coefficient D was calculated for crystalline quartz, and a very low value of about 10^{-20} cm^2 s^{-1} (Tolstikhin et al., 1974b) was obtained.

Quartz and topaz crystals, which preserve helium well and are poor in radioactive elements, are excellent detectors of the natural neutron flux. These detectors were installed by nature itself and have been working for billions of years. With the assumption that the neutron flux is identical for all parts of one large crystal, we calculated that a flux, F = 8 neutrons cm^{-2} day^{-1}, had been irradiating samples *27* and *28* (Fig. 8.9) during 1.7 b.y. This value is in perfect agreement with the average flux in granites (Gorshkov et al.,1966); it also testifies that there has never been any excess neutron flux in the crust (the Oklo-phenomenon is the only known exception).

TABLE 8.11

Concentrations of radioactive elements and ratios of (U + Th)/K and ^{3}He/^{4}He in country rocks of chambered pegmatites (after Prasolov, 1972)

Rock (see Fig. 8.9)	Concentrations (ppm)			Ratios	
	U	Th	K (10^3)	(U + Th)/K (10^{-3})	^{3}He/^{4}He (10^{-6})
17	3.3	6.2	44	0.215	0.05
16	7.2	30.4			0.04
5	8.3	8.3	42	0.275	0.04
*	3.0	17	42	0.475	—
40	1.7	1.1	6.3	0.445	0.23
41	1.1	3.5	7.6	0.605	0.23
**	1	4	8.3	0.600	—

* "Average" granite; ** "average" basic rock (after Turekian and Wedepohl, 1961).

162

This model is of paramount importance for helium isotope geochemistry because it provides a mechanism capable of generating helium with a high ^{3}He/^{4}He ratio in minerals poor in lithium.

Another result of this investigation also seems to be interesting and beneficial. While the ratios of ^{3}He/^{4}He in the trapped helium (average $2.2 \cdot 10^{-6}$) are about one order of magnitude higher than those in enclosing granites ($3 \cdot 10^{-8}$), they are very close to the same ratios in gabbrolabradorities which surround the granitic massif. Very high ratios of $(^4$He/^{40}Ar$)_{rad} = 10^2$ to 10^3 observed in included gases (compare with the average value of ~ 10) also disprove the idea that the granites (with an average ratio of U/K) are a source of the trapped gases. The authors postulated that the volatiles had been released by rocks under comparatively low-temperature metamorphism. The gabbrolabradorites (see Table 8.11) appear to be more probable candidates. It should be noted that the above-mentioned supposition was independently confirmed by δ^{13}C measurements (Mamchur et al., 1975).

8.5. Distribution of helium isotopes in terrestrial gases; relationships with geotectonics and heat flow

Volatile elements (helium isotopes inclusive) released from rocks and minerals occur as components of terrestrial fluids.

Fluids are complex mixtures of volatile components contributed by different sources whose chemical and isotopic compositions are levelled out by migration processes. Such an approach promises to be beneficial for the study of the principal regularities in the isotope geochemistry of rare gases as well as for outlining the main properties of the earth's sources of noble gas isotopes. Many authors have worked in the field of the helium isotope distribution in fluids during the last decade (Mamyrin et al., 1969a, b; Tolstikhin et al., 1969, 1972a, 1977b; Bennet and Manuel, 1970; Kamensky et al., 1971, 1974, 1976; Kononov et al., 1974; Polyak et al., 1976, 1979a, b; Craig and Lupton, 1976; Craig et al., 1978a, b; Wakita et al., 1978; Matveeva et al., 1979; Nagao et al., 1979, 1980a, b, 1981; and others). Extremely wide variations in ^{3}He/^{4}He ratios in fluids at large and a constant ratio in fluids of the same regional tectonic structure appear to be the most important tendencies in the geochemistry of terrestrial volatiles (see Fig. 7.1).

On the other hand, the development of the geoenergy approach to geotectonics has led to the discovery of regular differences between various structures possessing unequal magnitude of heat flow. It has been established that this parameter is related to the age of the tectono-magmatic activity, increasing from ancient structures to younger ones (Lee and Uyeda, 1965; Polyak and Smirnov, 1966, 1968; Smirnov, 1972; and others). Attempts at interpreting these relationships have led to the development of various geothermal models which imply the existence of deep-seated sources of heat localized in space and time (Smirnov, 1972; Kutas and Gordienko, 1972). However, the actual character of these sources has not been quite clear so far. At present new possibilities have emerged in this field, following achievements in the geochemistry of helium isotopes.

A careful interpretation of the helium isotope distribution in terrestrial gases and the relationships between $^3He/^4He$ ratios, tectonic structures and the heat flow has been carried out by Polyak et al. (1976, 1979b) and Polyak and Tolstikhin (1983); the results obtained are used as a basis for sections 8.5.1—8.5.3. In section 8.5.4 some geochemical implications are described.

8.5.1. The data used and the analytical method

The available data on the helium isotopic composition of various underground fluids, such as petroleum, natural gas, formation waters, hot springs and volcanic emanations, have been systematized on geotectonical principles (Table 8.12). Within one and the same geotectonic region all these types of fluids normally show the same $^3He/^4He$ ratio, irrespective of their hydrochemical and other features. This is due to the fact that silicate matter is the main source of helium isotopes in all types of fluids. This important peculiarity enables one to compare the $^3He/^4He$ ratios without any restrictions on the type of fluids; hence, more than 1100 measurements of the ratio in samples from 650 fluid manifestations have been examined.

For the geotectonic division of the regions the *International Tectonic Map of Europe* (Chernook, 1964), the *Tectonic Map of Eurasia* (Yanshin, 1966), and the *Tectonic Basement Map of the USSR* (Anonymous, 1974) have been used. Data from various structural units in each region as well as from areas of tectonic activation following the main folding stage were subdivided into special categories.

In analyzing the relationships between the helium isotopic composition and terrestrial heat flow, the values of the latter were taken from the *Heat Flow Map of the USSR* (Smirnov, 1982) and from published regional maps (Makarenko et al., 1968; Avetis'yanz et al., 1975; Lysak and Zorin, 1976; Kononov and Polyak, 1977). These maps are based on observed conductive heat losses (over 1000 measurements) regardless of the possible effects of topography, evolution of climate, sedimentation and other factors.

Data sets of $^3He/^4He$ values for different areas selected on geotectonic grounds were statistically processed. A check of the distribution models shows that, by Pearson's criterion χ^2, the distribution of values in sufficiently large data sets is controlled by the lognormal rule. Our analysis affords an objective evaluation of the average $^3He/^4He$ in various structures, within the 95% tolerance limits. The uniformity of the data sets and the range of the differences in their average values were determined using the nonparametric Kruskal-Walles (H) and Wilkinson (W) criteria, as well as the

TABLE 8.12

Isotopic helium compositions in underground fluids of different tectonic provinces

Index of regional data set	Tectonic provinces, structural unit	Regional values of ^{3}He/^{4}He (10^{-8}) at the 95% confidence level		average	References[a]
		tolerance limits (in parentheses: observed values)			
		$\min_{0.05}$	$\max_{0.05}$		
1	*Ancient (pre-Riphean) platforms*				
1.1	East European Platform				
1.1.1	stable segments	0.55	4.96	1.94	1
1.1.2	activated segments	2.30	15.9	6.82	1
1.2	Siberian Platform				
1.2.1	southwestern segment	0.77	14.3	4.19	1
1.2.2	Vilyuy syneclise	1.64	36.8	9.94	1
2	*Baikalides*	0.36	18.0	2.73	1
3	*Caledonides*	(1.5)	(3.2)		
4	*Hercynides*				
4.1	West Siberian plate	1.09	11.0	3.73	1
4.2	Scythian plate	2.54	30.0	11.0	1
4.2.1	Azov—Kuban and Terek—Kuma Basins	3.37	16.7	8.20	1
4.2.2	Stavropol arch	21.9	68.9	39.5	1
4.3	Turanian plate	3.32	26.6	10.7	1
4.4	Middle Asian zone of epiplatform orogeny				
4.4.1	areas of uplifting (Pamir)	(5.2)	(12.0)		1
4.4.2	areas of sinking	1.51	19.5	7.03	1
5	*Structures of the Mediterranean belt (Alpides)*				
5.1	foredeeps	2.42	16.0	6.94	1
5.1.1	Indol—Kuban	2.62	10.8	5.63	1
5.1.2	Terek—Caspian	2.42	27.6	9.25	1
5.2	intermontane (inner) depression				
5.2.1	Po depression	(8.9)	(10.0)		2
5.2.2	west Turkmenian depression	2.8	72.0	19.2	1
5.2.3	Trans-Caucasus depression	2.72	329	55.8	1, 3
5.2.3a	Kura	3.09	219	44.6	1, 3
5.2.3b	Rioni	5.12	968	108	1, 3
5.3	meganticlinoria				
5.3.1	Greater Caucasus	7.0	887	159	1, 3
5.3.2	Lesser Caucasus	48.5	1300	341	1, 3
5.3.3	central Apennines	(12.0)	(250)		2

TABLE 8.12 (*continued*)

Index of regional data set	Tectonic provinces, structural unit	Regional values of $^3He/^4He$ (10^{-8}) at the 95% confidence level			References[a]
		tolerance limits (in parentheses: observed values)		average	
		$min_{0.05}$	$max_{0.05}$		
5.4	areas of recent volcanic activity				
5.4.1	Phlegrean fields (Campi Flegrei)	(380)			2
5.4.2	Eolian Islands	(600)	(750)		4
5.4.3	eastern Sicily (nearby Etna)	(920)			2
6	*Structures of the Pacific belt*				
6.1	northwestern segment				
6.1.1	Sakhalin and western Kamchatka	11.5	1280	209	1, 5, 6
6.1.1a	Sakhalin	13.2	2260	215	1, 5
6.1.1b	western Kamchatka	5.0	1860	175	1, 6
6.1.2	central and eastern Kamchatka	178	2360	730	1, 6
6.1.3	Japanese Island Arc	(<30)	(1540)		7—13
6.1.4	Kuril Island Arc	640	1640	1040	1, 14
6.1.5	Marianas Island Arc	(420)	(920)		7
6.2	southwestern segment (New Zealand)	(560)	(1240)		15
6.3	eastern segment				
6.3.1	Rocky Mountains				
6.3.1a	Yellowstone Park	(390)	(2180)		16, 17
6.3.1b	Bueyeros Valley			439	18
6.3.2	Cascade Range	(450)	(1130)		16
6.3.3	central Mexican plateau	(85)	(230)		19, 20
6.3.4	trans-Mexican volcanic belt	(185)	(880)		19, 20
6.3.5	San Andreas fault zone	(14)	(1070)		21, 22
6.3.5a	Imperial Valley	(112)	(1070)		21
6.3.5b	Mexicali Valley	(710)	(890)	850	22
7	*Active rifting zones*				
7.1	continental				
7.1.1	Baikal Rift	0.51	2100	257	1, 23
7.1.2	Jordan Rift	(70)	(280)		15
7.1.3	West African Rift	42			15
7.1.4	East African Rift				
7.1.4a	Kenya segment	(490)			15
7.1.4b	Ethopia segment	(420)	(1960)		15
7.2	oceanic				
7.2.1	Red Sea Rift	(1200)	(1220)		24
7.2.2	Galapagos Rift	(1100)			25
7.2.3	Gulf of California Rift	(1110)			26
7.2.4	East Pacific Rise	(1600)			27, 28
7.2.5	Iceland	674	4230	1830	29, 30, 31

TABLE 8.12 (*continued*)

Index of regional data set	Tectonic provinces, structural unit	Regional values of ^{3}He/^{4}He (10^{-8}) at the 95% confidence level		References[a]
		tolerance limits (in parentheses: observed values)	average	
		$\min_{0.05}$ $\max_{0.05}$		
8	*"Hot spots" of oceanic plates (Hawaii)*	2100		32

[a] References:

1 = Polyak et al. (1979b)	12 = Nagao et al. (1980b)	23 = Lomonosov et al. (1976)
2 = Polyak et al. (1979a)	13 = Nagao et al. (1981)	24 = Lupton et al. (1977a)
3 = Matveeva et al. (1978)	14 = Tolstikhin et al. (1972)	25 = Lupton et al. (1977b)
4 = Polyak et al. (1980)	15 = Craig and Lupton (1978)	26 = Lupton (1979)
5 = Kamensky et al. (1974)	16 = Craig et al. (1978b)	27 = Craig et al. (1975)
6 = Kamensky et al. (1976)	17 = Torgersen and Jenkins (1979)	28 = Lupton et al. (1980)
7 = Craig et al. (1978a)		29 = Kononov et al. (1974)
8 = Matsubayashi et al. (1978)	18 = Phinney et al. (1978)	30 = Polyak et al. (1976)
	19 = Prasolov et al. (1982)	31 = Kononov and Polyak (1977)
9 = Wakita et al. (1978)	20 = Polyak et al. (in prep.)	
10 = Nagao et al. (1979)	21 = Welhan et al. (1978a)	32 = Craig and Lupton (1976)
11 = Nagao et al. (1980a)	22 = Welhan et al. (1978b)	

parametric criterion (Rodionov, 1964), regarded as the strongest one under conditions of lognormal distribution. When $|\,t\,| > 1.96$, the data sets compared by their average values are statistically heterogeneous; when $|\,t\,| < 1.96$, they are statistically indistinguishable (homogeneous), and the corresponding structures may be regarded as similar in their average isotopic composition of helium. It should be kept in mind, however, that the inference of similarity, on whatever criteria, is always conditional and may be modified when the size of any of the data sets increases.

The relationship between the helium isotopic composition in fluids and the value of conductive heat flow at the same points was studied by means of the usual methods of mathematical statistics (Miller and Kahn, 1962; Bondarenko, 1970). This enables one to establish the existence of a correlation between the parameters compared, to evaluate its closeness, and to approximate it graphically and analytically.

8.5.2. Helium isotopes and heat flow in various regions[1]

1. Ancient plates

The lowest average ^{3}He/^{4}He value (1.94 ± 0.38) is associated with the

---

[1] Numbers in this section correspond to those in Table 8.12 and Figs. 8.11—8.13.

underground fluids of the East European Platform. The majority of data bearing on this statistically uniform (by the H criterion) data set belong to the Volga—Ural anteclise with its Karelian folded basement. The remaining structural units of the platform are characterized by one or a few determinations per each unit so that differences between them cannot be traced. The data set differs greatly ($| t | = 7.37$) from those of regions with subsequent tectono-magmatic activation (the Dnieper—Donets Trough and the Rostov Height of the Ukrainian Shield) where the average $^3He/^4He$ value is almost three times greater. There is a similar pattern in the distribution of heat flow, whose mean value is also known to be minimum in the tectonically oldest structures. The heat flow in the Cis-Uralian region is always less than one HFU (unit of heat flow, $1 \cdot 10^{-6}$ cal cm^{-2} s^{-1} = 41 868 mW m^{-2}), locally it is 0.8, while its range in activated segments of the East European Platform is 1.0—1.6 HFU.

A similar differentiation of $^3He/^4He$ has been observed in the Siberian Platform (see Table 8.12). Here the mean value of $^3He/^4He$ is also somewhat higher in the activated Vilyuy syneclise and the southern part of the Anabar Shield than in the relatively stable southwestern part of the Platform: the Angara—Ilim region ($| t | = 2.41$). In the latter, however, this value is twice as high as in stable regions of the East European Platform ($| t | = 3.53$). This is consistent with the fact that tectonic processes were manifested later in this region. The distribution of heat flow over the Siberian Platform, in the same way as over the East European Platform, corresponds to variations in the isotopic composition of helium and makes apparent the same tendency, although not as vividly expressed: higher values are observed in the Vilyuy syneclise, up to 1.3 HFU, but lower in the Angara—Ilim region, less than 1.2 HFU.

2, 3. Baikalides and Caledonides

For the regions of the Baikalian folding only comparatively little data on fluids in the Pechora syneclise and northern areas of western Siberia were available. This data set is statistically similar to those from the stable areas of both ancient platforms. Curiously, its mean $^3He/^4He$ value is intermediate between structures of the East European Platform and the Angara—Ilim region. The heat flow in the Baikalian structures is higher than in the older ones, and amounts to about 1.5 HFU.

There are only four $^3He/^4He$ determinations for the region of the Caledonian folding, from two localities in south-central Kazakhstan. They are inadequate for a reliable estimation of helium isotopic composition.

4. Hercynides

The total body of data from the epi-Hercynian plates differs greatly from the data sets of the stable regions of the East European Platform ($| t | = 13.2$), the difference being less pronounced as compared to the data set of the Angara—Ilim area ($| t | = 2.64$). At the same time, it is similar to the data sets for the recently activated segments of the ancient platforms and for regions of the Baikalian folding ($| t | = 0.73$ and 1.72, respectively). It should be kept in mind, however, that the overall data set for the epi-Hercynian plates is not uniform. The west Siberian plate differs greatly from the Turanian and Scythian plates in the average isotopic helium composition, $(4.17 \pm 0.90) \cdot 10^{-8}$; it is statistically very similar to the Angara—Ilim region ($| t | = 0.23$) and particularly to regions of the Baikalian folding ($| t | = 0.002$). These relationships are corroborated by the results of some regional studies of another isotopic characteristic of subsurface fluids: the ratio of concentrations of radiogenic isotopes of helium and argon (section 8.1). With reference to $(^{4}\mathrm{He}/^{40}\mathrm{Ar})_{\mathrm{rad}}$, the west Siberian plate is likewise different from the Scythian and Hercynian, but similar to the East European Platform (Voronov et al., 1974). An analysis of the isotopic composition of methane carbon in the oil and gas regions of the USSR leads to the same conclusion (Prasolov and Lobkov, 1977). Thus, the isotopic data disprove the view about the similarity of all epi-Hercynian plates of the USSR based upon the fact that the observed average heat flow (1.35 HFU) is the same in all of these structures. This discrepancy warrants a special analysis.

The average $^{3}\mathrm{He}/^{4}\mathrm{He}$ values are almost the same, slightly over $1 \cdot 10^{-7}$ for the Turanian and the Scythian platforms ($| t | = 0.11$). Some heterogeneities in the $^{3}\mathrm{He}/^{4}\mathrm{He}$ ratios resulting from geotectonic peculiarities are also observed there; for instance, the ratio in the Stavropol Uplift sharply differs from the average ratio of the total data set for the Scythian plate. It is notable that the Stavropol Uplift is also distinguished by its heat flow value. These data indicate the presence of juvenile helium in the fluids of the Stravropol Uplift and link it, as demonstrated by further correlations, with the development of the Greater Caucasus.

Of importance in this connection is the isotopic composition of helium in segments of the Central Asian Hercynides known to have undergone Mesozoic tectonic activation. Judging from the rather scanty data at present on hand, this composition in the Pamir and Tien Shan mountain chains is indistinguishable from that observed in stable segments of the Turanian plate. This similarity indicates that the helium isotopic composition reflects not simply the age and intensity of tectonic movements but primarily the age of regional magmatic activity. Indeed, the mean $^{3}\mathrm{He}/^{4}\mathrm{He}$ value in fluids of superimposed neotectonic basins is significantly lower ($| t | = 3.22$) than in stable segments of the Turanian platform. A similar situation is observed in regions of the Alpine folding (for instance, compare items 5.2.3 and 5.3.1 and 5.3.2 in Table 8.12). A possible explanation is that fluids of a rising region contain helium which is (partially) derived from more or less deep-seated sources (rocks, magma chambers, etc.), whereas helium of basin fluids is released by sedimentary rocks eroded from the slopes of the rises.

The heat flow pattern is not uniform in structures developed during the epi-platform stage. It is as high as 1.6 HFU and higher in mountain structures, which is substantially higher than the average on epi-continental plates, in contrast with an average of 1.2 HFU for superimposed basins, and locally 0.8 HFU in the Fergana basin, Central Asia. The possible causes for such a relationship between the geochemical and geothermal data are given below.

5. Alpides

In these regions the distribution of heat flow is just as differentiated as in the zone of epi-platform orogeny. However, here the isotopic helium composition also differs from one structural unit to another, and the $^3He/^4He$ ratios vary even within geothermally similar groups of negative and positive structures.

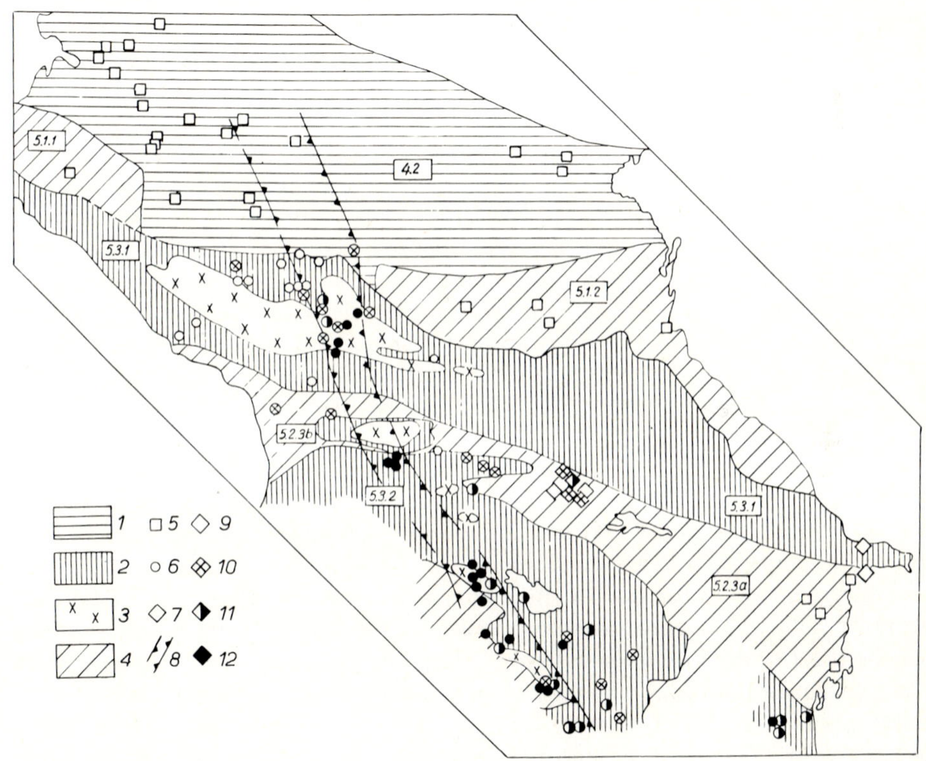

Fig. 8.11. Isotopic composition of helium in underground fluids of the Caucasus region (Matveeva et al., 1978).
$1-4$ = geological environment: 1 = epi-Hercynian Scythian plate; 2 = Alpine meganticlinoria; 3 = crystalline massifs; 4 = Alpine foredeeps and inner depressions. $5-7$ = types of fluids; 5 = bed waters and gases recovered from boreholes; 6 = cold and thermal mineral fluids; 7 = mud volcanoes; 8 = zone of high $^3He/^4He$ ratios. $9-12$ = values of $^3He/^4He$ ratios in fluids ($\times 10^{-6}$): 9 = $\leqslant 0.5$; 10 = $0.51-1.50$; 11 = $1.51-3.0$; 12 = > 3.0. For numbers see Table 8.12.

The lowest ^{3}He/^{4}He values in the Alpine belt, $(6.94 \pm 2.15) \cdot 10^{-8}$, mark the Cis-Caucasian foredeeps (Fig. 8.11). The Cis-Caucasian data set is statistically uniform and quite similar to that for superimposed basins of Central Asia ($|\,t\,| = 0.29$). The latter also differs from the data set for the stable regions of the Turanian plate by a lower average ^{3}He/^{4}He value. The relationship between the average ^{3}He/^{4}He in the Cis-Causasus and in stable and activated segments of the southern epi-Hercynian plates suggests similar tectono-magmatic conditions reflected in the helium isotopic composition in fluids of the Alpine foredeeps and the superimposed neotectonic basins. Such a similarity is readily accounted for if one takes into account the Hercynian age of the basement, at least in the exterior zones of the marginal troughs and folded structures, the source of their terrigenous fill, and also the absence of any evidence of Cenozoic magmatism in the Cis-Caucasus, except for the Mineralovodsk Height.

The average ^{3}He/^{4}He value of $(0.184 \pm 0.064) \cdot 10^{-6}$ in fluids of the west Turkmenian Basin is much higher ($|\,t\,| = 3.64$) than in the Cis-Caucasian Basin, and an even higher value, $(0.558 \pm 0.027) \cdot 10^{-6}$, is typical of the intermontane troughs of the Caucasus.

The heat flow field of these structures is marked by lower values. Its regional variations have a pattern different from that of the ^{3}He/^{4}He distribution; one may trace an opposite trend here. For example, the observed average heat flow is 1.1 HFU in the Terek—Kuma trough, 1.0 in the west Turkmenian and the Rioni troughs and a mere 0.8 HFU in the Kura trough.

The negative structures in the Caucasian segment of the Alpine belt have a prominent feature in common, namely mud volcanicity. The average ^{3}He/^{4}He value in gases of mud volcanoes decreases from the center of the segment to its periphery: from $47.2 \cdot 10^{-8}$ in the Kura Valley to $6.16 \cdot 10^{-8}$ in the Kerch Peninsula and $7.12 \cdot 10^{-8}$ in Cheleken Island. The isotopic helium composition from mud volcanoes coincides with the regional value for formation fluids. This implies the same source of helium in all types of fluids and rules out the possibility that the mud volcanoes are related to local flows of juvenile mantle emanations.

In the positive structures in the Caucasian segment of the Alpine belt the values of helium isotopic ratios and the observed heat flow are much higher than in the negative ones (see Table 8.12). However, although the heat flow in the meganticlinoria of the Greater and Lesser Caucasus is almost the same (2 HFU on the average), their helium isotopic composition differs substantially. In the Greater Caucasus the ^{3}He/^{4}He value is lower, averaging $(1.59 \pm 0.75) \cdot 10^{-6}$; its highest values $(0.8 \cdot 10^{-5})$ have been observed in the Elbrus area, which shows a weak activity at present (Masurenkov, 1971).

In the Lesser Caucasus the isotopic ratios of ^{3}He/^{4}He in subsurface waters are much higher than in the Greater Caucasus (Fig. 8.11); the average being $(3.40 \pm 0.96) \cdot 10^{-6}$, locally it may be as high as $1.5 \cdot 10^{-5}$. A comparison of the ^{3}He/^{4}He data sets for the Greater and Lesser Caucasus shows that they are statistically heterogeneous ($| t | = 2.18$). These data suggest that the regional isotopic composition of helium in areas of nearly contemporaneous volcanicity reflects not only the time of magmatic activity but also its scope, thereby making evident the regional intensity of mantle differentiation and degassing.

The first measurements of the $^3He/^4He$ ratio in thermal fluids of the central Appenines, Sicily and the Eolian Islands (see items 5.2.1, 5.3.3, 5.4.2 and 5.4.3 in Table 8.12) are close to the data obtained in the Caucasus region. Fluids of the Po Basin contain helium with a $^3He/^4He$ ratio quite similar to that in fluids of the Cis-Causasus foredeeps (Fig. 8.12). In the

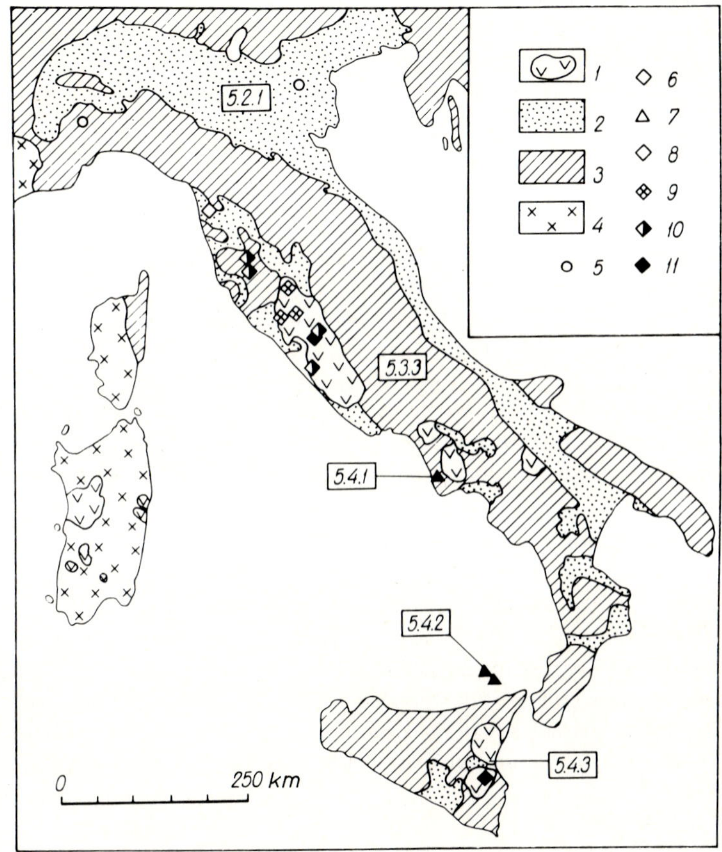

Fig. 8.12. Isotopic composition of helium in thermomineral fluids of Italy (Polyak et al., 1979a, 1980). *1—4* = geological environment: *1* = Late Cenozoic volcanics; *2* = post-orogenic successions; *3* = orogenic complex; *4* = European Hercynian foreland. *5—7* = typical gas component of fluids: *5* = N_2; *6* = CO_2; *7* = $H_2S—CO_2$. *8—11* = value of $^3He/^4He$ ratio ($\times 10^{-6}$): *8* = < 0.2; *9* = 0.2—0.8; *10* = 0.8—3.2; *11* = > 3.2. For numbers see Table 8.12.

central Appenine segment of the volcanic zone the ratio increases up to $2.4 \cdot 10^{-6}$, and in the Solfatara of the Flegreifield it reaches $3.8 \cdot 10^{-6}$. The highest ^{3}He/^{4}He value, $9.2 \cdot 10^{-6}$, is observed in Mofette Palichi, east Sicily. It is notable that the age of the latest manifestations of volcanic activity decreases from northern to southern Italy (Polyak et al., 1979a).

6. Structures of the Pacific Belt

It has been shown (Tolstikhin et al., 1972a; Kamensky et al., 1974, 1976; Craig et al., 1978a, b; Nagao et al., 1981; and others) that the ^{3}He/^{4}He ratios in thermal fluids of the Circum-Pacific belt vary as a rule from 10^{-6} to $1.5 \cdot 10^{-5}$, being noticeably higher than in any other region discussed above. These variations depend on the stage and type of structures in which the fluids circulate (see Table 8.12). In particular, the fold structures of western Kamchatka differ greatly in their average ^{3}He/^{4}He value from other areas of the peninsula which have been affected by recent volcanism (Fig. 8.13). On the other hand, they are similar to the structures of east Sakhalin, which belong to the same Hokkaido—west Kamchatka fold system, believed to be a zone of a comparatively early development of continental crust. The heat flow in this zone is also uniform and comparatively low (1.2—1.3 HFU).

The structures in east Kamchatka also make up part of the zone recently developed. Tectono-magmatic activity is fairly high there, and intense Quaternary volcanicity occurs not only in the east Kamchatka fold zone but also in some areas of the Median Range in central Kamchatka. Consequently, the average ^{3}He/^{4}He value is far higher here, $(7.4 \pm 1.1) \cdot 10^{-6}$, than in Sakhalin and western Kamchatka. As to the heat flow, it differs substantially from structure to structure, owing to the character of recent tectonics. Its mean value for central Kamchatka amounts to 2.2 HFU, but it is much lower in the graben-synclines of the eastern volcanic zone, where it approaches the value observed in the west of the peninsula.

The highest average value (see Table 8.12) for the transitory zone, ^{3}He/^{4}He $= (1.04 \pm 0.18) \cdot 10^{-5}$, is associated with hot springs in the Kuril Islands: Paramushir, Iturup, and Kunashir. The isotopic helium composition observed there is statistically different from that typical of volcanic areas of Kamchatka. This difference is quite consistent with the tectonics of the island arc where the crust is least "continentalized", being locally even close to the oceanic type. On the map of the observed heat flow the Kuril arc is bounded by the 2.0 HFU isoline; this figure rises to 2.5—3.0 HFU in the zone adjoining the South Okhotsk Deep.

The regularity thus established is supported by the data recently obtained for other structures of the northwestern segment of the Pacific belt, namely for the Japanese and Marianas Island Arc (see Table 8.12). From the geostructural point of view the former arc is similar to the Kamchatka Peninsula, whereas the latter is very much like the Kuril Islands. The same conclusion follows from analysis of the helium isotope composition. Statistical process-

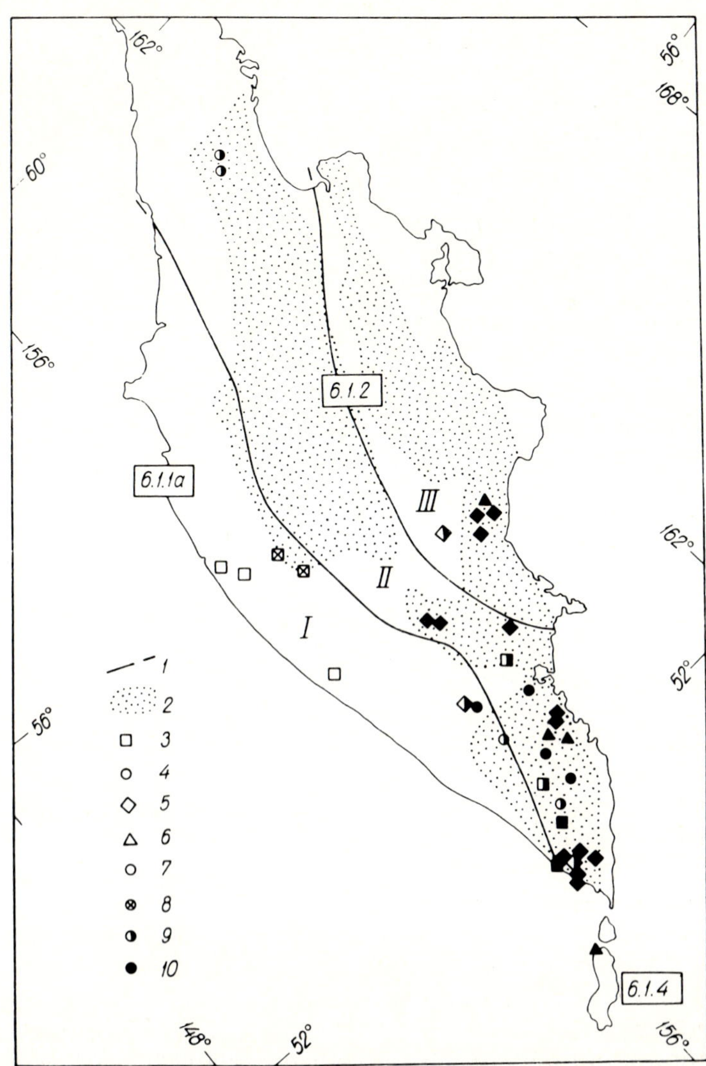

Fig. 8.13. Isotopic composition of helium in fluids of the Kamchatka region (after Kamensky et al., 1976).
1—2 = geological environment; 1 = boundaries of structural—facial zones: Kamchatka—Koryakian (I), central Kamchatka (II), eastern Kamchatka (III); 2 = areas of Quaternary volcanic accumulation. 3—6 = typical gas component of fluid: 3 = CH_4; 4 = N_2; 5 = CO_2; 6 = H_2S—CO_2. 7—10 = value of $^3He/^4He$ ratio ($\times 10^{-6}$): 7 = < 1.75; 8 = 1.75—3.5; 9 = 3.5—7.0; 10 = > 7.0. For numbers see Table 8.12.

ing of the available data showed that the isotopic helium composition for Japan and central—eastern Kamchatka can be represented by a single data set with an average ^{3}He/^{4}He value of $(7.7 \pm 0.8) \cdot 10^{-6}$, and for the Kuril and the Marianas Islands by a data set with average ^{3}He/^{4}He value of $(9.5 \pm 1.0) \cdot 10^{-6}$.

Results of the isotope analysis of helium from thermal fluids of some other parts of the Circum-Pacific belt have also been recently published (see Table 8.12 and references therein). The southwestern segment is characterized by data from New Zealand. The fluids of the eastern frame of the Pacific Ocean have been studied in the United States and Mexico (the Yellowstone and Lassen Parks, the San Andreas Fault Zone, the Mexican volcanic belt and some other areas). As a whole, these data are in good agreement with those for the northwestern Pacific island arc, although some fluids of the Yellowstone Park stand out against the usual values of the ^{3}He/^{4}He ratio, amounting to $2.18 \cdot 10^{-5}$. This record reported by Craig et al. (1978b) seems to be very important; it proves that under some areas of the Pacific belt there occur helium sources with ^{3}He/^{4}He ratios higher than the average value for oceanic basalts. It is interesting that exactly the same value was obtained for dacite lava from the Karymsky Volcano, Kamchatka (Tolstikhin et al., 1976); in both cases very high ^{3}He/^{4}He ratios were found in helium derived from acid magmas enriched in U and Th! In this connection it is worthwhile to recall that the Yellowstone Park area is now considered a continental "hot spot" caused by the upward mantle plume, which, according to helium isotope data, originated from an enriched reservoir (Smith et al., 1977; see section 7.5).

A comparison of the Pacific zone structures with the Alpine structures shows that on the basis of average ^{3}He/^{4}He the Hokkaido—west Kamchatka fold system is formally similar to the Caucasian meganticlinoria, whereas the volcanic areas of Kamchatka (and even more so those of the Kurils) are statistically quite different from both the Greater and the Lesser Caucasus ($| t | = 4.79$ and 3.90, respectively).

7. Contemporary rift zones

Rifting modifies the permeability of the crust and stimulates the escape of the deepest fluids.

Continental rift zones are marked by the greatest variation of ^{3}He/^{4}He values in subsurface fluids. These values are locally just as high as in volcanic regions of the island arcs, or as low as in ancient plates. In the Baikal rift the ^{3}He/^{4}He ratios vary along its axis, reaching $0.89 \cdot 10^{-5}$ in formation waters and hot mineral springs associated with Quaternary volcanism in the Tunkin trough (Lomonosov et al., 1976). The scatter in ^{3}He/^{4}He ratios in this zone led to a very broad range of average values $(2.57 \pm 2.51) \cdot 10^{-6}$. In the heat flow field the same trend is perceived: there is a belt of high values along the rift axis, while the highest values (2.5 HFU) are associated with the southwestern part of Lake Baikal.

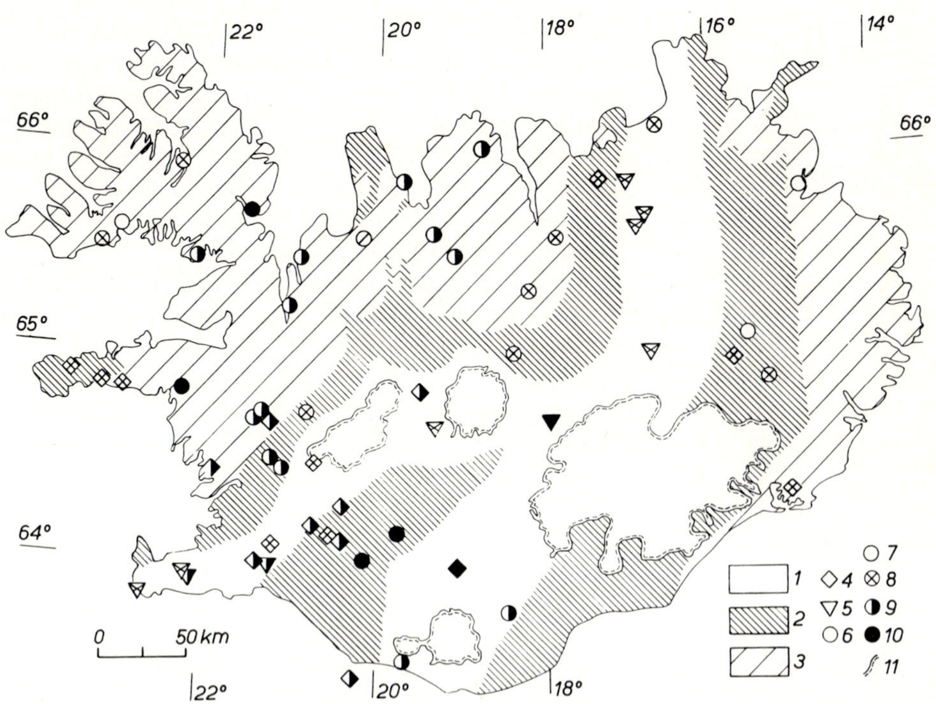

Fig. 8.14. Isotopic composition of helium in thermomineral fluids of Iceland (Kononov et al., 1974; Polyak et al., 1976; Kononov and Polyak, 1977).
1—3 = geological environment: *1* = active zone of rifting and volcanism; *2* = Quaternary flood basalts; *3* = Tertiary flood basalts. *4—6* = typical gas component of fluids: *4* = N_2; *5* = CO_2; *6* = H_2. *7—10* = value of $^3He/^4He$ ratio in fluids ($\times$ 10^{-5}): *7* = < 0.9; *8* = 0.9—1.8; *9* = 1.8—2.7; *10* = > 2.7. *11* = boundaries of glaciers.

Oceanic rifts and such specific structures as the Hawaiian Archipelago are marked by the highest and more or less homogeneous $^3He/^4He$ values. These regions contain mostly mantle helium (see section 7.2). Here we must note that the distribution of Icelandic $^3He/^4He$ ratios is consistent with the heat flow distribution, whose maximum (2.2 HFU) is associated with the zone of the highest isotopic ratios, while both parameters tend to decrease in the older parts of the island. Such relationships imply that both the helium isotopic composition and the heat flow are affected by more permanent (more "inert") factors than those controlling the surface volcanic activity (Fig. 8.14).

8.5.3. Relationship between isotopic helium composition and heat flow

$^3He/^4He$ values in various geological structures clearly indicate that in terrestrial fluids the ratio varies with time, reflecting the general trend of

176

the tectonic process (Fig. 8.15). The ^{3}He/^{4}He is at its highest in areas of recent tectonic-magmatic activity; it decreases as the continental crust becomes older and more consolidated. This decrease is particularly noticeable in the early stages of crustal development. The decrease ceases completely in structures of pre-Baikalian age and the like, which always show the same level of ^{3}He/^{4}He values, $(2 \pm 1) \cdot 10^{-8}$ (Fig. 8.15). Being the absolute regional minimum, this level obviously constitutes an objective feature of the isotopic composition of radiogenic helium for the ancient continental crust.

It is clear from the text and Fig. 8.15 that the trend of changes in the isotopic composition of helium and in heat flow is the same; in the course of geological time, both parameters had reached the absolute regional minimum in the most ancient tectonic units of the crust. In determining the relationships between these two parameters, their values were compared at the same localities in various geotectonic regions (398 pairs of figures). Correlation analysis shows a positive relationship between the parameters $(r_{x/y} = r_{y/x} = 0.72)$, approximated by a straight line and analytically described by the expression:

$$(^3He/^4He) \cdot 10^{-8} = \exp 6\, q - (5.2 \pm 0.2) \tag{8.4}$$

where q is the observed conductive heat flow, in HFU. This relation is graphically illustrated in Fig. 8.16.

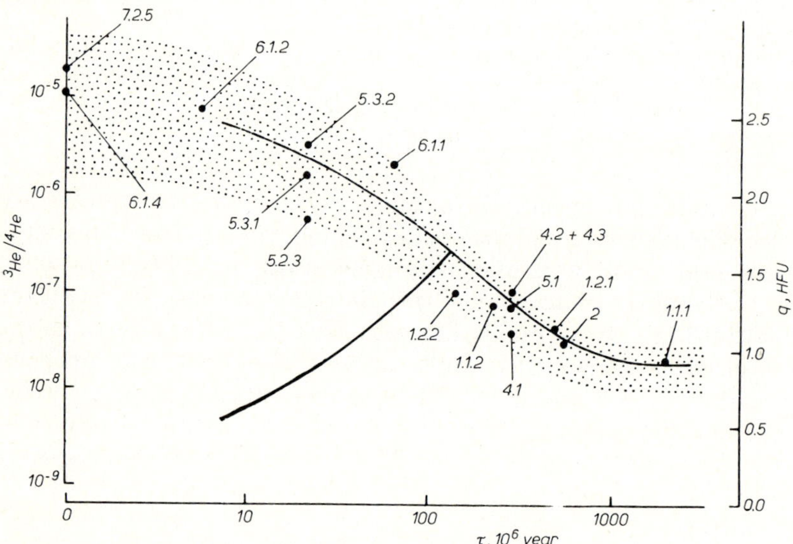

Fig. 8.15. Decrease of the ^{3}He/^{4}He ratio in helium of terrestrial fluids (dotted area) and heat flow (solid line) as a function of time since the end of tectono-magmatic activity of regions. Numbered dots refer to data set listed in Table 8.12 (Polyak et al., 1979b).

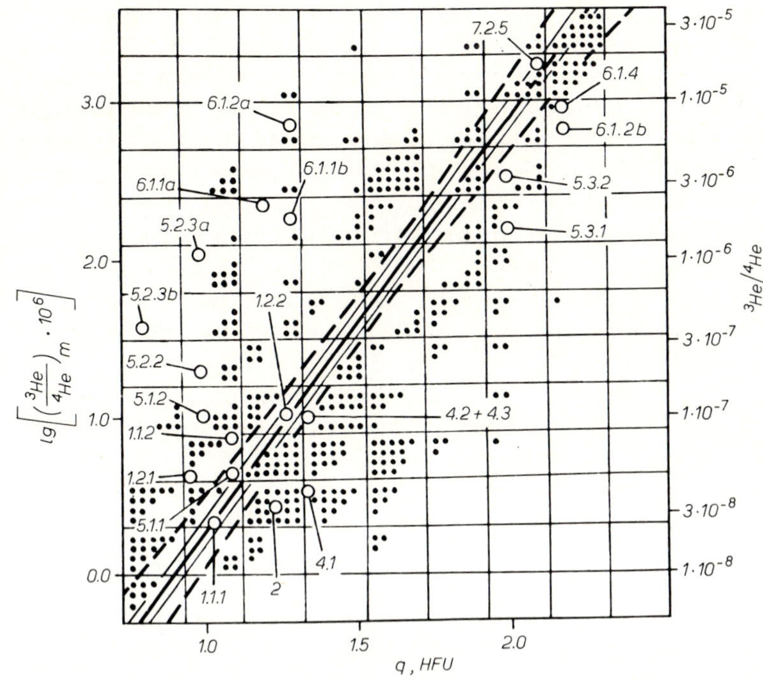

Fig. 8.16. Relationship between isotopic helium composition in terrestrial fluids and heat flow (Polyak et al., 1979b).
Solid line: regression curve ("abridged main axis", after Miller and Kahn, 1962); dashed lines: boundaries of the confident zone for the regression curve. The number of dots in each square of the correlation field corresponds to the number of observations of the ^{3}He/^{4}He ratio and heat flow in the given range. Numbered dots refer to data sets listed in Table 8.12.

It is interesting to compare the average values of the observed heat flow with the "helium" values of the flow given by formula 8.4. In stable regions, the two values are close and sometimes even coincide (see Table 8.12; Fig. 8.16). They differ, however, in mobile belts. The average observed heat flow values, both in the Alpine belt and in the west Pacific transitory zone, are lower than the "helium" values in negative structures, but higher in positive ones. This might be the effect of non-stationary disturbances of the geothermal field, due to vertical displacement of terrestrial masses. By these processes the surface heat flow decreases in regions of downwarping and sedimentation and increases in uplifted and eroded crustal blocks, producing two branches in the curve of observed heat flow (see Fig. 8.15). The helium isotopic composition cannot be affected by these processes or, for that matter, by any of the other factors controlling the scatter of the observed heat flow for one and the same geotectonic region (form of relief, climatic changes, differential heat conductance, the circulation of subsurface water). For this reason the "helium" estimates of heat flow (calculated from formula 8.4 with an average regional ^{3}He/^{4}He ratio) can be even closer to the "true" value of deep heat flow than the observed average flow. The true value of the latter can be determined by direct geothermal measurements allowing for corrections connected with the rate of displacement of silicate mass, the heat properties of rocks and their heat generation, morphology, etc. Such corrections, when available, lead to a far better agreement between the mean values of heat flow and the ^{3}He/^{4}He ratios.

It is apparent that the region where relation 8.4 holds true is bounded by the range of the following possible ^{3}He/^{4}He values in terrestrial objects: $(3 \pm 2) \cdot 10^{-5}$ in mantle helium and $(2 \pm 1) \cdot 10^{-8}$ in purely crustal helium. By the same token, the relation must then in fact be limited by the second parameter: the value of heat flow. Fig. 8.16 shows that the heat flow values corresponding to such limits are roughly equal to 1.0 ± 0.1 and 2.2 ± 0.1 HFU.

Some of the observed values, however, surpass these limits. Evidently they are related not to the general geothermal and geochemical (helium isotopic) trend but rather to the particularly strong effect of the factors cited above, which produce the scatter in the observed values of the heat flow. For example, the convective transfer of heat by subsurface waters can reduce the conductive heat flow down to zero, or increase it considerably.

Thus, the coincidental change of the ^{3}He/^{4}He ratio in subsurface waters and the regional heat flow in the course of geological time clearly indicates that their values characterize two different aspects — the geochemical and the geophysical — of one and the same process — tectogenesis. Consequently, any analysis of the character of the forces causing this process must make provisions for the established relation, this being an objective complex constraint. Its geochemical aspect indicates that the tectonic process is related to mass flow from the mantle, which carries along the residual primordial ^{3}He. On the other hand, the geothermal data rule out any possibility of this mass flow consisting solely of mantle helium; the outflow of the latter is estimated as a mere $4.0 \cdot 10^6$ atoms cm^{-2} s^{-1}, on the average (see Table 10.1). Despite the very high heat capacity of helium, this amount is absolutely incapable of producing adequate energy for the observed heat flow, and the only agent able to transport helium and transfer heat on the necessary scale must be some silicate mantle material. The relationship between the helium and the strontium isotopic ratios (see Fig. 8.7) favours this supposition.

Our proposal is a model which is consistent with the data discussed above and postulates impulses of heated mantle matter enriched in light isotopes of helium. Such impulses are supposed to occur within mobile belts where they convey the mantle melts and heat energy that had activated the tectonic process to the upper layers of the earth. Naturally, each impulse lasts as long as the period of regional tectono-magmatic activity. When the mantle impulse comes to an end, the excess of its contributed heat is gradually evacuated by conductive heat flow. At the same time, the ^{3}He/^{4}He ratio decreases in rocks and fluids, because residual mantle ^{3}He is released without compensation whereas radiogenic ^{4}He is being continuously generated.

After the mantle impulse ceases, its geothermal and isotopic helium parameters simultaneously reach a stable minimum level, which can be regarded as the continental background. According to current data, such background values of ^{3}He/^{4}He and the heat flow are roughly $(2 \pm 1) \cdot 10^{-8}$ and 1.0 ± 0.1 HFU, respectively.

Such a model — an impulsive mass heat flow from the mantle as a factor of tectogenesis — fully accounts for the observed distribution of ^{3}He/^{4}He in subsurface fluids as well as for conductive heat losses across the earth's surface; it also explains their correlative relation. It provides a new approach to evaluating the geo-energetic effect of mass heat flow from the mantle in structures of the continents and the transitory zones. According to earlier views, the magnitude of the heat flow in these areas is controlled solely by the actual volcanic and hydrothermal activity, whose products cover no more than 2% of the total deep heat loss. It has now become obvious that the mantle mass heat flow in such structures is responsible for the excess of regional conductive heat flow over the continental background. If so, over 15% of the conductive heat losses observed in the area of these crustal structures must come from the mantle.

The data on helium isotopic composition make it possible to re-evaluate the relative contribution of mantle and crustal sources of energy to terrestrail heat losses. According to a preliminary calculation, the proportion of mantle and crustal heat outflows was found to be similar to the earlier and independently determined proportion of mantle and crustal helium fluxes (approx. 2).

8.5.4. Helium isotopes and other volatiles in terrestrial fluids

In sections 8.3 and 8.4 it has been established that juvenile outgassing of the earth is a global process which is continuing at the present geological period. On their way to the earth's surface juvenile volatiles mix with volatile components of the crust and the atmosphere. Geochemical and isotopic investigations of volatile elements, such as H, O, C, S, and N, show that the contribution of mantle volatiles is generally smaller than that of crustal and atmospheric ones, with the only exception of helium.

Now we turn to estimations of the contributions of crustal, mantle and atmospheric helium to the helium of terrestrial fluids (1) and to a possible approach to a similar estimation of active volatile elements (2).

180

(1) *Contribution of mantle, crustal and atmospheric helium to helium of terrestrial fluids*

Keeping in mind the average ^{3}He/^{4}He ratios (R) in the mantle ($R_M \approx 3 \cdot 10^{-5}$, see section 7.5), the crust ($R_C = 2 \cdot 10^{-8}$, see section 8.1) and the atmosphere ($R_A = 1.4 \cdot 10^{-6}$, see section 10.1), we can estimate the contribution of helium from these three reservoirs to the helium observed in a sample (Kamensky et al., 1976). The only necessary assumption is that non-radiogenic isotopes of other noble gases in the same sample (^{20}Ne$_S$ or ^{36}Ar$_S$, for example) are derived from the atmospheric reservoir, viz. ^{20}Ne$_S \equiv {}^{20}$Ne$_A$. This assumption appears to be consistent with recent notions in isotope geochemistry of light noble gases (see section 7.4 and 7.5). ^{4}He$_A$/^{20}Ne$_A$ is known to be 0.32 in the atmosphere and 0.25 in the air dissolved in fresh water (assuming the water temperature to be $10°$C). Then the average ratio of these atmospheric components may be accepted as equal to 0.285.

Now a system of three simple equations may be written:

$$^4\text{He}_A = 0.285 \; {}^{20}\text{Ne}_S$$
$$^4\text{He}_A + {}^4\text{He}_C + {}^4\text{He}_M = {}^3\text{He}_S \qquad\qquad (8.5)$$
$$R_A \cdot {}^4\text{He}_A + R_C \cdot {}^4\text{He}_C + R_M \cdot {}^4\text{He}_M = R_S \cdot {}^4\text{He}_S$$

where ^{4}He$_A$, ^{4}He$_C$ and ^{4}He$_M$ are the unknown concentrations of atmospheric, crustal and mantle helium, respectively; ^{4}He$_S$ and ^{20}Ne$_S$ denote the measured concentrations of the isotopes in a sample.

The desired proportion of mantle helium can be found by solving system 8.5:

$$^4\text{He}_M/{}^4\text{He}_S = [(R_S - R_C) - (R_A - R_C) \cdot {}^4\text{He}_A/{}^4\text{He}_S]/(R_M - R_C) \qquad (8.6)$$

Similar to 8.6 a formula can be obtained for the contribution of crustal helium, and the proportion of atmospheric helium is simply defined as:

$$^4\text{He}_A/{}^4\text{He}_S = 0.285 \cdot {}^{20}\text{Ne}_S/{}^4\text{He}_S \qquad\qquad (8.7)$$

The result of the estimation by formula 8.6 is directly dependent on the R_M value, which is not precisely known (see section 7.5). Kamensky et al. (1976) used a high R_M value ($3 \cdot 10^{-5}$) and calculated on this ground ^{4}He$_M$/^{4}He$_S \approx 0.35$ for most of the hot springs of Kamchatka enriched in ^{3}He. At present more recent estimations of R_M are available, and if one assumes $R_M = 1.1 \cdot 10^{-5}$ (which is typical of oceanic basalts, thermal fluids of the ocean floor and probably of the depleted mantle reservoir), then the contribution of mantle helium in many thermal manifestations of the Circum-Pacific volcanic belt approaches 1. However, in any case Kamensky and his co-workers proved that the contribution of atmospheric helium is small or even negligible in the vast majority of terrestrial fluids; this is in full agreement with considerations presented in section 10.2.

As a rule the proportion of mantle, crustal and atmospheric components is more or less constant, at least within a period of direct observations, lasting about ten years or so. This applies to gas and oil deposits and gases dissolved in deep waters of platform regions, which are very conservative systems. Repeated measurements of the helium isotope composition in thermal fluids of Iceland and Kamchatka also confirm the validity of this statement. But in some cases the $^3He/^4He$ ratios in volcanic areas can change notably during a decade, as was reported for fumarolic gases from the Showa-shinzan Volcano, Japan (Nagao et al., 1980a). Seven samples of gases collected between 1958 and 1977 were analyzed; the $^3He/^4He$ and $^4He/^{20}Ne$ ratios decreased from $7.6 \cdot 10^{-6}$ and 130 to $3.1 \cdot 10^{-6}$ and 0.61, respectively, during this period, in harmony with a decline in volcanic activity and an increase in atmospheric contamination. The latter occurred due to flows of underground waters, which are the main carriers of air components. Similarly, Craig et al. (1978b) observed a disequilibrium between the dissolved and gaseous light noble gases in Acid Spring, Lassen Park, California: the $^3He/^4He$ and $^4He/^{20}Ne$ ratios in the gas phase were twice and 1000 times higher than those in the liquid phase, respectively. The above observations also clearly show the unique possibility of the helium isotope method: the isotope composition of helium sampled very close to the earth's surface (in soil, swamp or lake waters, etc.) does not undergo essential contamination by atmospheric helium and thus preserves the information about the origin of helium and other volatiles.

(2) *Isotope composition of helium and chemically active gases*

It is clear that the helium flux to the earth's surface is accompanied by the flux of other active volatiles, but an identification of their mantle component, even a qualitative one, appears to be a very complex problem. In rare cases such an identification is possible if based on the isotope composition of the volatiles. An example is the juvenile isotope composition of sulphur, $\delta^{34}S = 0$, in thermal fluids of internal areas of Iceland; this is in agreement with the mantle isotope composition of helium (see section 7.2.1). As a rule, the proportion of mantle volatiles and those of the earth's crust and atmosphere is too small to identify the volatiles on the grounds of isotope data. A comparative analysis shows the absence of any direct relationship between the helium isotopic abundance in terrestrial fluids and their chemical composition (Table 8.13). High $^3He/^4He$ values were found in the highest temperature fluids, namely the nitrogen-carbon dioxide hot springs of the geyseric type and the hydrogenous hot springs of the Icelandic type. This may serve as additional evidence that the isotopic ratio of $^3He/^4He$ is determined by the tectono-magmatic activity of a region as a whole, whereas the chemical composition of fluids is connected with more local and shallow geological units.

TABLE 8.13

Range of ^{3}He/^{4}He values in various types of thermal fluids (Polyak et al., 1976)

Gas manifestation	^{3}He/^{4}He (10^{-6})	Temperature (°C)	He + Ne	Ar	N$_2$	CH$_4$	CO$_2$	H$_2$S	H$_2$	O$_2$
Methane underground waters of regional bedded systems										
Liman, western Kamchatka	0.0092		—	4.1	—	95.9	0.0	0.0	0.0	0.0
Pinachevo, eastern Kamchatka	− 4.9		0.01	0.26	24.2	74.4	11		0.0	0.0
Tunka depression, Baikal region	8.9		0.242	0.31	23.3	74.5	1.60	0.0	0.02	0.0
Nitrogenous hot springs										
Gorjyachinsky spring, Baikal region	0.42	57	0.078	1.03	98.2	0.0	0.40	0.0	0.0	0.0
Malka hot spring, central Kamchatka	8.7	82.7	0.026	1.47	98.5	—	—	—	—	—
Hveravik, north western Iceland	33	77	0.0169	2.61	97.0	0.03	0.35	0.0	0.001	0.1
Carbon-dioxide mineral waters										
Yamarovka, Baikal region	0.55	2	0.018	1.03	—	0.02	99.0	0.0	0.0	0.0
Malka cold spring, central Kamchatka	4.6	5.6	—	1.57	—	0.0	98.5	0.0	0.0	0.0
Furubrekka, western Iceland	12.5	7	0.001	0.03	0.92	0.02	99.0	0.0	0.0003	0.01
Carbon dioxide hot springs										
Yermuk, Lesser Caucasus	4.1	54	0.0093	0.167	2.2	0.016	97.8	0.0	0.0	0.0
Nalychevo, eastern Kamchatka	8.0	75	0.002	5.80		0.74	93.4	0.0	0.0	0.0
Lysuholl, Western Iceland	13.0	57.5	0.0022	0.071	4.6	0.01	95.0	0.0	0.0	0.38

TABLE 8.13 (*continued*)

Gas manifestation	^{3}He/^{4}He (10^{-6})	Temperature (°C)	Gas components (% vol.)							
			He + Ne	Ar	N_2	CH_4	CO_2	H_2S	H_2	O_2
Nitrogen—carbon dioxide hot springs of "geyseric type"										
Geyser Valley, eastern Kamchatka	9.0	$t_{boil.}$	0.0024	0.60	26.9	3.85	66.0	0.0	2.67	0.0
Hveravellir—Kjölur, central Iceland	21	$t_{boil.}$	0.001	0.34	21.61	1.06	76.0	0.0	0.54	0.45
Great Geyser area, southwestern Iceland	26.5	$t_{boil.}$	0.0118	1.26	62.2	1.08	26.2	0.0	0.282	8.56
Hveravellir—Reykjahverfi, northern Iceland	17.0	$t_{boil.}$	0.004	2.84	85.0	0.35	11.7	0.0	0.007	0.1
Arhver, western Iceland	23.5	$t_{boil.}$	0.0167	2.3	95.1	0.3	2.23	0.0	0.062	0.01
Hydrogenous hot springs of "Icelandic type"										
Kerlingarfjöll, central Iceland	2.0	$t_{boil.}$	0.001	0.03	1.64	0.14	81.3	7.09	9.78	0.0
Idem	18.0	$t_{boil.}$	0.001	0.1	9.48	1.01	53.5		35.6	0.26
Landmannlaugar, Hver, southern Iceland	25.0	$t_{boil.}$	0.0005	0.084	3.0	0.35	86.5		9.79	0.27

184

In a general way the high ^{3}He component is associated with the CO_2 component: both most typically occur in regions of recent high tectono-magmatic activity (Kamensky et al., 1971, 1976; Tolstikhin et al., 1972a; Polyak et al., 1976; Craig et al., 1978a; Matveeva et al., 1978; Nagao et al., 1981; and others). However, it is most unlikely that this CO_2 component is of mantle origin; on the contrary, carbon dioxide appears to be mostly of metamorphic origin. High ^{3}He/^{4}He ratios are also observed in nitrogenous hot springs characterized by comparatively high helium concentrations (Table 8.13); nitrogen in these manifestations is mainly of atmospheric origin (Kamensky et al., 1976) and in some cases high N_2/Ar_{atm} ratios (as compared with that of the atmosphere) point to an addition of N_2 from sedimentary materials (Matsuo et al., 1978). Matsuo et al. (1978) suggested that the mantle ratio of N_2/Ar is equal to 12, but they are not explicit as to the ways of identification of the argon mantle component. A wide range of ^{3}He/^{4}He ratios is typical of the methane underground waters widespread in some active regions. However, high ^{3}He/^{4}He ratios have been observed in gas pools of these regions; for example, a ^{3}He/^{4}He value of $6.3 \cdot 10^{-6}$ has been reported for the Tungor gas pool, Sakhalin (Kamensky et al., 1974). Interpretation of helium and carbon isotopic ratios, the latter being determined in both CO_2 and CH_4 components, does not lead to the estimation of the contribution of the carbon mantle component (Kamensky et al., 1976; Nagao et al., 1980a). Welhan and Craig (1979) observed a large flow of abiogenic methane accompanying ^{3}He in hydrothermal vents of the East Pacific Rise; these ressults are the only ones which might be interpreted as evidence of a mantle flux of juvenile methane (see below).

Two approaches to the problem appear to be possible. The first consists of an investigation of samples containing only mantle components (Tolstikhin, 1978); the samples may be chosen on the basis of the Ar/He isotope correlation (see section 7.5, Fig. 7.8).

The second approach (Polyak et al., 1976; Tolstikhin, 1979) starts with a comparison of the ratio:

$$C_s/^3He_s \eqno(8.8)$$

in fluid or gas samples from different regions, where C_s denotes the concentration of a volatile component in a sample (carbon, sulphur or nitrogen, for instance), and ^3He_s is the concentration of ^{3}He in the same sample. We assume that all ^3He_s in ratio 8.8 is of mantle origin. If the ^{4}He/^{20}Ne ratio in the sample is similar to the atmospheric one, then a correction for atmospheric ^{3}He is to be introduced by the above method and the absence of tritigenic ^{3}He is to be testified.

The constant ^{3}He/^{4}He ratio in the upper mantle may be considered as evidence for a more or less homogeneous mantle with respect to concentrations of the most volatile elements. Then, other conditions being similar, the lowest $C_s/{}^3He_s$ ratios indicate a mantle source of an active component. Of course, only a very rough guess of C_s is possible because of a difference between helium and other volatiles in solubility in liquids and migration from/into solids. Moreover, fixation of active volatiles might also have changed their initial (juvenile) proportion. On the other hand, a detailed study of the $C_s/{}^3He_s$ ratio in fluids may be useful for the understanding of the role of the above-mentioned processes. In addition it is possible to take into consideration the $C_\Sigma/{}^3He_\Sigma$ ratio, where C_Σ and ${}^3He_\Sigma$ denote an active component and a light helium isotope, respectively, which have been hitherto released from deep earth into the atmosphere. If the residence time of any volatile component in the mantle is approximately the same (see Chapter 11), then a mantle sample should yield a $C_s/{}^3He_s$ value approximately equal to $C_\Sigma/{}^3He_\Sigma$, provided the above-mentioned complications are negligible.

This approach was applied to the problem of carbon origin in hot and cold underground carbon dioxide and nitrogen-carbon dioxide waters. Some authors suggest a large contribution of juvenile carbon dioxide in terrestrial fluids, others do not even mention this source. The total ratio of $C_\Sigma(C)/{}^3He_\Sigma$ amounts to about $1.8 \cdot 10^7$; $C_\Sigma(C)$ denotes the amount of carbon released by the earth, $C_\Sigma(C) \approx 4 \cdot 10^{45}$ atoms (Galimov et al., 1975), and ${}^3He_\Sigma \approx 2.2 \cdot 10^{38}$ atoms (Tolstikhin et al., 1977; see Chapter 10). In Iceland, the Kuril—Kamchatka volcanic belt and the Caucasus, this total ratio was compared with those in the highest-temperature vapor jets, where a totally different solubility of He and CO_2 in water cause no significant changes in their proportion. In some high-temperature springs of Iceland $CO_2/{}^3He$ values of $\approx 1.7 \cdot 10^7$ are observed. This indicates the juvenile origin of their carbon dioxide, which is in agreement with the results obtained for helium and sulphur. A low ratio of $CH_4/{}^3He \approx 0.6 \cdot 10^{-7}$ is observed in East Pacific waters. On the other hand, most hot springs of Kamchatka and the Caucasus are characterized by high $CO_2/{}^3He$ ratios, reaching 10^{10}; this confirms the idea of a crustal, metamorphic origin of CO_2.

Due to a lack of experimental data, this method cannot be further specified at present, and needs additional investigations.

8.6. The ${}^3H$—${}^3He$ method; its application in physical hydrology and limnology

The progress in accurate measurement techniques of both helium isotope concentrations in water samples has allowed the introduction of a new method, based on radioactive tritium, ${}^3H$, β-decay and radiogenic helium, ${}^3He_{tri}$ accumulation. This method can be used to estimate the effective time since water mass isolation from the atmosphere. The method supplies the time parameter necessary to compute a set of physical parameters in hydrologic and limnologic systems, such as gas exchange rates, gas renewal, vertical diffusivity, ${}^4He$ fluxes in a water reservoir, etc.

In the first publication on this subject by Mamyrin et al. (1969a), an attempt was made to simultaneously interpret the ^{3}H and ^{3}He/^{4}He data. Later Tolstikhin and Kamensky (1969) outlined the ^{3}H—^{3}He method of water age determination, and Devirts et al. (1971) reported the first results of ^{3}He and ^{3}H measurements in hot springs. Later on considerable progress was made in this field by a joint Canadian—American team (W.B. Clarke and Kugler, 1973; W.B. Clarke et al., 1976, 1977; Torgersen et al., 1977, 1979, 1981; Top and Clarke, 1981).

In the following section we will give a general description of the method (8.6.1) and some examples of its utilization (8.6.2—8.6.4).

8.6.1. The ^{3}H—^{3}He method

To obtain the ^{3}H—^{3}He "age" of a water sample, radiogenic ^{3}H$_{tri}$ (produced by ^{3}H *in situ* decay) must be separated from dissolved atmospheric and injected ^{3}He (Torgersen et al., 1981):

$$^3He_{tri} = {}^3He_{tot} - {}^3He_{sat} - [{}^4He_{tot} - {}^4He_{sat}] \cdot ({}^3He/{}^4He)_{in} \qquad (8.9)$$

where the subscripts "tot", "sat" and "in" refer to the total, saturated and injected components, respectively. The values of ^{3}He$_{tot}$ and ^{4}He$_{tot}$ are supplied from measurements and ^{3}H$_{sat}$ and ^{4}He$_{sat}$ have been reported by Weiss (1970, 1971). The $(^3He/^4He)_{in}$ ratio in helium injected from wall rocks can be estimated from independent data; in many cases the ratio which is typical of deep waters of the region is applicable (see section 8.5).

When the ^{3}He$_{tri}$ and ^{3}H concentrations are determined, the water mass age can be calculated from the following equation (Torgersen et al., 1979, 1981).

$$t = 17.69 \ln \left\{ \frac{4.01 \cdot {}^3He_{tri}}{{}^3H} \cdot 10^{14} + 1 \right\} \qquad (8.10)$$

where t is age, years; ^{3}He$_{tri}$ is the tritigenic ^{3}He concentration, cm^3 STP g^{-1}; and ^{3}H is the tritium concentration in tritium units, T.U. (1 T.U. = 1 ^{3}H/10^{18} ^{1}H). Because the waters are of a mixed character, the ages t obtained from eq. 8.10 should not be taken at face value, but interpreted in the context of a particular system.

8.6.2. ^{3}H and ^{3}He in hot springs

In order to shed additional light on the origin of ^{3}He excess Devirts et al. (1971) and Tolstikhin et al. (1972a) compared high ^{3}He/^{4}He ratios and ^{3}H concentrations in thermal waters. The concentrations of ^{4}He and the ^{3}He/^{4}He ratios were measured in gas bubbling through water, whereas the ^{3}H con-

TABLE 8.14

Concentrations of ^{3}H, ^{4}He and ^{3}He/^{4}He ratios in hot springs of the South Kuril Islands (Devirts et al., 1971)

No. Location	Concentration		^{3}He/^{4}He (10^{-6})
	^{3}H in water (T.E.)	^{4}He in gas (% vol.)	
Kunashir Island			
1 Stolbovsky spring	< 5	0.01	7.6
2 east part of fumarole field, Mendeleyev Volcano	< 20	0.002	6.2
3 upper Mendeleyev springs	< 20	0.0013	8.8
Iturup Island			
4 Goryachy Klyuch	< 20	0.008	8.0

tent was estimated directly in the water (Table 8.14). However, these data allowed only a rough comparison between the juvenile and tritigenic concentrations of ^{3}He. Assuming that the partial pressure of gases in bubbles is related to their contents in waters in accordance with Henry's Law[1], it is possible to estimate the total ^{3}He concentration in waters. Near the earth's surface the total gas pressure in bubbles approaches the atmospheric value, which means that the partial pressure of ^{3}He can be regarded as equal to $\sim 5 \cdot 10^{-7}$ Torr or somewhat lower (Table 8.14). Multiplying this pressure with the solubility coefficient $S = 7 \cdot 10^{-3}$ cm^3 He/cm^3 H$_2$O for the Stolbovsky hot spring of the South Kuril Islands, one will find the ^{3}He concentration, ^{3}He$_{tot}$:

$$^3\text{He}_{tot} = p \cdot S = 3.5 \cdot 10^{-11} \text{ cm}^3 \, ^3\text{He/cm}^3 \, \text{H}_2\text{O} = 0.94 \cdot 10^9 \text{ atoms} \, ^3\text{He/cm}^3 \, \text{H}_2\text{O}$$

Somewhat lower ^{3}He$_{tot}$ concentrations are estimated for other thermal springs listed in Table 8.14.

On the other hand, β-decay of ^{3}H in these waters has produced a maximum of $\sim 10^6$ atoms ^{3}He/cm^3 H$_2$O. Thus, a comparison of ^{3}He concentrations calculated from ^{3}He excess in bubbling gas and the ^{3}H content in water has led to the conclusion that practically all ^{3}He in thermal fluids is of juvenile origin and the contribution of tritigenic ^{3}He is negligibly small

---

[1] In the light of recent data this assumption is not always correct (see, for example, Craig et al., 1978b), and the problem of an equilibrium of bubble and dissolved gases appears to be very interesting.

(Craig and Lupton, 1976; Craig et al., 1978b). The conclusion is valid for thermal waters characterized by a high enough concentration of excess ^{3}He.

8.6.3. The ^{3}H—^{3}He method in physical limnology

The ^{3}H—^{3}He method gives a unique possibility for investigating the physical mechanisms of mixing and migration in lakes, such as the rate of gas exchange at the lake surface, vertical diffusion in the epilimnion, mixing and gas renewal, etc. (Torgersen et al., 1977, 1979, 1981). Fig. 8.17 illustrates a considerable difference in the distribution of δ^3He in June and August for Lake Huron, in spite of the fact that the ^{3}H concentrations are practically constant for both months and throughout the depth. The similarity in ^{3}H—^{3}He ages within the limits of the hypolimnion shows that the total water column is subject to mixing. After a more stable summer period the surface ages equal to a few days were observed. They were lower than those of the June profile; near the thermocline the ^{3}H—^{3}He ages increase sharply and reach values of about 100 days in the hypolimnion. Such a distribution of δ^3He values or ^{3}H—^{3}He ages indicates frequent mixing and contact with the atmosphere for the epilimnion and long periods of isolation throughout the hypolimnion.

The ^{3}H—^{3}He data allow us to estimate the gas exchange rates (Torgersen et al., 1977, 1979). The theory of dissolved gas exchange with a gas reservoir

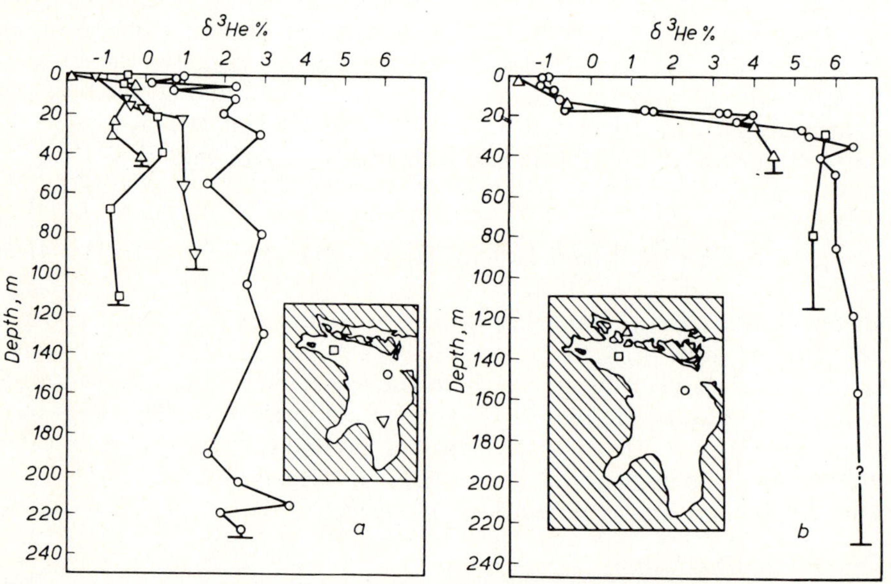

Fig. 8.17. δ^3He profile for Lake Huron taken during "*Martin Karlsen*" cruises: a, 22—28 June, 1974; b, 2—29 August, 1974. Stations as indicated. (Torgersen et al., 1977.)

states that the exchange rate is restricted by some resistance of a thin surface layer, z, through which gas must diffuse at molecular rates. The flux
equation across z is:

$$\text{Flux} = k_{He} \{ [^3He]_0 - [^3He]_{equil} \} \tag{8.11}$$

where k_{He} is the gas exchange coefficient, m day^{-1}; $k_{He} = D_{He}/z$, D_{He} being
the diffusion coefficient of helium in water, equal to $5.42 \cdot 10^{-2}$ cm^2 s^{-1};
$[^3He]_0$ is the ^3He concentrations in the surface layer (square brackets denote
concentrations); and $[^3He]_{equil}$ is the dissolved atmospheric ^3He. On the other
hand, assuming the ^3He concentration from the thermocline to be negligible,
the flux can be defined with (steady-state conditions implied) as:

$$\text{Flux} = A_0 \cdot Z_{epi} \tag{8.12}$$

where A_0 is the production rate of ^3He per unit volume, and Z_{epi} denotes the
depth of the epilimnion. From eqs. 8.11 and 8.12, k_{He} can be determined as
follows:

$$k_{He} = \frac{A_0 Z_{epi}}{([^3He]_0 - [^3He]_{equil})}$$

Note that the ratio of $([^3He]_0 - [^3He]_{equil})/A_0$ constitutes the mean residence
time of ^3He at $Z = 0$ ($\tau_Z = \tau_0$); therefore:

$$k_{He} = Z_{epi}/\tau_{Z=0} \tag{8.13}$$

For a calculation of the gas exchange coefficient the average epilimnion age
can be used provided the epilimnion is a well-mixed layer. For example, the
average age of the Lake Erie epilimnion is 4.2 days and the depth of the
epilimnion is 15 m. In accordance with eq. 8.13, the gas exchange coefficient
k_{He} is equal to 3.6 m day^{-1}. From eq. 8.11 it is also possible to estimate z:

$$z = D_{He}/k_{He} = 1.3 \cdot 10^{-2} \text{ cm}$$

The same operation with Rn concentrations in ocean waters resulted in
$k_{Rn} = 2.4$ m day^{-1} and $z = 4 \cdot 10^{-3}$ cm. Thus, estimations of k and z from
ocean Rn data are consistent with those obtained for lakes from the $^3H-$
^3He system.

The $^3H-^3He$ method can be used to assess vertical diffusivity in surface
waters. In the simplest case, assuming vertical advection and horizontal
gradients to be negligible, a single water layer has two ^3He inputs (*in situ* ^3H
β-decay and the diffusive input) and one output (the diffusive loss-out of the

190

layer). For a constant vertical eddy diffusion, K_v, the equation describing
this model is:

$$\frac{[^3\text{He}]_{tri}}{t} = A_0 + K_v \frac{\partial^2 \, [^3\text{He}]_{tri}}{\partial \, Z^2} \qquad (8.14)$$

where $[^3\text{He}]_{tri} = [^3\text{He}]_0 - [^3\text{He}]_{equil}$ (see eq. 8.11). For steady-state conditions,
$\partial \, [^3\text{He}]_{tri}/\partial \, t = 0$ and eq. 8.14 can be solved:

$$[^3\text{He}]_{tri} = [^3\text{He}]_{tri,\,0} + \frac{F + A_0 Z_m}{K_v} - \frac{A_0}{2 \, K_v} Z^2 \qquad (8.15)$$

with boundary conditions:

$$K_v \frac{[^3\text{He}]_{tri}}{Z} = A_0 \, (Z_m - Z) + F \qquad (8.16)$$

and:

$$[^3\text{He}]_{tri} = [^3\text{He}]_{tri,\,0} \text{ at } Z = 0$$

where Z_m is the depth of the layer; F is the input flux; and $[^3\text{He}]_{tri,\,0}$ is the
surface excess ^3He concentrations. Eq. 8.16 enables us to compute concentrations of excess $[^3\text{He}]_{tri}$ and plot them in coordinates of concentration versus
depth, where K_v is a parameter. A comparison of calculated and observed
curves in these coordinates gives the value of K_v. Thus, for Lake Huron $K_v =$
10 m² day⁻¹ = 1.2 cm² s⁻¹ was estimated (Torgersen et al., 1977).

   Ample data (including $^3\text{H}-^3\text{He}$ water mass ages) have been obtained for
Fayetteville Green Lake, New York, by Torgersen et al. (1981). The authors
have found that an upper (16—32.5 m, average $^3\text{H}-^3\text{He}$ age equal to 3.1
years) and a lower (32.5—52 m, average age 5.8 years) monimolimnions, separated by a secondary chemocline, can be observed in the lake. A decrease
of the ^3H concentration in the deep waters, with constant $^3\text{H}-^3\text{He}$ ages, is
interpreted as a result of ground-water input; enrichment of deep waters
with N^+, Cl^{-1}, ^4He and CO_2 agrees with such an interpretation.

8.6.4. The $^3H-^3He$ method and ^4He fluxes through lakes

   The $^3\text{H}-^3\text{He}$ "ages" or the residence time of a helium atom was successfully applied for the estimation of ^4He fluxes (W.B. Clarke and Kugler, 1973;
W.B. Clarke et al., 1977; Top and Clarke, 1981). The purpose of these works
was to localize zones of high ^4He fluxes with the view of prospecting ura-

nium and thorium. Radiogenic ^{4}He excess in lake water depends not only on the ^{4}He input but also on helium degassing rates — that is, the residence time of the helium atom. Thus, a low ^{4}He input rate can produce a large excess of ^{4}He if the water preserves helium long enough (high residence time); a relatively large ^{4}He input rate can produce a negligibly low ^{4}He excess if the residence time of a helium atom is very short. In fact the residence time, τ, varies from about two weeks to four years judging from ^{3}H—^{3}He measurements in water samples from 107 lakes and ponds of central Labrador (Top and Clarke, 1981).

Taking into account the above consideration, W.B. Clarke et al. (1977) introduced the so-called "prospecting" index I^x:

$$I^x = \Delta\ ^4\text{He} \cdot V/\tau \tag{8.17}$$

where $\Delta\ ^4$He is the ^{4}He excess in a water sample relative to saturated concentrations:

$$\Delta\ ^4\text{He} = \left\{ \frac{[^4\text{He}]_{\text{obs}}}{[^4\text{He}]_{\text{sat}}} - 1 \right\} \cdot 100\%$$

$[^4\text{He}]_{\text{obs}}$ is the observed ^{4}He concentration; $[^4\text{He}]_{\text{sat}}$ is the concentration calculated for the case of equilibrium between the atmospheric and the dissolved helium; τ is the residence time obtained from ^{3}H—^{3}He data; and V is the volume of the lake.

One of the lakes, Anna, shows a very high $I^x = 8.7 \cdot 10^5$ cm^3 yr^{-1}. The amount of uranium, which is necessary to supply a radiogenic helium flux through the lake, is estimated to be $7 \cdot 10^9$ kg if ^{4}He is released into the lake at the present-day production rate ($1.2 \cdot 10^{-7}$ cm^3/1 g U yr). A deposit of this size is about ten times larger than the total known reserves in the Western World. To avoid difficulties with the size of the hypothetical deposit, the authors (W.B. Clarke et al., 1977; Torgersen and Clarke, 1978) assumed that the helium (accumulated in a uranium body of a reasonable size during $\sim 10^9$ years) was released from uranium minerals due to leaching by ground water (during the last 10^4 years) which carried the helium from the body to the lake. On the other hand, the authors noted that there was no correlation between excess helium and uranium concentrations in lakes.

This explanation is by no means the only possible one and is not very convincing either. Probably it is not necessary to take into account the uranium deposits. Deep waters of the earth's crust (circulating among rocks with ordinary contents of uranium and thorium) normally contain a large excess of radiogenic helium (Yakutseny, 1968). The concentration of excess ^{4}He in the water (injected into Lake Anna) may be assumed, other conditions being similar, equal to 10^{-4} cm^3/cm^3 H$_2$O; this value is somewhat lower

192

than the average one, typical of deep ground waters of ancient plates (Ya-kutseny, 1968). Such a concentration may be expected in water circulating among ordinary granitic rocks (U = 3 ppm; Th = 10 ppm; porosity $\sim$ 3%) during $2 \cdot 10^6$ years. To supply the excess of $\Delta\,^4\mathrm{He} = 3 \cdot 10^{-8}$ cm^3/cm^3 H$_2$O observed in Lake Anna, about $3 \cdot 10^{-4}$ cm^3 of deep ground water should be mixed every year with 1 cm^3 of the lake water; this gives approximately 10^{10} cm^3 deep water per year for the whole lake. These data allow us to esti-mate the velocity of the ground-water flow as $\sim$ 0.3 cm yr^{-1}. Actually, the velocity must be lower because a high I^x value is found only in two samples (among more than twenty) taken from Lake Anna, and the total $\Delta\,^4\mathrm{He}$ is essentially lower than that adopted for the calculations. It should be noted that the velocity of deep waters in ancient plates varies, as a rule, from 0.1 to 1.0 cm yr^{-1}.

In a more recent work Top and Clarke (1981) proposed another formula for estimating the "prospecting index":

$$I = \Delta\,^4\mathrm{He}/\tau \tag{8.18}$$

where the notation is the same as in eq. 8.13. Comparing the I values de-termined for a large number of lakes and ponds in central Labrador, Top and Clarke (1981) suggested the zones with high I values to be the most promis-ing ones for U and Th prospecting. A more accurate interpretation is that samples with high I values correspond to a fault zone providing the uplift of deep waters to the surface.

Another presentation of index I appears also to be suitable, namely:

$$I^* = \Delta\,[^4\mathrm{He}]\,H/\tau \tag{8.19}$$

where $\Delta\,[^4\mathrm{He}]$ is the average absolute concentration of excess ^{4}He, atoms cm^{-3}, along depth H, cm; τ is the residence time, s. Note, that eq. 8.15 represents the ^{4}He flux through a lake bottom. Assuming that depth H in station 11, Lake Anna, is 55 ft = $2 \cdot 10^3$ cm and taking into account Δ $[^4\mathrm{He}] = 3.1 \cdot 10^{-8}$ cm^3/cm^3 H$_2$O = $0.83 \cdot 10^{12}$ atoms cm^{-3} and $\tau = 1.07$ year = $3.37 \cdot 10^7$ s, one can calculate the highest value of the ^{4}He flux through the lake bottom, $I^* \approx 5 \cdot 10^7$ cm^2 s^{-1} — that is, only about twenty times higher than the total terrestrial flux. Averaging the I^* values for all stations in Lake Anna appears to result in a flux lower than the terrestrial flux; thus, it is not necessary to imply any uranium enrichment of rocks near the lakes showing a high "prospecting index".

Chapter 9

HELIUM ISOTOPES IN SEAWATER

Attention to helium dissolved in seawater was drawn by Revelle and Suess (1962), who pointed out that refined analytical methods might make it possible to detect excess helium produced by the outgassing of the sea floor. If such excess helium were to be found, it would enable us to study the helium flux through the oceanic crust and application of the results for an evaluation of such problems as the helium balance in the atmosphere, the contents of radioactive elements in the sub-oceanic mantle and heat flow, the tectonic structure of the sea floor and oceanic streams. Several authors (König et al., 1964; Bieri et al., 1964, 1967, 1968; Craig et al., 1967) determined the concentrations of light noble gases in seawater. Despite a sufficient degree of accuracy of their measurements there were still doubts concerning the presence of excess helium in seawater.

To shed more light on the origin of oceanic helium, W.B. Clarke et al. (1969) measured the ^{3}He/^{4}He ratio in dissolved helium and discovered that helium from deep Pacific water was substantially enriched in ^{3}He compared to the atmospheric helium. Later it was found that excess ^{3}He and ^{4}He were supplied by injection of volatiles through the oceanic crust and that the ^{3}He/^{4}He ratio in injected helium was similar to that in thermal fluids of continental and island hot springs.

Intensive research by the Canadian—American group not only produced a difinition of the source of helium isotopes in the oceans (section 9.1) but also led to a description of regional regularities in the ^{3}He and ^{4}He distribution in the Pacific and Atlantic Oceans, in the Red and the Caribbean Seas, etc. It was shown that ^{3}He is a unique tracer for detecting mantle-derived volatiles injected into seawater (section 9.2).

9.1. Origin of helium isotopes in seawater

There are three processes responsible for the concentration of helium isotopes in seawater: (1) solution of atmospheric helium; (2) injection of helium through the oceanic crust; and (3) decay of radioactive nuclides which produces helium isotopes in water (first of all ^{3}H β-decay).

The purpose of this section is to discuss these processes.

194

9.1.1. Rare gases dissolved in oceanic water

Earlier it was assumed that the noble gas content in seawater is provided by a thermodynamic equilibrium between the atmosphere and the ocean surface. In this case the measured concentrations C should be equal to the calculated C^* (cm^3 kg^{-1}) from the formula:

$$C^* = \frac{X_i^{\mathrm{A}} \cdot \beta_i}{\rho} (P - P^*) \tag{9.1}$$

where X_i^{A} denotes the volume fraction of gas i in dry air; β is the Bunsen absorption coefficient; ρ is the density of seawater; P denotes atmospheric pressure; and P^* is the equilibrium vapor pressure of water.

However, when the isotope dilution method sufficiently improved the precision of measurements (Craig et al., 1967; Bieri et al., 1968) and accurate data of the solubility of light noble gases became available (Weiss, 1970, 1971), a significant difference (named saturation anomaly) between the calculated and the observed contents unexpectedly emerged.

Craig and Weiss (1971) pointed out three main processes causing the saturation anomaly: (1) pressure variations; (2) temperature changes that are achieved after the equilibrium; (3) injection of bubbles containing air gases in the surface water layer.

The saturation anomaly (Δ, %) for the "i" component is calculated from the approximative formula (Craig and Weiss, 1971):

$$\Delta_i\,(\%) = \left(\frac{C_i}{C_i^*} - 1 \right) \cdot 100\% = \Delta_P - 100 \left[\frac{\mathrm{d}\,\ln C_i^*}{\mathrm{d}T} \right] \Delta T + \frac{X_i^{\mathrm{A}} \cdot a}{C_i^*} \tag{9.2}$$

where C_i^* is the equilibrium concentration given by eq. 9.1; C is the measured concentration; Δ_P is the deviation of atmospheric pressure, %; ΔT is the observed potential temperature minus the saturation temperature; a denotes the injected air component, cm^3 kg^{-1}. C_i^* and d $\ln C_i^*$/dT values are given by Weiss (1970, 1971).

Concentrations of three noble gases (other than helium) are required to define the unknown parameters Δ_P, ΔT and to evaluate their effects on helium isotopes. Hitherto precise measurements have been reported for two atmospheric noble gases (neon and argon); as for helium isotopes, only rough estimates of their saturation anomalies are available. Nevertheless, the estimations show that the ^{4}He excess, $\Delta(^4\mathrm{He}) \approx 10$—12%, is formed by both the saturation anomaly ($\approx 8\%$) and the helium influx ($\approx 3\%$). The latter value was used for the calculation of the ^{3}He/^{4}He ratio in injected helium and the flux of helium isotopes through the oceanic crust.

The discovery of ^{3}He excess in the sea had drawn attention to the importance of determining the effect of isotope fractionation in the ^{3}He—^{4}He mixtures dissolved in water and seawater. This effect was studied by Weiss (1970). The isotopic fractionation factor α is given as:

$$\alpha = \frac{(^3\text{He}/^4\text{He})_{\text{aqueous phase}}}{(^3\text{He}/^4\text{He})_{\text{gas phase}}} = \frac{\beta_3}{\beta_4} \tag{9.3}$$

where β is the measured Bunsen solubility coefficient.

It follows that $(\alpha - 1) \cdot 10^2$ is the single-stage percentage enrichment of the ^{3}He/^{4}He ratio in the aqueous phase relative to that in the gas phase (Table 9.1). Careful micro-gasometric measurements enable one to define a linear least-square regression line:

$$(\alpha - 1)\% = (-1.16 \pm 0.18) - (0.011 \pm 0.007) \cdot T \tag{9.4}$$

where T denotes temperature, $^\circ$C.

These data are in agreement with Clarke et al.'s (1969) measurements of the ^{3}He/^{4}He ratio dissolved in surface seawater; they prove that high excess of ^{3}He amounting to 30% cannot be accounted for by isotope fractionation under solubility.

TABLE 9.1

Experimental solubility values of helium isotopes in water and seawater (Weiss, 1970)

Temperature (°C)	Bunsen coefficient, β ($\times 10^{-3}$)		$\alpha - 1$ (%)
	^{3}He	^{4}He	
Distilled water			
0.60	9.254 ± 0.026 (4)	9.355 ± 0.004 (4)	−1.1
20.11	8.620 ± 0.016 (4)	8.724 ± 0.024 (6)	−1.2
40.14	8.574 ± 0.019 (5)	8.713 ± 0.036 (4)	−1.6
Seawater (salinity 36.425°/$_{oo}$)			
0.07	7.655 ± 0.012 (4)	7.771 ± 0.025 (5)	−1.5
20.13	7.339 ± 0.009 (4)	7.420 ± 0.029 (4)	−1.1
40.46	7.346 ± 0.028 (4)	7.488 ± 0.015 (4)	−1.9

Notes: β is given in cm^3 gas (under standard pressure, 760 Torr) per cm^3 water; errors ± 1 σ; in brackets number of replicate trials are given.

9.1.2. *Helium isotopes injected in seawater*

Both helium isotopes are injected into seawater due to the circulation of fluids through the ocean floor. Depending on the source of helium the ^{3}He/^{4}He ratio in the fluid can be higher (W.B. Clarke et al., 1969) or lower (Top et al., 1980) than the atmospheric one. However, in most cases seawater is enriched in light isotopes due to mantle helium injection in rift zones, oceanic rises, etc (W.B. Clarke et al., 1969, 1970; Jenkins and Clarke, 1976; Lupton, 1976, 1979; and others). To compare the ^{3}He/^{4}He ratios in thermal fluids of continents (or islands) and in excess helium of seawater, Devirts et al. (1971) and Tolstikhin et al. (1972a) assumed the ^{3}He/^{4}He ratios in deep and surface waters of the Pacific Ocean to be equal to $1.55 \cdot 10^{-6}$ and $1.38 \cdot 10^{-8}$, respectively, and the fraction of juvenile injected ^{4}He $\approx 5\%$ (data from Craig and Clarke, 1970). Assuming that helium observed in deep water (He$_0$) includes dissolved air helium (He$_A$) and injected juvenile helium (He$_j$), the authors solved the simple equations:

$$\frac{^3\text{He}_A + {}^3\text{He}_j}{^4\text{He}_A + {}^4\text{He}_j} = \left(\frac{^3\text{He}}{^4\text{He}} \right)_0 = 1.55 \cdot 10^{-6}$$

$$\frac{^4\text{He}_j}{^4\text{He}_A + {}^4\text{He}_j} = 0.05 \tag{9.5}$$

and produced an approximate value of $(^3\text{He}/^4\text{He})_j \approx 0.5 \cdot 10^{-5}$, which was in reasonable agreement with the ^{3}He/^{4}He values in the Kuril—Kamchatka hot springs. Later the ^{3}He/^{4}He ratio in helium injected in deep water of the East Pacific was determined more precisely by Craig et al. (1975), who took into account the mean saturation anomaly, Δ He = 11.9%, for this region and estimated the contribution of non-atmospheric helium, Δ_jHe = 3.5%. Substituting these values for those of eq. 9.6 they calculated the fraction of injected juvenile helium X_i:

$$X_i = \Delta_j\text{He}/(1 + 10^{-2}\,\Delta\,\text{He}) = 3.1\% \tag{9.6}$$

The solubility fractionation being neglected, the ratio of $(^3\text{He}/^4\text{He})_j$ was found from the formula:

$$(^3\text{He}/^4\text{He})_j = \frac{1 + \delta\,(^3\text{He})/X_i}{(^3\text{He}/^4\,\text{He})_A} = 1.6 \cdot 10^{-5} \tag{9.7}$$

where $\delta\,(^3\text{He}) = \left[\dfrac{(^3\text{He}/^4\text{He})_0}{(^3\text{He}/^4\text{He})_A} - 1 \right] \cdot 100$

and $(^3He/^4He)_0 = 1.85$ is the ratio measured in deep ocean waters.

By utilizing data from Craig et al. (1975) and the approximate eq. 9.5, we may obtain $(^3He/^4He) \approx 1.5 \cdot 10^{-5}$.

The best approach so far to the determination of the $(^3He/^4He)_j$ ratio was suggested by Lupton (1979), who used the percentage deviation of $^3He/Ne$ and $^4He/Ne$ from the solubility ratios:

$$\Delta \left[\frac{^3He}{Ne} \right] \% = \left[\frac{C\,(^3He)/S\,(^3He)}{C\,(Ne)/S\,(Ne)} - 1 \right] \cdot 100 \qquad (9.8)$$

where $C\,(^3He)$ and $C\,(Ne)$ denote the measured concentrations of ^3He and Ne, respectively, and $S\,(^3He)$ and $S\,(Ne)$ are the expected concentrations for air-saturated water at the potential temperature and salinity of the sample. Such a normalization of results eliminates errors in sample weight, variations

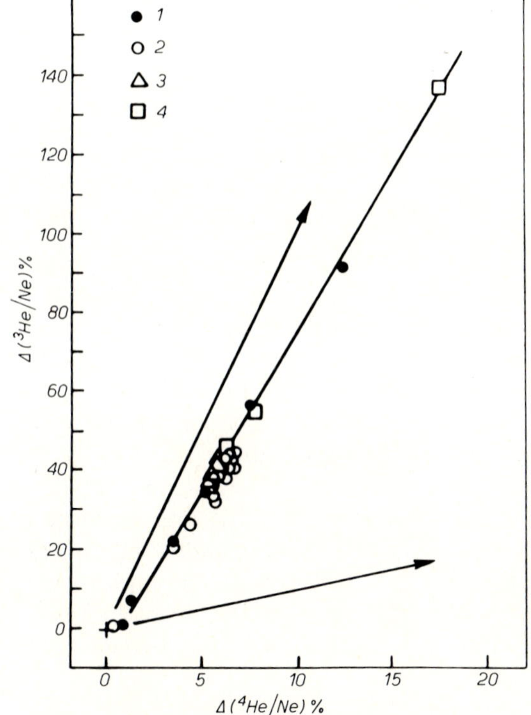

Fig. 9.1. $\Delta\,(^3He/Ne)$ versus $\Delta\,(^4He/Ne)$ for the Gulf of California: *1* = Guaymas Basin; *2* = Mazatlan Basin and Galapagos Rift; *3* = normal bottom waters; *4* = hydrothermal plumes. Samples from both areas fall on a line whose slope corresponds to the $^3He/^4He$ ratio of injected helium of $1.1 \cdot 10^{-5}$. Lower and upper arrows indicate the $^3He/^4He$ ratio in air and this ratio multiplied by ten, respectively. Cross refers to air-saturated water. (Lupton, 1979).

198

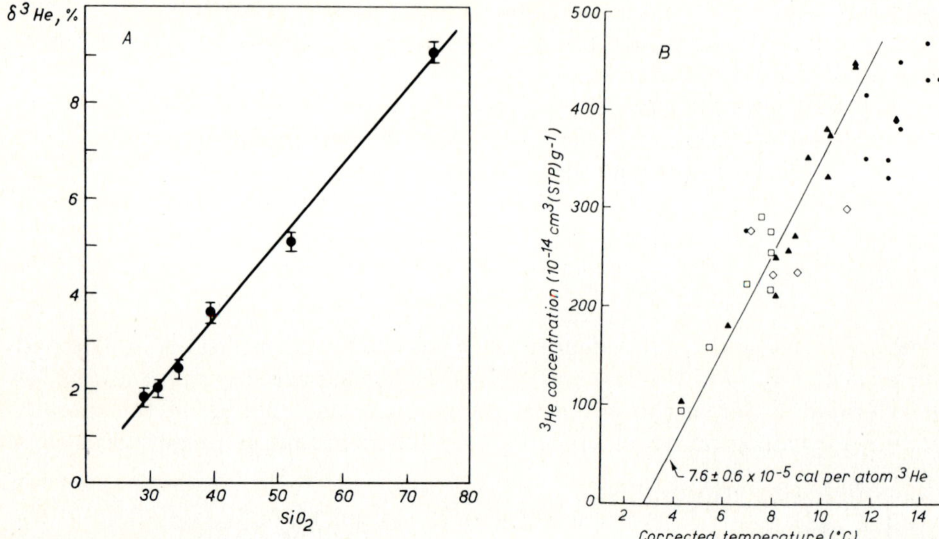

Fig. 9.2. Left: correlation between δ ^{3}He (%) and SiO_2 content (μmol/kg H_2O) in the core of lower intermediate water in the South Atlantic (Jenkins and Clarke, 1976). Right: correlation between excess ^{3}He and temperature for Clambake (▲), Oysterbed (□) and Dandelions (◇) vent areas of the Galapagos Spreading Center; samples from Garden of Eden area (●) show significant bias from the correlation (Jenkins et al., 1978).

in original atmospheric components, etc. The $(^3He/^4He)_j$ ratio was determined from the slope of Δ (^{3}He/Ne) versus Δ (^{4}He/Ne). Data for the Gulf of California and the Galapagos Rift fall on a single line indicating a $(^3He/^4He)_j$ ratio of $1.1 \cdot 10^{-5}$ (Fig. 9.1). Extremely important correlations between ^{3}He excess and: (1) the SiO_2 content in South Atlantic waters (Fig. 9.2, left), and (2) the corrected temperature of seawater in the Galapagos Spreading Center (Fig. 9.2, right) have been established by Jenkins and colleagues (Jenkins and Clarke, 1976; Jenkins et al., 1978). The former confirms an idea on the transport of juvenile volatiles by silicate melts (compare with Fig. 8.7); the latter is in good agreement with the $^3He/^4He$: heat flow relationships discussed in Section 8.5.3. Both correlations may be considered as key points to future investigations.

9.1.3. Radiogenic ^{3}He in seawater

Fairhall (1969) attempted to explain ^{3}He excess in seawater due to ^{3}H β-decay, but Craig and Clarke (1969) showed that Fairhall had mistakenly used the ^{3}He/^{4}He enrichment in relation to the ratio in atmospheric helium as a measure of absolute ^{3}He enrichment. In fact, this latter is defined by the saturation anomaly Δ_3:

$$\Delta_3 = \left\{ \delta\,^3\text{He} + \Delta_4 + 10^{-2}\left[\delta\,^3\text{He}\cdot\Delta_4\right] - (\alpha - 1)\right\}\alpha^{-1} \qquad (9.9)$$

Substituting into eq. 9.9 the numerical values discussed in this section, namely, $\delta\,^3$He = 30%, Δ_4 = 12%, $(\alpha - 1)$ = 1.2%, α = 0.988, we obtain Δ_3 = 47%, whereas the maximum ^{3}H concentration in ocean waters, equal to 2 T.E. (1 T.E. = 1 atom ^{3}H/10^{18} atoms H = $6.7 \cdot 10^{17}$ atoms ^{3}H/1 kg H$_2$O) provides a ^{3}He enrichment of about 8.7%, which makes up only about one-fifth of the real ^{3}He excess.

Yet, in surface waters as well as in young waters driven into deep layers by oceanic streams the ^{3}He excess is really produced by the ^{3}H decay. Measurements of tritium concentrations and ^{3}He excess enable determination of the water mass "ages", which is of paramount importance for the study of oceanic streams and mixing processes. The application of this method is in some cases complicated because of the difficulties in identifying the ^{3}He excess (which is either the result of an injection of mantle helium or of ^{3}H decay or both). A possible approach to the identification problem includes:

(1) accurate measurements of ^{3}He, ^{4}He, Ne, Ar and Kr concentrations in seawater;

(2) calculation of the saturation anomaly for ^{4}He and a comparison of its observed and calculated concentrations: if the ^{3}He excess in seawater is not accompanied by ^{4}He excess then ^{3}He comes from ^{3}H.

In cases when a sufficiently reliable identification was achieved, the ^{3}H—^{3}He method was used to measure stream velocity, study mixing processes, etc. (Jenkins et al., 1972; Jenkins and Clarke, 1976; Top et al., 1980). Age determination of water parcels obtained by this method for various regions of the Atlantic (from 0 to 25 years) may serve as an example.

9.2. Distribution of helium isotopes in oceanic waters

The concentration of helium isotopes in Pacific waters was reported by W.B. Clarke et al. (1969, 1970), Craig et al. (1975), Weiss (1977), Lupton et al. (1977b, 1980) and Lupton (1979). Fig. 9.3 gives the $\delta\,^3$He values for some profiles mostly located in the region of the East Pacific Rise.

One of the most striking features of the profiles is the presence of a mid-depth maximum of $\delta\,^3$He near the top of the rise crest where an injection of juvenile volatiles takes place. Interpreting this feature, Craig et al. (1975) pointed out that at a low depth dissolved helium is in equilibrium with the atmosphere whereas waters underlying the mid-depth maxima mix with the

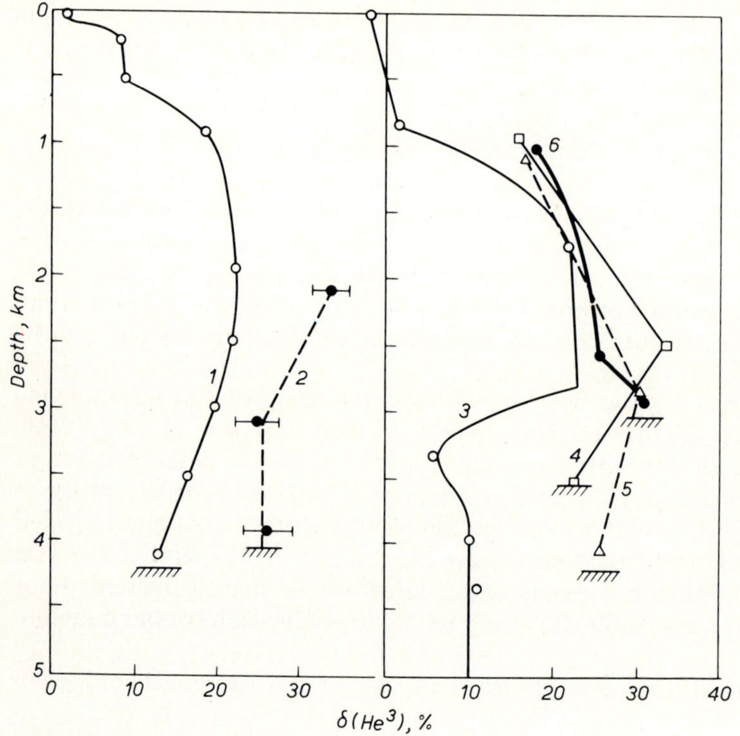

Fig. 9.3. ³He profiles in the North (left) and South (right) Pacific (Craig et al., 1975). Location: *1* = 28° N, 121° W; *2* = 8° N, 113° W; *3* = 31° S, 177° W; *4* = 6° S, 110° W; *5* = 7° S, 102° W; *6* = 6° S, 107° W.

northward flowing Antarctic water, which is clean of excess helium. Thus, mid-depth waters do not only receive volatiles as a result of hydrothermal manifestations in the upper parts of the rise, but also preserve them intact because they are better isolated from the atmosphere. This is confirmed by some other characteristics of these waters, viz. SiO_2 data, etc.

In some cases bottom waters also show a very high enrichment in ³He. For instance, water taken along the central axis of the Galapagos spreading center is characterized by a comparatively high δ ³He anomaly, the mean value being equal to 30.4%. The value is consistent with other measurements over the East Pacific Rise. In samples selected from near the hydrothermal plumes through the oceanic crust, which are characterized by potential temperature spikes, δ ³He approaches 99%, which is the highest value observed in the oceans (Lupton et al., 1977b). A very high enrichment in injected helium, δ ³He = 68%, was observed by Lupton (1979) in the Guaymas Basin, Gulf of California, which suggests a recent hydrothermal activity in the basin.

Geochemical studies of deep waters of the East Pacific Rise have confirmed that ^{3}He is an extremely sensitive tracer for detecting submarine hydrothermal fluids diluted with seawater and that the Mn content in seawater can serve as another tracer (Weiss, 1977; Lupton et al., 1980). ^{3}He may be considered a conservative tracer which is the clearest indicator of a mantle input and which is integrated by the ocean over a 10^3-yr time scale. In view of this, ^{3}He is useful for estimating the global fluxes of associated species which are due to submarine hydrothermal activity; it is also useful for investigating ocean currents. Manganese is thought to be a non-conservative tracer removed from seawater by scavenging, with a residence time of about fifty years (Weiss, 1977); it is applicable for elucidating local heterogeneities near rising plumes of hot vent water. If the residence times of Mn and ^{3}He, as well as their ratio in a plume, are known, then the Mn/^{3}He ratio may be used to define a time scale within horizontally spreading plumes (Lupton et al., 1980).

A record value of excess helium of $14 \cdot 10^{-6}$ cm^3 g^{-1} (He/Ne $\approx$ 100!) was found by Lupton et al. (1977a) in Red Sea brines; a ^{3}He/^{4}He ratio in helium of $1.2 \cdot 10^{-5}$ implies that the helium is derived from a depleted mantle reservoir.

Careful study of helium isotope distribution (ca. 20 profiles, ca. 350 measurements of helium concentration and ^{3}He/^{4}He ratio) was carried out for Atlantic waters (Jenkins et al., 1972; Jenkins and Clarke, 1976; Lupton,

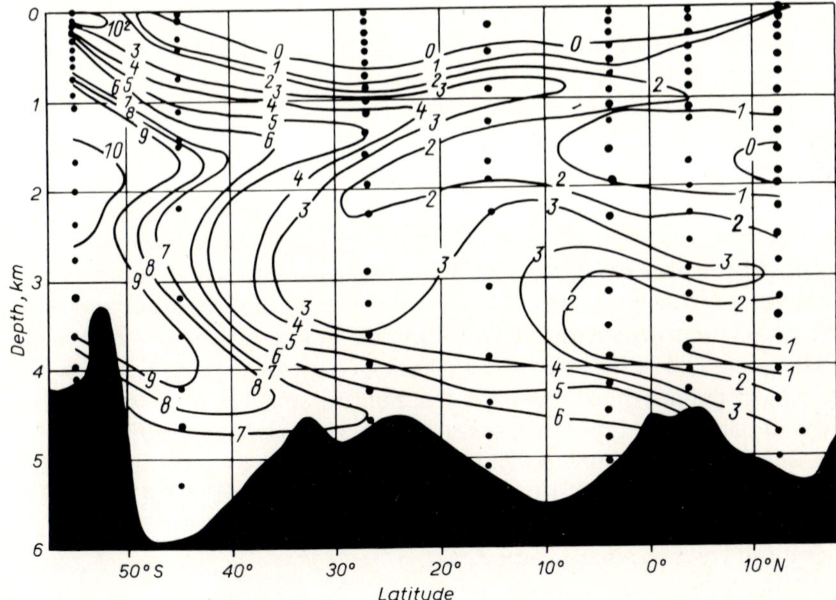

Fig. 9.4. The meridional distribution of δ ^{3}He in the South Atlantic. Dots show sampling positions (Jenkins and Clarke, 1976).

202

1976). The measured ^{3}He enrichment is far lower than that in the Pacific, the maximum values of $\delta\,^3$He being about 10% — that is, three times less than the average $\delta\,^3$He in the East Pacific. Some profiles show a complex structure with several maxima observed at different depths. Excess ^{3}He in the upper maxima is produced by technogenic tritium decay. As for the mid-depth maxima, in many cases it is difficult to decide whether they are also radiogenic or caused by a juvenile helium injection.

Deep waters of the North Atlantic contain mantle helium injected through the Gibbs Fracture Zone. The $\delta\,^3$He can be used here to trace the westward sea current (Jenkins and Clarke, 1976).

In the Atlantic (30°N, from 26° to 56°W) there is no obvious ^{3}He anomaly in deep water as indicated by four profiles across the Mid-Atlantic Ridge (Lupton, 1976).

A meridional system of profiles in the South Atlantic (Fig. 9.4) may serve as an excellent illustration of the possibilities provided by the use of $\delta\,^3$He for sea-floor tectonics and ocean currents. A perfect correlation between $\delta\,^3$He and the SiO_2 content is observed in the South Atlantic (see Fig. 9.2). Because of the different origins of these components, such a correlation means that either both components were injected by a single active hydrothermal system (Fig. 9.4) or else they were supplied by some underwater eruptions. To decide between the two alternatives, further comprehensive experimental data are required.

Jenkins and Clarke (1976) and Top et al. (1980) discussed the advantages of the ^{3}H—^{3}He method. It has been shown that the method enables estimation of the share of deep and surface water as well as the rate of water flows. Thus, Jenkins and Clarke (1976) reported a continuous increase in age of the Denmark Strait overflow water. ^{3}H—^{3}He ages of water estimated from four profiles in the North Atlantic exhibit a linear correlation with a latitude consistent with the rate of 0.7 cm s^{-1}, which corresponds with estimates obtained from other data.

Using the ^{3}H—^{3}He system Top et al. (1980) constructed a two-box model of the Baffin Bay water exchange, which yielded a deep water renewal time of 77—455 years. The result is close to estimates based on other methods. Profiles disposed in the central Baffin Bay are characterized by high $\delta\,^3$He and ^{3}H—^{3}He ages. These data show the effect of a shallow pycnocline, which is common to regions where seasonal ice melting results in a thin surface layer of low density. The layer decreases greatly the exchange rate of dissolved gases with the atmosphere and, consequently, increases the life time of ^{3}He in water. Top et al. (1980) noted that in some cases the He, Ne and ^{3}H data revealed information which cannot be obtained by applying conventional oceanographic tracers, such as Tp, S, O_2.

Chapter 10

HELIUM ISOTOPES IN THE EARTH'S ATMOSPHERE

In the course of time helium isotopes are released by solid earth and pass into the atmosphere which is the main gas reservoir of our planet. The isotope ratio of $^3He/^4He$ and the helium content in the atmosphere depend on three main processes: earth outgassing, entrance of solar wind and interstellar gas into the atmosphere, and helium isotope escape from the atmosphere.

We shall consider some experimental observations of helium isotope distribution in contemporary atmosphere (10.1) and some geochemical results inferred from a hypothetical "nondissipated" atmosphere (10.2). Then we intend to review recent data on ^4He and ^3He fluxes into the atmosphere and calculate the life time of helium isotopes, which proves that helium escape actually exists (10.3). In the last section (10.4) a new model of ^4He and ^3He escape from the earth's atmosphere will be presented (Pudovkin et al., 1981) which is in reasonable agreement with both helium fluxes and the temperature of the thermosphere.

10.1. Distribution of helium isotopes in the atmosphere

It has been firmly established that the helium content in the atmosphere

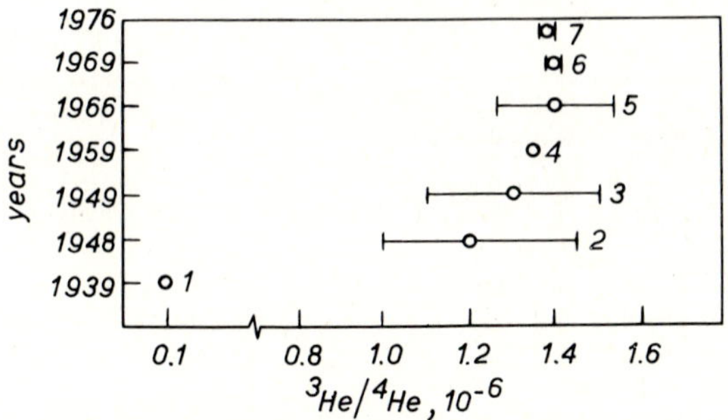

Fig. 10.1. Isotope compositions of helium in the terrestrial atmosphere. Data are taken from: *1*, Alvarez and Cornog (1939a); *2*, Aldrich and Nier (1946); *3*, Coon (1949); *4*, Hoffman and Nier (1958); *5*, Alimova et al. (1966); *6*, Mamyrin et al. (1970a, b); *7*, W.B. Clarke et al. (1977).

below the heights of about 100 km is constant (within the accuracy limits of measurements) and equal to $(5.239 \pm 0.005) \cdot 10^{-4}\%$ by volume (Cook, 1961; Kockarts and Nicolet, 1962; Bieri et al., 1971).

The ratio of ^{3}He/^{4}He in the atmosphere has been known very approximately before we made our measurements, although it was used for some geochemical and geophysical calculations, as well as a standard for mass-spectrometric measurements. Alvarez and Cornog were the first to observe ^{3}He in nature. According to their estimation the ^{3}He/^{4}He ratio in the terrestrial atmosphere amounted to $1 \cdot 10^{-7}$. Later some other authors reported their estimations of the ratio (Fig. 10.1) though they did not describe their

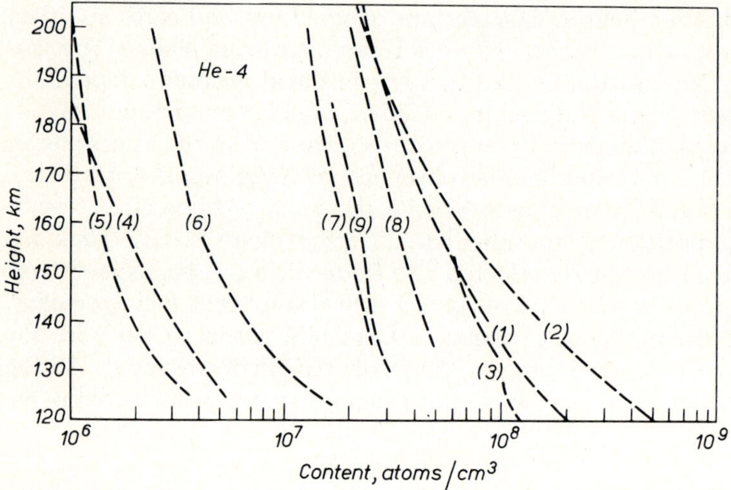

Fig. 10.2. Mass-spectrometric observations of the ^{4}He distribution in the thermosphere (Kockarts, 1973). Numbers near the curves refer to:

No.	Launching site	Date	Zone time	Reference
1	White Sands (32° 24N, 106° 21W)	Nov. 30, 1966	04 : 45	Kasprzak et al. (1968)
2	Idem	Dec. 2, 1966	14 : 09	Idem
3	Fort Churchill (58° 44N, 93° 49W)	Dec. 12, 1966	13 : 05	Müller and Hartman (1969)
4	White Sands (32° 24N, 106° 21W)	June 21, 1967	12 : 49	Krankowsky et al. (1968)
5	Idem	July 20, 1967	02 : 00	Idem
6	Idem	Idem	12 : 24	Idem
7	Salto di Quirra (40° N, 9° 30E)	Oct. 4, 1967	14 : 23	Bitterberg et al. (1970)
8	Idem	Oct. 10, 1967	15 : 33	Idem
9	Fort Churchill (58° 44N, 93° 49W)	Feb. 4, 1969	08 : 35	Hickman and Nier (1972)

methods and equipment. The large dispersion of ^{3}He/^{4}He values due to insufficient accuracy of measurements stimulated more careful determinations of the isotope composition of air helium. This has been carried out by Mamyrin et al. (1970a, b) and resulted in determining a ^{3}He/^{4}He value of $(1.399 \pm 0.013) \cdot 10^{-6}$. Using the same method Clarke et al. (1976) made another attempt to determine the ratio and obtained (within the limits of accuracy) the same value of ^{3}He/^{4}He $= (1.384 \pm 0.005) \cdot 10^{-6}$. Hereafter a ^{3}He/^{4}He ratio of $(1.39 \pm 0.01) \cdot 10^{-6}$ is used for air helium; this has been shown to be constant in the lower atmosphere (Mamyrin et al., 1970b). A simple calculation leads to the conclusion that the β-decay of technogenic tritium produces a very small amount of ^{3}He in the atmosphere (Leventhal and Libby, 1968; Vinogradov et al., 1968) contributing less than 0.01% of the atmospheric ^{3}He.

As to the ^{4}He distribution in the upper atmosphere, this is established by three different means: infrared and ultraviolet optical observations, satellite drag data and mass-spectrometric measurements. So far no direct way has been suggested to find out the distribution of ^{3}He because of its very low abundance. Mass-spectrometric measurements of the ^{4}He content within the range of 100—200 km are shown in Fig. 10.2. The ^{4}He concentration varies between wide limits (Fig. 10.2), whereas the major atmospheric constituents and even atomic hydrogen (another minor component) show considerably smaller variations. A comparison of data shown in Fig. 10.2 and in its caption suggests that the helium content depends on the season: in winter helium is accumulated. Thus, the problem is to explain how a ^{4}He mixing ratio of $5.24 \cdot 10^{-6}$ in the biosphere can increase by an order of magnitude between approximately 90 and 120 km.

As the main source of atmospheric ^{4}He is solid earth, an increase in the ^{4}He content can only be due to transport processes which start from a height of about 100 km. The helium continuity equation (Kockarts, 1973) is:

$$\frac{\partial n_1}{\partial t} + \mathrm{div}\,(n_1 \vec{w}_1) = 0 \tag{10.1}$$

where n_1 denotes the content of the minor component (^{3}He or ^{4}He); t is time and $\vec{w}_1$ is the transport velocity of ^{3}He or ^{4}He. The transport velocity arises from three effects:

$$\vec{w}_1 = \vec{w}_D + \vec{w}_E + \vec{u} \tag{10.2}$$

where $\vec{w}_D$ is the molecular diffusion velocity; $\vec{w}_E$ is the eddy diffusion velocity; and $\vec{u}$ is the velocity associated with any fluctuations, winds, etc. The vertical component of the molecular diffusion velocity $\vec{w}_D$ for the minor constituent n_1 at height z is given by Nicolet (1957), MacDonald (1963) and Kockarts (1973):

206

$$w_D = -D \left(\frac{1}{n_1} \cdot \frac{dn_1}{dz} + \frac{1}{H_1} + \frac{1-\alpha}{T} \cdot \frac{dT}{dz} \right) \tag{10.3}$$

where D is the molecular diffusion coefficient; α is the thermal diffusion factor, $\alpha = 0.38$; T is temperature; H_1 is the scale height:

$$H_1 = \frac{kT}{m_1 g}$$

where k is the Boltzmann constant; m_1 is the atomic mass; and g is the gravity acceleration. Two other members of eq. 10.2, $\vec{w}_E$ and $\vec{u}$, denote processes which lead to the mixing of atmospheric components; so the maximum increase in helium content and $^3\text{He}/^4\text{He}$ ratio due to transport processes can be estimated only by formula 10.3.

From 10.3 we derive for a stationary atmosphere:

$$\frac{dn_1}{dz} = -n_1 \left[\frac{1}{H_1} + \frac{(1-\alpha)}{T} \cdot \frac{dT}{dz} \right] \tag{10.4}$$

If $n_1 \ll n$ and $n_1 + n \approx n$, where n denotes the content (atoms cm^{-3}) of the atmospheric major component, the solution of 10.4 can be presented as (MacDonald, 1963):

$$\frac{n_1(Z_1)}{n_1(Z_2)} = \left[\frac{T(Z_2)}{T(Z_1)} \right]^{1-\alpha-\beta} \cdot \left[\frac{n(Z_1)}{n(Z_2)} \right]^{\beta} \tag{10.5}$$

where β is the mass ratio of minor m_1 and major m components, $\beta\,(^3\text{He}) = 0.114$ and $\beta\,(^4\text{He}) = 0.153$. The following parameters of the atmosphere are assumed: (a) $Z_1 = 100$ km; $n(Z_1) = 0.2 \cdot 10^{14}$ cm^{-3}; $m(n) = 26.2$; $T(Z_1) = 200°$K. (b) $Z_2 = 220$ km; $n(Z_2) = 0.29 \cdot 10^{10}$ cm^{-3}; $m(n) = 14.5$; $T(Z_2) = 1000°$K. For height $Z_1 = 100$ km the helium content is similar to that of the low atmosphere, namely $5.24 \cdot 10^{-4}\%$, so $^4\text{He}(Z_1) = 1.05 \cdot 10^8$ atmos cm^{-3} and $^3\text{He}(Z_1) = 147$ atoms cm^{-3}.

Substituting the numerical data in eq. 10.5, we can calculate the helium content and $^3\text{He}/^4\text{He}$ ratio for height $Z_2 = 220$ km: $^4\text{He} = 0.45\%$ vol. or $0.13 \cdot 10^8$ atoms cm^{-3}; $^3\text{He} = 24$ atoms cm^{-3}; $^3\text{He}/^4\text{He} = 1.8 \cdot 10^{-6}$. The calculated ^4He content is in reasonable agreement with the experimental observations presented in Fig. 10.2. The isotopic effect of the transport process does not appear to be large; as will be shown in section 10.4, thermal escape from the atmosphere results in a far greater fractionation of the helium isotopes.

10.2. Helium isotopes in a "nondissipated" atmosphere

For some geochemical purpose it seems to be useful to estimate the ^{3}He and ^{4}He contents in a hypothetical "nondissipated" atmosphere (Tolstikhin et al., 1977b).

To restore the abundance of primordial isotopes, we can use the well-defined linear relationship between helium and neon isotopes that exists in practically all materials of the solar system: $(^4He/^{20}Ne)_{prim} = 500$ (see Fig. 6.6). The ^{20}Ne content in the earth's atmosphere is $66.4 \cdot 10^{18}$ cm^3, and all ^{20}Ne is primordial. The isotopic composition of primordial helium, ^{3}He/^{4}He $= 2.4 \cdot 10^{-4}$, is well known from the isotope cosmochemistry of inert gases (see Chapter 6). The relative abundance of helium isotopes in a "nondissipated" atmosphere is easily found from these data:

$$^4He_{A,prim} = {}^{20}Ne_A \, (^4He/^{20}Ne)_{prim} = 33 \cdot 10^{21} \text{ cm}^3$$

$$^3He_{A,prim} = {}^4He_{A,prim} \, (^3He/^4He)_{prim} = 0.8 \cdot 10^{19} \text{ cm}^3$$

To estimate the upper limit of the total ^4He_A content in a "nondissipated" atmosphere, an analogy between the isotope geochemistry of helium and argon may be utilized. Atmospheric argon, like helium, contains radiogenic (^{40}Ar) and primordial (^{36}Ar) isotopes, and both have appeared in the atmosphere as a result of degassing of the earth until now. It is known from the isotope geochemistry of argon and from K-Ar dating that ^{40}Ar predominates in the earth's crust. In the earth's mantle the $^{40}Ar/^{36}Ar$ ratio is not lower than in the atmosphere (see section 7.5). It is unlikely that the migration process of helium isotopes from the solid earth into the atmosphere is different from the migration of argon isotopes, and it may be assumed, therefore, that the distribution of argon and helium isotopes is the same, and that the ratio of $^4He/^3He$ is higher in the earth's mantle than in a "nondissipated" atmosphere. In other words, if it is considered as established that $(^{40}Ar/^{36}Ar)_M > (^{40}Ar/^{36}Ar)_A$, then it is logical to assume that $(^4He/^3He)_M > (^4He/^3He)_A$. Then the upper limit of the ^4He content in a "nondissipated" atmosphere, ^4He_A, must be, considering an accepted bulk value of $(^4He/^3He)_M = 3 \cdot 10^{-5}$:

$$^4He_{A,\Sigma} \lesssim {}^3He_A \, (^4He/^3He)_M \lesssim 3.3 \cdot 10^{23} \text{ cm}^3, \text{ and } {}^4He_{A,rad} \lesssim 3.0 \cdot 10^{23} \text{ cm}^3$$

Another method of calculating the content of radiogenic ^4He in a "nondissipated" atmosphere is based on the observed $(^4He/^{40}Ar)_{rad}$ ratio in natural gases ("sources" of atmospheric inert gases). The constancy of this ratio in different natural gases has been repeatedly emphasized. The ratio has been used, for example, to estimate the U/K ratio in the earth and the magnitude of the $^4He_{rad}$ flux from the earth into the atmosphere (Wasserburg et

208

al., 1963). In the early works (Zartman et al., 1961; Gerling et al., 1967a), the average $(^4He/^{40}Ar)_{rad}$ ratio was taken as ≈ 10 and it was recognized that during the early stages of the evolution of the earth it must have been slightly lower. At present there are good reasons to believe that $(^4He/^{40}Ar)_{rad} \approx 10$ is not the average ratio but is close to its upper limit in terrestrial gases. It has been shown, for example, that "young" gases of tectonically active regions, where degassing is most intensive, have considerably lower $(^4He/^{40}Ar)_{rad}$ ratios, such as 3 or even 1 (Voronov et al., 1974). Comparisons of these ratios in rocks and gases (Tolstikhin and Drubetskoy, 1975; see section 8.1) have confirmed the view that in natural gases of tectonically passive platform regions the $(^4He/^{40}Ar)_{rad}$ ratio is substantially higher than in rocks because of the selective loss of ^4He from the latter (cf. section 8.1). It may be assumed, therefore, that the ratio $(^4He/^{40}Ar)_{rad} = 10$ is definitely higher than the average, either in time or space, and may be considered as the upper limit. Taking these considerations into account and knowing that the content of radiogenic ^{40}Ar in the atmosphere is $3.7 \cdot 10^{22}$ cm^3, it is easy to estimate the upper limit of the amount of radiogenic helium in a "nondissipated" atmosphere: $\lesssim 3.7 \cdot 10^{23}$ cm^3. The total amount of helium, considering the $He_{A,prim}$ content of the atmosphere, should not exceed $4.0 \cdot 10^{23}$ cm^3.

It should be emphasized that the two estimates of the maximum abundance of ^4He in a "nondissipated" atmosphere were obtained independently from reliable data on the concentrations and isotope ratios of inert gases, repeatedly reviewed in the literature. The results of the calculations are very close, practically identical, and the following figures will be used, therefore, in our discussion of a "nondissipated" atmosphere: maximum content of $^4He = 3.3 \cdot 10^{23}$ cm^3; the content of $^3He = 1.0 \cdot 10^{19}$ cm^3, and the ratio of $^3He/^4He \gtrsim 3.0 \cdot 10^{-5}$.

Several important conclusions follow from the values obtained for the upper bound of the ^4He amount in a "nondissipated" atmosphere and the fact of agreement between the two independent calculations.

(1) The agreement of results may be considered a convincing argument in favor of the idea that the excess ^3He in the earth is a result of participation in its accretion of meteoritic matter containing primordial noble gases.

(2) The agreement of the results obtained by the two different methods speaks in favor of the continuous accumulation and preservation of the atmosphere during the earth's history. The first estimate of the helium isotopic amount is based on the amount of primordial ^{20}Ne, which was able to migrate into the atmosphere most actively in the early stages of evolution of the planet and would have been largely lost if the ancient atmosphere had dissipated into space. The second estimate is based on the ratio of radiogenic isotopes, which migrated gradually into the atmosphere in the course of continuous production by radioactive decay of U, Th, and K. Catastrophic dissipation of the ancient atmosphere could not have a significant effect on

the present-day amount of ^{40}Ar in the atmosphere or on the amount of radiogenic helium in a "nondissipated" atmosphere.

(3) Knowing the age of the earth (4.6 billion years) and its mass ($5.89 \cdot 10^{21}$ g), assuming that Th = 3U, and using the calculated upper limit of radiogenic ^{4}He in a "nondissipated" atmosphere, it is possible to estimate the concentration of uranium. This turns out to be $\lesssim 3 \cdot 10^{-8}$ g g^{-1}, in good agreement with the estimates made on the basis of direct cosmo- and geochemical data and of thermal models of the earth (MacDonald, 1959; Wasserburg et al., 1964; Larimer, 1971). As noted above, this estimate was obtained by a new method, independently from those used before. The estimate is correct only if the greater part of helium has been released from the solid earth (see Chapter 11).

(4) The calculated value of ^{4}He$_{A,\Sigma}$ is equivalent to a concentration of helium in a "nondissipated" atmosphere of about several percent by volume. Such a concentration would result in a very strong contamination of the accessible parts of the crust (gases, waters, rocks) with air helium. For example, in most thermal springs of the Kamchatka—Kurils volcanic zone the relationship between measured contents of helium isotopes (^{4}He$_{E}$, ^{3}He$_{E}$) derived mostly from solid earth (Kamensky et al., 1976) and of atmospheric argon (^{40}Ar$_{A}$) is as follows:

$$^{4}\text{He}_{E}/^{40}\text{Ar}_{A} \lesssim 10^{-3}; \quad ^{3}\text{He}_{E}/^{40}\text{Ar}_{A} \lesssim 10^{-8}$$

In the case of a "nondissipated" atmosphere the corresponding ratios would be:

$$^{4}\text{He}_{N}/^{40}\text{Ar}_{A} \approx 3; \quad ^{3}\text{He}_{N}/^{40}\text{Ar}_{A} \approx 10^{-4}$$

where N denotes the helium concentration in a "nondissipated" atmosphere. Utilizing the above-mentioned values it is easy to show that, in the absence of helium dissipation, the ratios between helium from the solid earth and that from a "nondissipated" atmosphere would be very low, ^{4}He$_{E}/^{4}$He$_{N} \lesssim 3 \cdot 10^{-3}$, ^{3}He$_{E}/^{3}He_{N} \lesssim 10^{-4}$, and it would be impossible to distinguish the small helium component of crustal (or mantle) origin from the major component originating from a "nondissipated" atmosphere.

It is precisely because of the unique characteristic of helium in that it continuously escapes from the atmosphere, that the helium isotopes vary so widely and are so regularly distributed among the different geospheres (see section 8.5).

10.3. The problem of helium escape

The problem of helium escape arose more than twenty years ago (Nicolet,

1957; Bates and MacDowell, 1959). It was established that thermal escape could not explain the proportions of helium isotopes in the atmosphere. Later on, several papers and reviews were devoted to this problem (Bates and Patterson, 1962; Kockarts and Nicolet, 1962; MacDonald, 1963; Axford, 1968; Johnson and Axford, 1969; Holzer and Axford, 1971; Sheldon and Kern, 1972; Kockarts, 1973; and others), but no solution had been found.

The discovery of the high ^{3}He/^{4}He isotope ratio in the mantle (Chapter 7) and recent estimates of the ^{3}He flux from the solid earth into the atmosphere have changed the situation radically. In addition, more adequate models of the atmosphere have been suggested and a better understanding of the processes in its upper regions have been achieved (Alpert, 1972; Stickland, 1972). All this led to a revision of the problem of helium escape.

Reliable estimates of the influx of helium isotopes represent a very important part of the problem. There are several methods of estimating the ^{4}He flux ($^4F_{EA}$) from the earth (E) into the atmosphere (A).

One of these methods is based on helium content measurements in seawater (Table 10.1). If one knows the ^{4}He excess in seawater and the lifetime of ^{4}He atoms in the ocean, then it is possible to calculate the helium flux through the oceanic crust. Recent estimates (Craig et al., 1975) have led to very low values of $^4F_{EA} \approx 0.3$ atoms cm^{-2} s^{-1}, which may be con-

TABLE 10.1

Values of ^{4}He flux from the earth into the atmosphere in 10^6 atoms cm^{-2} s^{-1}.

No.	Value	Reference
1	2.0	Bieri et al. (1964)
2	1.9—3.2	Bieri et al. (1967)
3	0.77	Craig and Clarke (1970)
4	1—2	Bieri and Koide (1972)
5	0.3	Craig et al. (1975)
6	0.3	Tugarinov et al. (1975)
7	0.1	Naughton et al. (1973)
8	0.011 and 0.064	Barnes and Bieri (1976)
9	3.2	Wasserburg et al. (1963)
10	1.7	Nicolet (1957)
11	2.0	MacDonald (1963)
12	2.5 ± 1.5	Kockarts (1973)
13	$2.5 \pm {}^{2.5}_{1.5}$	Bieri et al. (1967)
14	4.0	Tolstikhin (1975b), Tolstikhin et al. (1975)

Values 1—5 were obtained from measurements of the ^{4}He content in ocean water and 7—8 were obtained by extrapolation of results of a few flux measurements for total earth. Values 8 are measurements of the flux at two points on the Pacific floor. Value 9 is a calculated value involving the (^{4}He/^{40}Ar)$_{rad}$ ratio. Others are calculated values.

sidered as a lowest bound to this flux for the whole earth. There is some evidence that the outgassing of the earth's interior in such regions as the Circum-Pacific volcanic belt, Iceland and Hawaii is considerably more intensive than in oceanic rift zones, rises, etc.

Some authors (Naughton et al., 1973; Tugarinov et al., 1975) estimated the total terrestrial flux by using very few measurements of $^4F_{EA}$. These estimates seem to be less reliable because the accuracy of extrapolation is rather low.

A method described by Wasserburg et al. (1963), which involves the ratio of $(^4He/^{40}Ar)_{rad.}$ in terrestrial fluids and the ^{40}Ar content of the atmosphere, is still very useful despite the fact that the value of $(^4He/^{40}Ar)_{rad.}$ looks less constant in the light of recent data than in earlier publications (see section 10.2). A value of $^4F_{EA} = 3.2 \cdot 10^6$ atoms cm^{-2} s^{-1} has been obtained by this method (Table 10.1).

The calculated values of the flux are deduced from *a priori* established contents of heavy radioactive elements in the earth. Various estimates of the U and Th contents of our planet have restricted the ^{238}U content to between $1.5 \cdot 10^{-8}$ g g^{-1} and $3.3 \cdot 10^{-8}$ g g^{-1} (Wasserburg et al., 1964; Taylor, 1979; and others). It is considered that practically all helium produced by radioactive decay is released by the solid earth. Therefore, $^4F_{EA}$ values may be calculated. In particular, a value of $4.0 \cdot 10^6$ atoms cm^{-2} s^{-1} has been widely accepted because this value is in agreement with (1) the total heat flux of the earth, (2) the ratio between the heat flux of the crust and that of the mantle, and (3) the abundance of primordial and radiogenic rare gases in the earth and in the atmosphere (Bieri et al., 1967; Kockarts, 1973; Tolstikhin et al., 1975). In subsequent calculations the above-mentioned value will be used, although one should bear in mind that it should probably be considered the upper bound of the flux (see Chapter 11).

Estimates of the average ^3He influx from various sources ($^3F_{EA}$) were presented by many authors (Nicolet, 1957; Damon and Kulp, 1958a; MacDonald, 1963; Kockarts, 1973; and others). The most detailed and careful investigations were published by Johnson and Axford (1969) and later by Bühler and co-workers (1976). Among other things, it was shown that the contributions from some natural sources of ^3He are uncertain and one has to be satisfied with approximate values. Nevertheless, Johnson and Axford (1969) evaluated the ^3He flux into the atmosphere as 4.83 atoms cm^{-2} s^{-1}; they believed that the main ^3He contributor was solar wind precipitation (4.0 atoms cm^{-2} s^{-1}). Bühler et al. (1976) reported an even higher influx of ^3He (Table 10.2).

During the last decade extremely high isotopic ratios of $^3He/^4He \sim 10^{-5}$ were discovered in terrestrial helium (Mamyrin et al., 1969a), and the distribution of both helium isotopes in the atmosphere, the crust and the upper mantle have been widely investigated (Chapter 7). The $^3He/^4He$ ratio in the earth's interior appears to be about $3 \cdot 10^{-5}$ which is about twenty times

TABLE 10.2

Estimates of various sources of atmospheric ^{3}He (Tolstikhin et al., 1975; Bühler et al., 1976) in atoms cm^{-2} s^{-1}

^{3}He source	Influx	References
Solar wind (auroral)	5	Bühler et al. (1976)
Interstellar gas	5	Holzer and Axford (1971)
Primordial	4	Craig et al. (1975)
Solar cosmic rays	1 ?	Lupton (1973)
Cosmic dust and meteorites	0.1	Johnson and Axford (1969)
Galactic cosmic rays	0.6	Johnson and Axford (1969)
Earth crust	0.1	Morrison and Pine (1955)
Total	16	
Earth outgassing	80	Tolstikhin (1975b); Tolstikhin et al. (1975)

higher than this ratio in the atmosphere, $(1.39 \pm 0.01) \cdot 10^{-6}$. Hence estimates of the ^{3}He flux from the solid earth have been changed radically; even the lowest estimates of the terrestrial flux (Craig et al., 1975) are similar to the total influx from all known helium sources. Models of earth degassing and differentiation (Tolstikhin, 1975b; Tolstikhin et al., 1975) resulted in a very high value of $^3F_{EA} \approx 80$ atoms cm^{-2} s^{-1}. In the following discussion this value, which can also be considered to be the upper bound to the ^{3}He flux into the atmosphere, will be used.

Comparison of the $^3F_{EA}$ and $^4F_{EA}$ fluxes adopted above with the atmospheric contents of ^{3}He and ^{4}He led to the conclusion that both helium isotopes escape from the terrestrial atmosphere. The total number of ^{4}He and ^{3}He atoms per vertical atmospheric column of 1 cm^2 cross-section is known to be $1.13 \cdot 10^{20}$ and $1.58 \cdot 10^{14}$ atoms, respectively. The ratio of the total number of atoms, N, to their influx is usually considered to be the lifetime, τ, of the atom in the atmosphere. Then the lifetime of a ^{4}He atom amounts to:

$$\tau_4 = \frac{1.13 \cdot 10^{20}}{4 \cdot 10^6} = 0.28 \cdot 10^{14} \text{ s} = 0.9 \cdot 10^6 \text{ years} \qquad (10.6)$$

For fluxes of the light helium isotope of 4.83 atoms cm^{-2} s^{-1} (Johnson and Axford, 1969) and 80 atoms cm^{-2} s^{-1} (Tolstikhin et al., 1975), the following lifetimes can be obtained, respectively:

$$\tau_3' = 1.0 \cdot 10^6 \text{ years} \qquad (10.7)$$

$$\tau_3 = 0.06 \cdot 10^6 \text{ years} \qquad (10.8)$$

It is also worthwhile to calculate the ratios of the ^{3}He and ^{4}He lifetimes according to the earlier and more recent data:

$$\tau'_3/\tau_4 = 1.12 \qquad (10.9)$$

$$\tau_3/\tau_4 = 0.07 \qquad (10.10)$$

It is obvious that the lifetimes are less than the earth's age, no matter what values of the fluxes were adopted. This proves that the escape of ^{3}He and ^{4}He from the earth's atmosphere into space actually exists.

The most probable thermal escape was considered in detail by Nicolet (1957). In his calculations, Nicolet used Jeans's formula (Jeans, 1925) for the escape rate of neutral atoms:

$$F_j = 2\pi n_c \int_{v_{esc}}^{\infty} f(v)v^3 \cos\Theta \sin\Theta \, d\Theta \, dv \qquad (10.11)$$

where $f(v)$ is a Maxwellian velocity distribution function. In accordance with Jeans's model, Nicolet supposed that escape takes place at the critical level, r_c, where the escaping atom possesses the velocity of $v > v_{esc} = 2GM/r_c$ and has a chance to pass through the upper part of the atmosphere without collisions. Using some numerical parameters of the upper atmosphere, Nicolet (1957) deduced formulae for the life time's (10.12) and escape rate, cm^{-2} s^{-1} (10.13) in the case of an isothermal atmosphere:

$$\tau_j = 2.6 \, \frac{n_{j,0}}{n_{j,a}} \left(\frac{H_{j,0,a}}{R_0} \right)^{\frac{1}{2}} \cdot \exp\left(0.95 \, R_0/H_{j,0,a}\right) \qquad (10.12)$$

$${}^jF_{AS} = 3 \cdot 10^5 \, n_{j,a} \left(\frac{R_0}{H_{j,0,a}} \right)^{\frac{1}{2}} \cdot \exp\left(-0.95 \, R_0/H_{j,0,a}\right) \qquad (10.13)$$

where j denotes the species of the escaping component; R_0 is the earth's radius; a is the geocentric height of the thermopause; $n_{j,0}$ and $n_{j,a}$ are the contents of the escaping atoms at sea level and the thermopause level, respectively;

$$H_{j,0,a} = \frac{kT_a}{m_j g_0}$$

g_0 is the acceleration due to gravity at sea level; and T_a is the thermopause temperature.

Moreover, the height of the critical level h_c is taken to be ≈ 500 km and R_0/R_a 0.95.

214

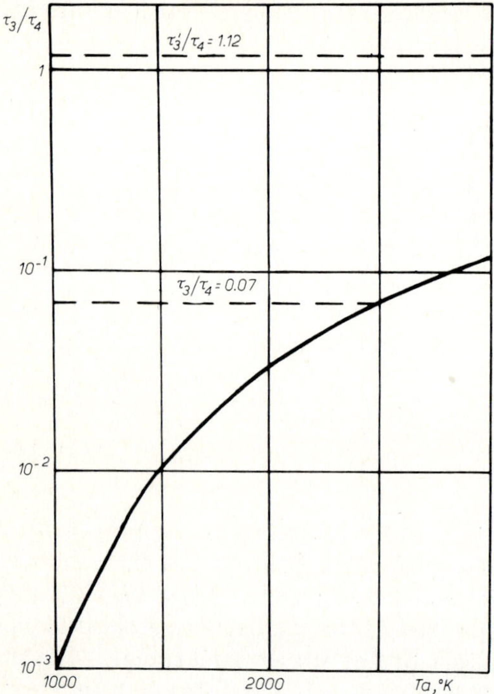

Fig. 10.3. Escape time ratio τ_3/τ_4 as a function of thermopause temperature, T_a (°K), in Nicolet's model (Pudovkin et al., 1981).

A theoretical dependence of the τ_3/τ_4 ratio on the thermopause temperature was established (Fig. 10.3) on the basis of eq. 10.12. With the assumption that there is a balance between gains and losses of helium isotopes in the atmosphere, Nicolet estimated T_a from this dependence; the numerical value of $\tau_3'/\tau_4 = 1.12$ from eq. 10.9 was used (Fig. 10.3). The results $T_a \gg 3000°$ K obtained in such a way seems improbable.

Now that new data concerning the flux of ^{3}He to the atmosphere have become available, it will be useful to adjust it to the thermal model of escape developed by Nicolet. Using the ratio $\tau_3/\tau_4 = 0.07$ (see eq. 10.10), one obtains $T_a = 2500°$K, which is also considerably higher than the average temperature of the thermopause assumed in recent models of the atmosphere (CIRA-72; Jacchia 1977). The thermopause temperature can also be estimated from the dependence of $^jF_{EA}$ (j = ^{3}He, ^{4}He) on T_a (eq. 10.13, Fig. 10.4). If dynamical equilibrium of helium actually takes place in the atmosphere ($F_{EA} \approx F_{AS}$), the thermopause temperatures estimated from the flux $^3F_{EA}$ and from the flux $^4F_{EA}$ must be the same or close to each other.

Using production values of $1.74 \cdot 10^6$ atoms cm^{-2} s^{-1} for ^{4}He and 2 atoms cm^{-2} s^{-1} for ^{3}He, Nicolet obtained T_a (^{4}He) = $3000°$K and T_a (^{3}He) =

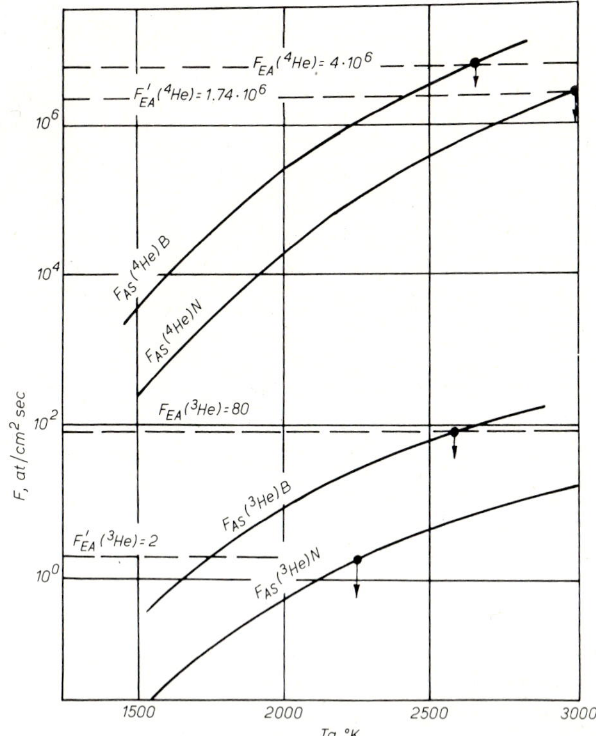

Fig. 10.4. Fluxes of helium isotopes from the critical level (Nicolet, 1957; curves N) and those from the finite atmosphere (Biutner, 1959; curves B) as a function of thermopause temperature, T_a (°K). Suffixes EA and AS indicate fluxes from the earth into the atmosphere and from the atmosphere into space, respectively (Pudovkin et al., 1981).

2250° K (see Fig. 10.4). The discrepancy between these estimates appears to be enormous. New estimates of the helium fluxes into the atmosphere (i.e. $^4F_{EA} = 4 \cdot 10^6$ atoms cm^{-2} s^{-1} and $^3F_{EA} = 80$ atoms cm^{-2} s^{-1}) have led to values of T_a (^{3}He), T_a (^{4}He) > 3000° K.

Thus, Nicolet's model, which employs Jeans's formula for the escape rate, is consistent with only very small escape fluxes; this does not agree with recent estimates. These low escape fluxes, as well as the high values of T_a, can be easily understood if the escape process takes place only from some fixed boundary level, while all the lower layers merely supply the atoms to this critical level. The upper layers are not taken into consideration.

10.4. Recent investigations of helium escape

It is possible to show that at any level there is a probability that escaping atoms will pass through all the atmospheric levels above without collision. As the height increases the escape occurs within a larger solid angle (Mitra, 1952). The contribution of the finite thickness of the atmosphere to the escape flux was estimated by Biutner (1958, 1959) and later her results were used by Lindenfeld and Shizgal (1979).

However, it must be pointed out that Biutner's model appears to be too inexact: (1) the concentration of the escaping component was assumed to be constant within the entire escape region; (2) the escape layer was considered to be isothermal and mono-componental; and (3) the boundaries of the escape layer were taken to be a function of height alone, without a direct calculation of the escape probability — that is, only approximate estimates were based upon insufficient data. Biutner considered the escape process only between 550 and 1000 km. The escape probability below 550 km was assumed to be negligible and the contribution of the atmosphere above 1000 km was not taken into account. Due to these limitations the escape fluxes in Biutner's model exceeded Nicolet's only by one or two orders of magnitude (see Fig. 10.4). Far higher escape fluxes may be obtained if the escape process within a wider height range is considered.

Let us assume (see Pudovkin et al., 1981) a spherically symmetrical earth and suppose that the atoms escape from a height of about 110 km (below which the probability of escape may be considered to be negligible) to an altitude of about $(0.5-1.0) \cdot 10^4$ km, where neutral helium is essentially absent[1]. Let us also assume (see Telesnin, 1973) that the flux of upward-going particles escaping from some fixed level decreases with distance above the layer (x) according to the relationship:

$$I = I_0 \exp\left(-x/\overline{\lambda}\right) \tag{10.14}$$

where $\overline{\lambda}$ denotes the mean free path determined by the collisions of the upward-going particle with the molecules of atmospheric gases. At last, let us represent the equivalent thickness of the atmosphere between level r and altitude r_1 (the upper limit of the integration) as H_r. Then the flux intensity of outward-going particles (from level r) which are passing through the thickness of the atmosphere and moving at angle Θ to the vertical is:

$$I_\infty(\Theta)\,\mathrm{d}\Theta = I_r(\Theta)\,\mathrm{d}\Theta \prod_{i=2}^{N} \exp\left\{-\left[\sqrt{r^2\cos^2\Theta + H_r^i(H_r^i + 2r)} - r\cos\Theta\right]\frac{1}{\overline{\lambda}_i}\right\} \tag{10.15}$$

---

[1] At these heights all helium is ionized. The escape fluxes of such helium are now being investigated by many authors (Axford, 1968; Banks and Holzer, 1969; Banks and Kockarts, 1973; Raitt et al., 1978; and others). In our model the escape flux of ionized helium is assumed to be considerably lower than the thermal flux of neutral helium.

where:

$$\frac{1}{\overline{\lambda}_i} = n_i \, \pi \, \left(\frac{\sigma_1 + \sigma_i}{2}\right)^2 \cdot \sqrt{1 + \frac{m_1}{m_i}} \qquad (10.15a)$$

(MacDaniel, 1964; Nicolet, 1964);

$$I_r(\Theta) = 2\pi n(r) \sin\Theta \, \cos\Theta \int_{v_{esc}}^{\infty} f(r, v) \, v^3 \, dv$$

σ_1 and m_1 denote the efficient diameter and the atomic mass of the respective helium isotope; σ_i and m_i are the diameter and molecular mass of the air constituent of species i (oxygen, nitrogen, or atomic oxygen); H_r^i is the atmospheric scale height at level r for the same atmospheric constituent.

Eq. 10.15 was derived from eq. 10.14 using the cosine theorem.

Then, by integrating eq. 10.15 with respect to angle Θ we can calculate the total number of atoms which escape from unit volume at level r per unit time:

$$I_r = 2\pi n(r) P(r) \int_{v_{esc}}^{\infty} f(r, v) \, v^3 \, dv \qquad (10.16)$$

where:

$$P(r) = \int_0^{\pi/2} \left[\prod_{i=2}^{N} \exp \left\{ -\left[\sqrt{r^2 \cos^2\Theta + H_r^i (H_r^i + 2r)} - r\cos\Theta \right] n_i \cdot \right. \right.$$

$$\left. \left. \pi \left(\frac{\sigma_1 + \sigma_i}{2}\right)^2 \sqrt{1 + \frac{m_1}{m_i}} \right\} \right] \sin\Theta \, \cos\Theta \, d\Theta \qquad (10.16a)$$

Recent data concerning the composition and the height distribution of the temperature in the upper atmosphere (CIRA-12; Jacchia, 1977) were used for the calculations of the escape probability, $P(r)$, and show that it actually approaches its maximum value (≈ 0.5) within the range of 600 km and 1000 km. However, within the range of 300 and 600 km the factor cannot be considered unimportant. Besides, in the model proposed by Pudovkin et al. (1981) the helium content is not supposed to be constant but changes with height according to the law (Mitra, 1952):

$$n(r) = n_0 \exp \left[-\frac{R_0}{r} \cdot \frac{(r - R_0)}{KT(r)/mg_0} \right] \qquad (10.17)$$

218

That is why the escape flux (see eq. 10.16) from heights of 300—600 km is comparable with that from the critical level.

In addition, the contribution of the part of the atmosphere ranging from 10^3 to 10^4 km to the escape flux has turned out to be significant because the critical escape velocity, v_{esc}, decreases with height.

Taking into account that concentration n diminishes with height (see eq. 10.17), and that due to ionization of helium the escaping flux decreases, we may multiply the result obtained by the area of the sphere of radius r and express the total number of helium atoms escaping from the terrestrial atmosphere as:

$$F_{AS} = 4\pi \int_{r_0}^{r_1} I(r)r^2 dr = 8\pi^2 \int_{r_0}^{r_1} dr\, r^2 P(r) \frac{n_0}{1 + \dfrac{n_i(r)}{n(r)}} \cdot$$

$$\exp\left[-\frac{R_0}{r} \cdot \frac{(r-R_0)}{KT(r)mg_0}\right] \int_{v_{esc}}^{\infty} f(r,v)\, v^3 dv \tag{10.18}$$

where n_i is the concentration of ionized helium; $P(r)$ is given by eq. 10.16a.

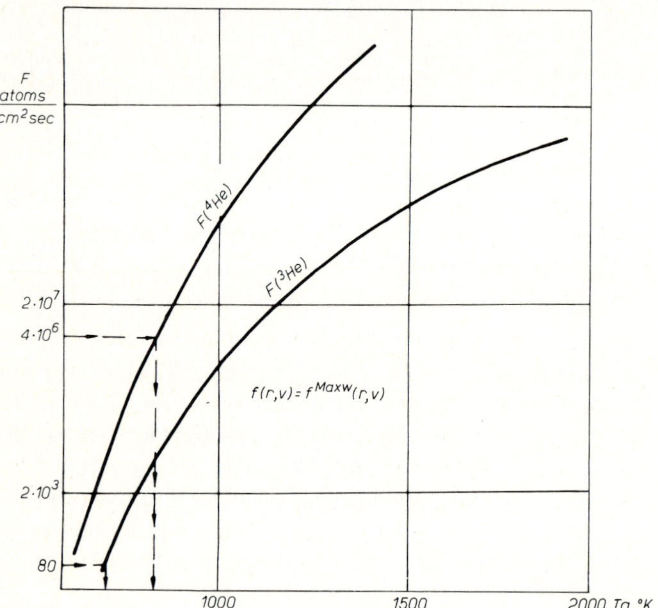

Fig. 10.5. F (^{3}He) and F (^{4}He) fluxes from the entire atmosphere (atoms cm^{-2} s^{-1}) as a function of thermopause temperature, T_a ($^\circ$K); the equilibrium distribution function was assumed for the helium atoms (Pudovkin et al., 1981).

The results of numerical integration of eq. 10.18 for different values of T_a have led to fluxes which exceed Nicolet's fluxes by about six orders of magnitude (Fig. 10.5); Maxwellian velocity distribution of helium atoms was assumed.

The thermopause temperature, at which there is a balance between production and loss of helium isotopes, in its turn decreases to T_a (^{3}He) $\approx 700°$K and T_a (^{4}He) $\approx 800°$K. These values appear to be rather low but the non-Maxwellian effect (associated with the thermal escape of helium) leads to an increase in T_a.

The distribution function $f(r, v)$ in eq. 10.18 differs from the Maxwellian due to the escape of high-velocity particles. This perturbation of the distribution function and the decrease in the escape flux resulting from this non-equilibrium effect was carefully investigated by Lindenfeld and Shizgal (1979). The non-Maxwellian distribution function for the escaping component was found in their model by means of the Boltzmann equation in which the isotropic loss term $L(r, v)$ was involved when $v > v_{esc}$.

$$(\frac{\partial}{\partial t} + v \nabla_r + \dot{v} \nabla_v) f(r, v, t) = (\frac{\partial f}{\partial t})_{coll} - L(r, v) \tag{10.19}$$

The collision operator and the loss term were used without any approximation. Lindenfeld and Shizgal (1979) set the time derivative at zero and the drift terms on the left-hand side of the Boltzmann equation and linearized eq. 10.19 by including the unknown $f(r, v)$ in the form of:

$$f(r, v) = f^{Maxw} [1 + \Phi(r, v)] \tag{10.20}$$

where f^{Maxw} is the equilibrium distribution function. Then they solved the integral equation for the perturbation $\Phi(r, v)$ and made corrections for the fluxes calculated in eq. 10.11.

Differences from the equilibrium escape flux F_{AS} can be represented as:

$$F_{AS} = F_y (1 - \eta) \tag{10.21}$$

Here the correction η for the helium flux depends on the temperature of the background gas and amounts to several percent (Fig. 10.6).

However, linearization of the Boltzmann equation by using $f(r, v)$ in eq. 10.20 is valid only when $\Phi(r, v) \ll 1$ (Cercignani, 1975), and the latter inequality does not seem to be plausible for the entire thickness of the atmosphere. Therefore, it is hardly possible to apply the results of Lindenfeld and Shizgal (1979) to the entire region of escape.

In the model by Pudovkin et al. (1981) the loss term which has been introduced on the right-hand side of the Boltzmann equation is less precise

220

compared to Lindenfeld and Shizgal's approximation but it is suitable for a larger range of height 100—1000 km.

Moreover, the τ-approximation for the collision operator is used (Krall and Trivelpiece, 1973) because the number of collisions of helium atoms is considered to be negligible compared to the number of collisions between helium and other constituents of the atmosphere. In this case eq. 10.19 can be rewritten as:

$$\left(\frac{\partial}{\partial t} + v \nabla_r + \dot{v} \nabla_v\right) f(r, v, t) = \frac{f^{\text{Maxw}}(r, v, t) - f(r, v, t)}{\tau(r)} -$$

$$L(r, v) + Q(r, v) \tag{10.22}$$

where f^{Maxw} is the equilibrium (Maxwellian) distribution function of gas particles; τ is the average time between collisions of atoms under consideration with atoms of other gas constituents; $\nu = 1/\tau$ is the average frequency of the collisions (see eq. 10.25); $L(r, v)$ is the rate of loss of particles; and $Q(r, v)$ is the intensity of the source.

It is necessary to note that $f(r, v, t)$ in eq. 10.22 is the total distribution function which takes into account all three kinds of particles (Biutner, 1958, 1959): (1) low-energy particles ($v < v_{\text{esc}}$) captured by the gravitational field of the earth; (2) "stationary" high-energy particles ($v > v_{\text{esc}}$) scattered by

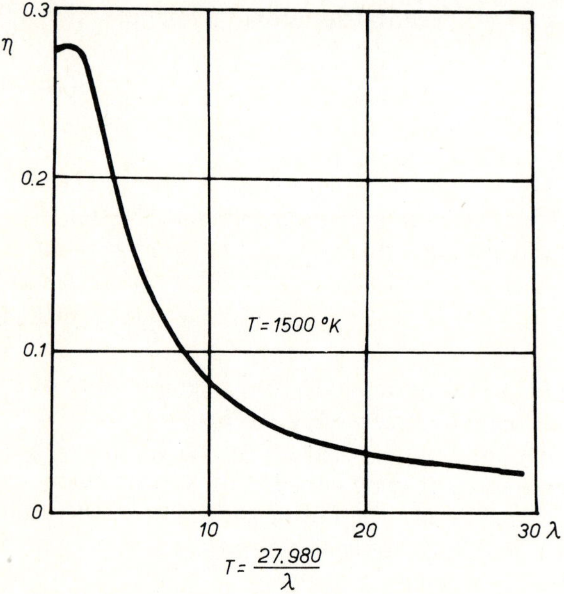

Fig. 10.6. Corrections η to Jeans's flux versus the temperature of the background gas (after Lindenfeld and Shizgal, 1979).

collisions with molecules of atmospheric gases; and (3) high-energy particles ($v > v_{esc}$) escaping from the atmosphere.

As for the first group of particles, we take into account a relatively large lifetime of helium within the earth's atmosphere and believe their distribution function to be close to the Maxwellian one. Concerning the other two groups, it is difficult, if not impossible, to distinguish between them. Nevertheless, in order to simplify the problem and taking into account the fact that there always exists a finite probability for a particle to escape from the atmosphere without collisions (see eqs. 10.16 and 10.16a), we shall suppose that escaping particles do not collide with "stationary" ones so that eqs. 10.19 and 10.22 are valid only with respect to the "stationary" particles.

At the same time the existence of escaping particles affects the distribution function of trapped particles that may be considered the source of the escaping particles.

The drift terms on the left-hand side of the Boltzmann equation are deleted due to the assumption that relaxation of the equilibrium is provided by local collisions whereas diffusion and conductivity play a negligible role in this process. So for the stationary case $\partial f/\partial t = 0$ and with the assumption that $Q(r, v)_{v > v_{esc}} = 0$, eq. 10.22 can be rewritten as:

$$\frac{f^{Maxw}(r, v) - f(r, v)}{\tau(r)} = \frac{f(r, v)}{\tau_{esc}(r, v)} \tag{10.23}$$

where $\tau_{esc}(r, v)$ is the mean time of escape from unit volume at level r for the outward-going atom.

$$\tau_{esc} = \frac{k}{vP(r)} \tag{10.24}$$

where $P(r)$ is given by eq. 10.16a; $k = 1$ cm is the dimensional coefficient.

From eq. 10.23 one can easily obtain:

$$f(r, v > v_{esc}) = \frac{f^{Maxw}(r, v)}{1 + \dfrac{P(r) v}{\nu(r) k}} \tag{10.25}$$

where (Alpert, 1972):

$$\nu(r) = \left(\frac{8RT(r)}{\pi M_{air}(r)}\right)^{\frac{1}{2}} \sqrt{2\pi} \sigma_{air}^2(r) n_{air}(r) \tag{10.26}$$

Analysis of eq. 10.25 shows: (1) $f(r, v)$ is close to the Maxwellian distribution due to the low values of $P(r)$ and the high values of $\nu(r)$ at low heights

222

($\lesssim$ 300 km), and therefore it is in full agreement with $f(r, v)$ calculated by Lindenfeld and Shizgal (1979); (2) the deviation of $f(r, v)$ from the Maxwellian distribution becomes considerable and approaches several orders of magnitude at heights of about 10^3—10^4 km.

The results of numerical integration of eq. 10.18 with $f(r, v)$ described by eq. 10.25 are illustrated in Fig. 10.7, in which the fluxes of ^{4}He and ^{3}He on a logarithmic scale are plotted against the temperature of the thermosphere T_a. Dotted horizontal lines show the assumed influx of helium isotopes into the earth's atmosphere. As indicated in Fig. 10.7, the balance between entry and escape is achieved at $T_a = 1300°$K for ^{4}He and $T_a = 1220°$K for ^{3}He.

It follows that the difference obtained in the thermosphere temperature does not seem to be very large, taking into account the approximate character of the calculations. As for the absolute value of temperature $T_a = (1260 \pm 40)°$K, this is in perfect agreement with recent data concerning the temperature of the thermosphere.

Although the above calculations (Pudovkin et al., 1981) have led to

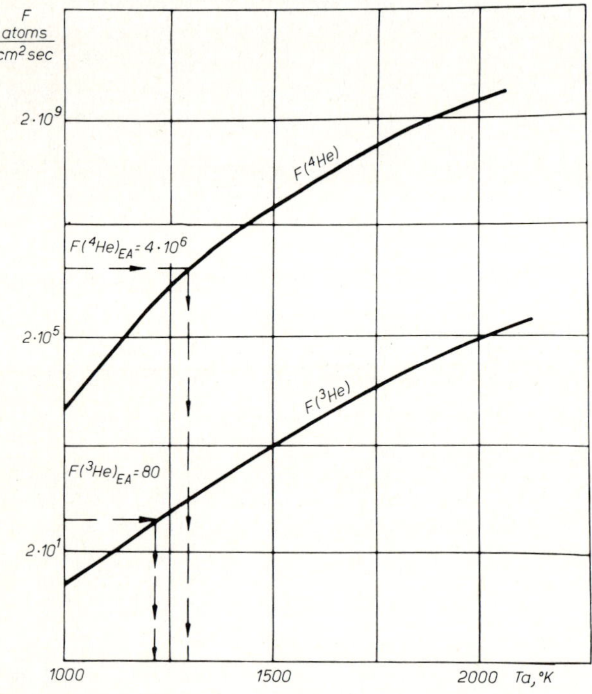

Fig. 10.7. Relationship between the F (^{3}He) and F (^{4}He) fluxes from the entire atmosphere (atoms cm^{-2} s^{-1}) as a function of the thermopause temperature, T_a ($°$K); the non-Maxwellian distribution function for escaping components was used (Pudovkin et al., 1981).

agreement between the temperatures of ^{3}He and ^{4}He escape as well as to a fairly reasonable average value of thermosphere temperature T_a, this investigation must be continued because of a possible decrease in values of $^4F_{\mathrm{EA}}$ and especially of $^3F_{\mathrm{EA}}$. Recent experimental results in the isotope geochemistry of helium show that probably the latter value is somewhat overestimated. If the flux of the light isotope decreases more than the heavy one then the difference between the estimates of the thermospheric temperatures T_a (^{3}He) and T_a (^{4}He) will be larger.

The following ways may be proposed to perfect the model.

(1) The average thermosphere temperature T_a could be somewhat lower than that obtained above (lower than $1260°\mathrm{K}$), and helium isotopes might escape mostly during the short intervals of high solar activity when $T_a \gg \overline{T_a}$. According to the thermal mechanism of dissipation, the difference between the lifetimes of ^{3}He and ^{4}He in the atmosphere (τ_3 and τ_4) decrease while the thermosphere temperature increases. Consequently, the model will have a solution if the relative contribution of the ^{3}He flux is lower than that adopted here.

(2) The search for an effective nonthermal escape which does not result in high mass differentiation of the isotopes may serve as another approach to the problem. The escape of charged atoms is not accompanied by a significant isotopic effect, but in the light of recent data it is necessary to check the contribution of these atoms in the entire helium flux from the atmosphere.

(3) Attention should be paid to more precise estimates of the ^{3}He flux from extraterrestrial sources of helium, such as solar wind or interstellar gases. It is possible that these sources can contribute more light helium to the atmosphere; in any case there is a tendency at present to increase the influx from such sources.

(4) It is also useful to bring together the models of earth degassing and helium dissipation because in this case the solution may be obtained directly by changing such input parameters as contents of radioactive elements and constants of degassing and differentiation. Moreover, such a model appears to be interesting because there is a connection between the hypothetical initial catastrophic degassing of the earth and the present process of helium dissipation from the atmosphere: if a significant proportion of primordial gases (^{3}He among them) was released by early degassing, the present ^{3}He flux would be smaller and the disagreement between T_a (^{3}He) and T_a (^{4}He) would increase (see Chapter 11).

OUTGASSING OF THE MANTLE AND CRUST; THE EVOLUTION OF NOBLE GASES IN THE ATMOSPHERE

In this chapter we make an attempt to simulate the process of earth degassing and differentiation using the most valid data of cosmochemistry and geochemistry of noble gases and radioactive elements. The simulation is based on the assumption that the process was continuous, but a possibility of an early catastrophic degassing event is also discussed (section 11.1). The results of such model calculations depend on the validity of the assumptions, the initial and boundary conditions, as well as on actual observations which are considered a decisive factor and used for comparison with calculated values. Considerable attention is given to all these input data; they are carefully examined in section 11.2. It is important that our model (Tolstikhin, 1975b; Tolstikhin et al., 1975) uses new data recently obtained in the isotope geochemistry of helium and argon in the mantle.

Such model calculations, despite a considerable simplification of natural processes, appear to be very useful because they enable one to interpret experimental results in more systematic quantitative terms. Thus, an agreement between new helium isotope data and the well-known data on the abundance of radioactive elements in the solid earth and noble gases in the atmosphere may be considered as a weighty argument in favour of the presence of primordial noble gases in the earth's interior (section 11.3). Moreover, some important parameters of the evolution of the earth, such as the residence time of atoms of noble gases or radioactive elements in the mantle, the flow of juvenile volatile components and their initial abundance in the solid earth, etc., may be defined only from the model calculations (section 11.3). Probably it is the only possible approach to the problem of the early degassing of the earth. The computed fluxes of light and heavy helium isotopes from the earth into the atmosphere agree with their escape fluxes from the atmosphere into space (section 11.4; see also Chapter 10).

11.1. The problem of the earth degassing

Two alternative hypotheses of the process of earth degassing and differentiation have been recently put forward. The first postulates that the earth had been initially outgassed during the period of its accretion (or very soon after). Some authors believe that this early period of the earth's history was characterized by intensive heating, by the melting of terrestrial material and by a more or less complete outgassing (Ringwood, 1966; Fanale, 1971). The strongest argument in favour of this hypothesis is the absence of "measurable" amounts of primordial isotopes in the solid earth and their presence in the atmosphere (Fanale, 1971). All terrestrial samples which had been investigated by that time (before 1970) may be subdivided into two groups:

(1) The first group comprises ancient rocks and minerals originating within the crust, such as beryls, cordierites, etc., containing excess noble gases (Damon and Kulp, 1958b; Gerling, 1961; Shukolyukov and Tolstikhin, 1965; Kirsten and Gentner, 1966; Gerling et al., 1967b, 1968; Ginzburg and Panteleev, 1971; later Kaneoka, 1974; and others). It was shown in Chapter 8 that crustal matter cannot be expected to contain any juvenile volatiles, since the transfer of material from the mantle into the crust is accompanied by its intensive outgassing. Metamorphism causes an even more complete removal of deep gases, whereas radioactive decay continuously renews the supply of radiogenic isotopes, ^{4}He and ^{40}Ar, in rocks and fluids. K-Ar dating may serve as clear evidence for this effect. So the study of crustal samples does not produce crucial data to explain the occurrence of juvenile volatiles in the earth's interior.

(2) The second group includes samples of probable mantle origin, such as young erupted rocks, oceanic basalts, ultrabasic inclusions in basalts, etc. These samples as a rule contain a small amount of noble gases, the greatest share of which is derived from the atmosphere and the crust, so that to detect the primordial component a considerable progress in analytical technique had to be achieved (compare papers by Fanale, 1971; Mamyrin et al., 1969a; W.B. Clarke et al., 1969).

The only fact provided by the isotope geochemistry of noble gases which might suggest an early degassing is the ^{129}Xe excess detected in a few terrestrial samples. It is not clear yet whether this excess is typical of the mantle or due to an unknown local process within the crust. The most substantial enrichment of ^{129}Xe was observed in a sample of CO_2-well gas from New Mexico (Phinney et al., 1978). The sample was also characterized by the highest ^{40}Ar/^{36}Ar ratio in all terrestrial underground gases and waters, which implies a long residence time of the gas in the crust. The ^{3}He/^{4}He in the sample is well below the mantle value. It might be worthwhile to include the ^{129}Xe excess in degassing models as soon as its origin becomes clear.

Though it was proven that the arguments based on the absence of primordial isotopes of noble gases in terrestrial materials are not valid as these isotopes are more or less evenly distributed in the mantle, some new arguments have emerged in favour of a high-temperature accretion of the earth and its very early degassing.

Kaula (1979) has shown that if only about 10% of the impact energy had been retained by the growing earth (the time of growing is assumed to be relatively continuous, i.e. about 50 m.y.), then the temperature of the planet would have been sufficient for the vaporization of silicate matter. Wetherill (1975) has also adopted a high-temperature model of the earth's formation. S.P. Clarke et al. (1976) believed that the accumulation of the central parts of the earth proceeded at a temperature of several thousand degrees.

Terrestrial rocks older than 3.8 b.y. do not exist, and even those whose age is > 3.0 b.y. are very rare. Moreover, low initial $^{87}Sr/^{86}Sr$ and $^{207}Pb/^{204}Pb$ ratios in the oldest rocks rule out the formation of the early granitic crust (Moorbath, 1977). In other words, the ratios show the non-existence of very ancient rocks whose origin involves low-temperature subsurface processes such as erosion and sedimentation. A possible explanation of these data is that the earth's accretion and early differentiation were high-temperature processes.

It is likely that the earth's magnetic field appeared more than 3.8 b.y. ago and, presumably, the core had been formed even earlier. During the segregation of the core a large amount of gravitational energy was transformed into heat and released by the earth. In section 8.5 it was shown that the heat outflow is accompanied by juvenile degassing of the earth. If the processes of losses of heat and volatiles which occurred during the early period of earth evolution were similar to those of the recent geological epoch, then a large amount of gases might have been released during this early period as well.

To summarize the first hypothesis it may be said that there is enough ground to discuss the early degassing event.

The second hypothesis which postulates a continuous degassing of the earth has been strongly confirmed by recent investigations of the helium isotope distribution in the earth's matter. It has become clear that the process of juvenile degassing is still going on. Moreover, direct observations as well as the relationships between $^3He/^4He$ and $^{87}Sr/^{86}Sr$ ratios (see Fig. 8.7) and between the $^3He/^4He$ ratio and the heat flow (see Fig. 8.16) lead to the conclusion that the differentiation of terrestrial matter in the mantle and the juvenile degassing of the mantle are two major manifestations of the same global process of the earth's evolution. These relationships show that it is advantageous to make use of the Sm-Nd, Rb-Sr and U-Pb isotopic systems to clarify the history of terrestrial volatiles. From this point of view, if continuous fractionation of the mantle and the formation of the crust are taken for granted, a continuous outgassing of the planet appears to be very probable.

Last but not least, the abundance of radiogenic nuclides in the atmosphere, mainly of ^{40}Ar, is also compatible with the theory of a continuous degassing of the earth.

Summarizing this section it should be emphasized that a gradual degassing and differentiation of the mantle and crust is now a certainty, and in the

228

following sections we shall present a model whose main objective is to show these processes: while radioactive elements are released from the mantle and accumulated in the crust, volatiles pass from the mantle into the atmosphere. The main purpose of the model computation is to define more precisely some important parameters of the earth and its evolution, such as the initial content of primordial noble gases in the earth, the content and distribution of radioactive elements, the residence time of a noble gas atom in the mantle and in the crust, etc. As mentioned above, a quantitative agreement of the distribution of helium isotopes in the earth's reservoirs with other well-established geochemical data throw additional light on the problem of the origin of the ^{3}He excess in the earth.

11.2. A model of the earth degassing and differentiation; description, assumptions, observations and parameters

Among a considerable number of works on earth outgassing published in the last three decades there are many that threat the problem in the light of noble gas geochemistry. The earliest works discuss the problem of migration of radiogenic isotopes, ^{4}He and ^{40}Ar, from the crust into the atmosphere (Shillibeer and Russell, 1955; Damon and Kulp, 1958a; Turekian, 1959). More recent works focus attention on the abundance of primordial isotopes in the atmosphere and other terrestrial reservoirs (Ozima and Kudo, 1972; Ozima, 1973; Schwartzman, 1973a, b; Tolstikhin, 1975b; Fisher, 1978; and others).

Turekian (1959, 1964) was the first to use first-order differential equations for the description of losses of radiogenic ^{40}Ar by the solid earth and its accumulation in the atmosphere:

$$d\,^{40}\mathrm{Ar}/d\,t = \lambda_e \cdot {}^{40}\mathrm{K_E} - \Lambda \cdot {}^{40}\mathrm{Ar_E} \tag{11.1}$$

where λ_e and Λ denote the constant of electron capture decay and the earth outgassing coefficient, respectively; $^{40}\mathrm{K_E}$ and $^{40}\mathrm{Ar_E}$ are the amounts of the isotopes in the earth. Such a presentation of the degassing process was later adopted by many authors (MacDonald, 1963; Russell and Ozima, 1971; Ozima and Kudo, 1972; Tolstikhin et al., 1975; Fisher, 1978; Hamano and Ozima, 1978; and others). It is essential that the first-order equation describes many natural processes, namely radioactive decay, the kinetics of chemical reactions, the migration of atoms in minerals, etc. The equation is also applied to the analysis of evolution of the U-Pb, Rb-Sr and Sm-Nd systematics.

Differential equations of the first order are used in the model presented here to describe three processes: the differentiation of the earth into the mantle and the crust (Λ_{MC}), the degassing of the mantle (Λ_{MA}) and the degassing of the crust (Λ_{CA}).

(1) The proportionality coefficients, Λ, in these equations are assumed to be constant.

(2) The second assumption is that during the earth's history there has been no loss of noble gases (except helium) from its atmosphere. At present no mechanism is known through which neon and heavier noble gases could escape from the atmosphere (see also section 10.2).

(3) The third assumption postulates the equality of the mantle degassing coefficients, Λ_{MA}, for all three light noble gases, He, Ne and Ar. As will be shown in section 11.3, the differentiation coefficients ($\Lambda_{MC} \approx 0.15 \cdot 10^{-9} \, yr^{-1}$) slightly differ from the outgassing ones ($\Lambda_{MA} \approx 0.4 \cdot 10^{-9} \, yr^{-1}$), but Λ_{MA} are always greater than Λ_{MC}. So, if we admit to a considerable difference between the degassing coefficient for helium and argon, Λ_{MA} (He) $\gg \Lambda_{MA}$ (Ar), then the latter will be similar to the differentiation coefficient for U, Rb and K; however, even in this case Λ_{MA} (Ar) would not be much lower than Λ_{MA} (He). Nevertheless, this is very unlikely because difference in the behaviour of noble gases and radioactive elements is far greater than that of helium and argon. Note that this assumption does not hold good for the crustal degassing coefficients, where Λ_{CA} (He) is always greater than Λ_{CA} (Ar); this is due to the fact that the behaviour of noble gases in the upper crust and in the mantle is governed by different laws: the crust looses its helium isotopes at low temperatures leaving its argon isotopes intact.

The model describes the following scenario: At the time of accretion, 4.5 b.y. ago ($t = 0$), the earth contained initial concentrations of radioactive isotopes $^{40}K_0$, $^{238}U_0$ and $^{232}Th_0$ as well as primordial noble gases $He_{p,0}$, $^{20}Ne_0$, $^{36}Ar_0$ (heavy noble gases are beyond the scope of the model). At the moment $t = 0$ some proportion of noble gases could have been released in the atmosphere; at that moment radiogenic isotopes, ^4He_r and $^{40}Ar_r$ began to be produced and accumulated. In the course of time noble gases continued to be released by the mantle and accumulated in the atmosphere (Λ_{MA}); meanwhile, the material of the mantle underwent differentiation which resulted in an accumulation of radioactive elements in the crust (Λ_{MC}). The radiogenic isotopes of noble gases emerged from the crust into the atmosphere (Λ_{CA}). In addition, the atmosphere went on producing ^3He at a rate P_3. ^3He and ^4He isotopes kept escaping from the atmosphere. The contribution of radiogenic ^3He and ^{36}Ar was not considered in the model because it was negligible in the total balance of these isotopes.

The system of equations (Table 11.1) representing the above-mentioned description has an unambiguous solution which can be obtained either by numerical integration of the equations or analytically. For example, the solution of eq. 11.9, which describes the evolution of the uranium content in the curst, is given below:

$$U_c = U_o \left\{ \exp\left(-\lambda_U t\right) - \exp\left[\left(\Lambda_{MC} + \lambda_U\right) t\right] \right\} \tag{11.15}$$

230

TABLE 11.1

Mathematical description of the model of earth degassing and differentiation

Mantle (A)

$$d^{40}K_M/dt = -(\lambda_{40} + \Lambda_{MC}) \cdot {}^{40}K_M \tag{11.2}$$
$$d^{40}Ar_M/dt = \lambda_{40} \cdot {}^{40}K_M - \Lambda_{MA} \cdot {}^{40}Ar_M \tag{11.3}$$
$$dU_M/dt = -(\lambda_U + \Lambda_{MC}) \cdot U_M \tag{11.4}$$
$$d^4He_M/dt = \lambda'_U \cdot U_M - \Lambda_{MA} \cdot {}^4He_M \tag{11.5}$$
$$d^{36}Ar_M/dt = -\Lambda_{MA} \cdot {}^{36}Ar_M \tag{11.6}$$

Crust (C)

$$d^{40}K_C/dt = -\lambda_{40} \cdot {}^{40}K_C + \Lambda_{MC} \cdot {}^{40}K_M \tag{11.7}$$
$$d^{40}Ar_C/dt = \lambda_{40} \cdot {}^{40}K_C - \Lambda_{MC} \cdot {}^{40}Ar_C \tag{11.8}$$
$$dU_C/dt = -\lambda_U \cdot U_C + \Lambda_{MC} \cdot U_M \tag{11.9}$$
$$d^4He_C/dt = \lambda'_U \cdot U_C - \Lambda_{CA} \cdot {}^4He_C \tag{11.10}$$

Atmosphere (A)

$$d^{40}Ar_A/dt = \Lambda_{MA} \cdot {}^{40}Ar_M + \Lambda_{CA} \cdot {}^{40}Ar_C \tag{11.11}$$
$$d^4He_A/dt = \Lambda_{MA} \cdot {}^4He_M + \Lambda_{CA} \cdot {}^4He_C - {}^4F_{AS} \tag{11.12}$$
$$d^{36}Ar_A/dt = \Lambda_{MA} \cdot {}^{36}Ar_M \tag{11.13}$$
$$d^3He_A/dt = \Lambda_{MA} \cdot {}^3He_M + P_3 - {}^3F_{AS} \tag{11.14}$$

(1) U_M, U_C denote concentrations of ${}^{238}U$, ${}^{235}U$ or ${}^{232}Th$ in the mantle and crust, respectively; $\lambda'_U = \lambda_U \alpha$, where λ_U is the decay constant of uranium isotopes or thorium, and α is the number of helium atoms produced by radioactive disintegration; $\lambda_{232} = 0.04947 \cdot 10^{-9}$ yr^{-1}, $\lambda_{235} = 0.9848 \cdot 10^{-9}$ yr^{-1} and $\lambda_{238} = 0.155 \cdot 10^{-9}$ yr^{-1}; $\alpha_{232} = 8$, $\alpha_{235} = 7$, $\alpha_{232} = 6$.

(2) ${}^{40}K = 1.19 \cdot 10^{-4} \cdot K \cdot \lambda_e/(\lambda_e + \lambda_\beta)$, where K is the potassium content in the mantle and the crust (the subscripts are the same as for uranium); λ_e and λ_β are constants of the K-capture and β-decay of ${}^{40}K$; $\lambda_e = 0.585 \cdot 10^{-10}$ yr^{-1}, $\lambda_\beta = 4.72 \cdot 10^{-10}$ yr^{-1}, $\lambda_{40} = \lambda_e + \lambda_\beta$.

(3) ${}^{40}Ar_M$, ${}^{40}Ar_C$ and ${}^{40}Ar_A$ denote concentration (cm^3 g^{-1}) of radiogenic argon in the mantle, the crust and the atmosphere, respectively; similar symbols are used for ${}^{36}Ar$ and He.

(4) For ${}^3He_M$ and ${}^{20}Ne_M$, equations similar to eq. 11.6 can be written; for ${}^{20}Ne_A$ we can write an equation similar to eq. 11.13.

It follows from eq. 11.15 that the uranium content in the crust is proportional to its initial content. So the results of calculations can be easily adopted for any initial contents of radioactive and primordial isotopes.

The choice of values used as parameters and observations is very approximate in models of the type described in the present chapter. We shall choose the values we believe to be the better known and call them observations; we shall use these values to make others, which we shall define as parameters, more pricise.

TABLE 11.2

Some observations on the cosmochemistry and geochemistry of noble gases

No. Observation	Value	No. Observation	Value
Meteorites		8 $(^4\mathrm{He}/^{40}\mathrm{Ar})_{\mathrm{rad}}^{\mathrm{gas}}$	5 ± 3
1 $(^3\mathrm{He}/^4\mathrm{He})_{\mathrm{prim}}$	$2.4 \cdot 10^{-4}$	$(^4\mathrm{He}/^{40}\mathrm{Ar})_{\mathrm{rad}}^{\mathrm{rock}}$	$\leqslant 1$
2 $(^4\mathrm{He}/^{20}\mathrm{Ne})_{\mathrm{prim}}$	500	9 $t\,(\mathrm{U}/\mathrm{He})/t\,(\mathrm{K}/\mathrm{Ar})$	< 1
3 $(^{40}\mathrm{Ar}/^{36}\mathrm{Ar})_{\mathrm{prim}}$	10^{-4}	10 $t\,(\mathrm{K}/\mathrm{Ar})/t\,(\mathrm{Rb}/\mathrm{Sr})$	< 1
		11 $\mathrm{U_C}$, ppm	$0.7{-}2.0$
Earth's mantle		*Atmosphere*	
4 $(^3\mathrm{He}/^4\mathrm{He})_{\mathrm{M}}$	$(2{-}4) \cdot 10^{-5}$		
5 $(^{40}\mathrm{Ar}/^{36}\mathrm{Ar})_{\mathrm{M}}$	$500{-}1000$	12 $(^3\mathrm{He}/^4\mathrm{He})_{\mathrm{A}}$	$1.39 \cdot 10^{-6}$
		13 $^4\mathrm{He}$, cm³ g⁻¹	$0.43 \cdot 10^{-8}$
Earth's crust		14 $(^{40}\mathrm{Ar}/^{36}\mathrm{Ar})_{\mathrm{A}}$	295.6
6 $^{36}\mathrm{Ar}$, $^3\mathrm{He}$, $^4\mathrm{He}_{\mathrm{prim}}$	0	15 $^{36}\mathrm{Ar_A}$, cm³ g⁻¹	$2.7 \cdot 10^{-8}$
7 $(^3\mathrm{He}/^4\mathrm{He})_{\mathrm{C}}$	$2 \cdot 10^{-8}$	16 $^{20}\mathrm{Ne_A}$, cm³ g⁻¹	$1.49 \cdot 10^{-8}$

Observations 1 and 2 are discussed in sections 6.3.2 and 6.2.2, respectively; 3 is adopted from Ozima and Kudo (1972); 4 and 5 are described in section 7.5; 6, 7 and 8 are discussed in section 8.1; 12 is from section 2.3; 14 is from Cook (1961); 13, 15 and 16 are given in section 6.4; 9, 10 and 11 are described in the text. Masses of mantle and crust are adopted as equal to $400 \cdot 10^{25}$ g and $2.27 \cdot 10^{25}$ g, respectively.

The observations are shown in Table 11.2; most of them have been discussed earlier and the corresponding sections are listed in the table as well. The inequality of ages determined by the Rb/Sr, K/Ar and (U + Th)/He methods (observations 9 and 10, Table 11.2) is well known from the experience of dating terrestrial rocks (Gerling, 1961; Hurley and Rand, 1969). The average ratio of radioactive elements in the earth, K/U, has been widely discussed in the literature (Wasserburg et al., 1963, 1964; Wakita et al., 1967; Lambert and Heier, 1968; Smyslov et al., 1979 and others). The value of $\mathrm{K}/\mathrm{U} \sim 10^4$ (observation 11) at one time appeared to be a good approximation of experimental data, but recently a somewhat higher proportion of these elements has also been adopted (Jacobsen and Wasserburg, 1979).

The parameters are listed in Table 11.3. The variation limits of the contents of radioactive elements are adopted on the grounds of Table 11.4, where data on meteorites, rocks of presumed mantle origin and thermal models of the earth are presented. The minimal potassium content, capable of producing the observed amount of radiogenic $^{40}\mathrm{Ar}$ in the atmosphere, is 110 ppm (Larimer, 1971) and its maximal content is assumed to be the same as in chondrites.

The minimal concentrations of primordial noble gases $^{36}\mathrm{Ar_0}$ and $^{20}\mathrm{Ne_0}$

were calculated (see section 6.4) as a ratio of their amount in the atmosphere and the mass of the silicate part of the earth.

The most probable rate of ^{3}He production in the atmosphere, P_3, is that obtained by Johnson and Axford (1969) and later by Bühler et al. (1976); limits shown in Table 11.3 are wider than their estimates. The possible values

TABLE 11.3

Parameters and limits of their variation

No.	Parameter	Dimension	Values	Reference
1	U_0	ppm	0.02—0.06	Table 11.4
2	Th_0	ppm	$3\,U_0$	
3	K	ppm	110—880	
4	$^{36}Ar_0$	10^{-8} cm^3 g^{-1}	> 2.7	Table 6.2
5	$^{20}Ne_0$	10^{-8} cm^3 g^{-1}	> 1.49	
6	P_3	atoms g^{-1} yr^{-1}	0.05—0.5	section 10.3
7	Λ_{MA}	10^{-9} yr^{-1}	> 0	
	Λ_{MC}	10^{-9} yr^{-1}		
	Λ_{CA}	10^{-9} yr^{-1}		
8	T_a	°K	1000—1500	section 10.4
9	L	—	0—1	

TABLE 11.4

Present-day contents of radioactive elements in silicate earth (ppm)

Materials	K	U	Th	References
Chondrites	880	0.011	0.043	Turekian (1964)
	850	0.005—0.06	0.039	Vinogradov (1961)
	—	0.010—0.020	—	Fisher (1972)
Carbonaceous chondrites	—	0.01	—	Krähenbühl et al. (1973)
	845	0.011	0.04	Wasserburg et al. (1964)
Achondrites	430	0.080—0.020	0.006—0.5	Wasserburg et al. (1964)
The earth	225	0.025—0.035	0.12	Wasserburg et al. (1964)
	130	0.012	0.035	Larimer (1971)
		0.03		Tolstikhin et al. (1977b)
Silicate portion of the earth	250	0.026	—	Jacobsen and Wasserburg (1979)
	200	0.020	0.08	O'Nions et al. (1979)
Samples of rocks of probable mantle origin	54	0.018	0.055	Wakita et al. (1967)
The earth	110—800	0.01—0.03	0.03—0.15	this work

of the degassing coefficients and the share of each volatile component lost in the early outgassing event as well as their initial amount in the solid earth do not call for an additional explanation.

Problems of measuring the thermopause temperature, T_a, are discussed by Jacchia (1977); the presented limits in Table 11.3 appear to be the most realistic ones.

The sets of parameters for which the calculated results agree with the observations are called solutions. The existence of solutions may be taken as proof of the potency of the proposed model. They can also be utilized in refining other parameters.

11.3. A model of the earth degassing and differentiation; evaluation of solutions

11.3.1. Degassing and differentiation coefficients and the contents of radioactive elements

The degassing coefficients of the crust, Λ_{CA} (Ar) and Λ_{CA} (He), have little influence on other parameters; they may be determined independently and for this reason we believe them to be constant. The best way to define Λ_{CA} (Ar) was proposed by Hamano and Ozima (1978). They compared results of the Rb/Sr and the K/Ar dating (Fig. 11.1), which had been carried out by Hurley and Rand (1969); they have found the parameter of the best fit

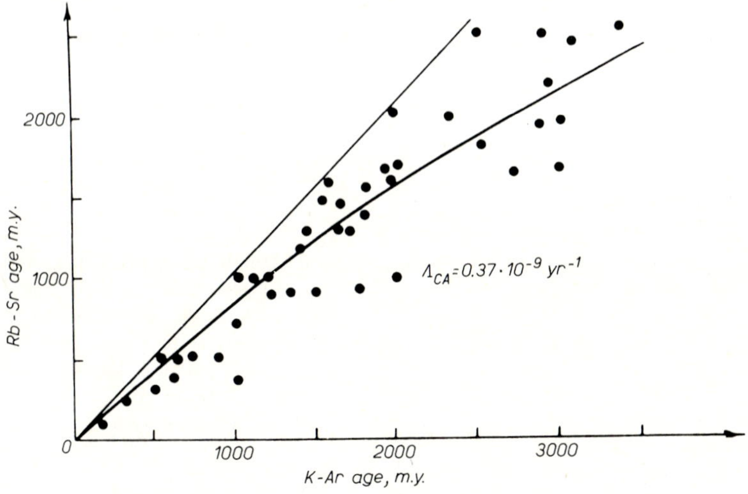

Fig. 11.1. Relationship between Rb-Sr whole rock and K-Ar mineral ages (Hurley and Rand, 1969) used by Hamano and Ozima (1978) to estimate the degassing constant Λ_{CA}.

234

curve, Λ_{CA} (Ar) = $0.37 \cdot 10^{-9}$ yr^{-1}. Similar estimations enable one to define Λ_{CA} (He) as equal to $\sim 10^{-8}$ yr^{-1}.

The solution of the model is illustrated in Fig. 11.2, where a U-He system is shown as an example. The content of radiogenic helium in the mantle as a function of the degassing coefficient Λ_{MA}, the differentiation coefficient Λ_{MC} and the concentration of uranium are shown in areas *1* and *2* on Fig. 11.2. Substituting the bound contents of U_c in eq. 11.15 (which are natu-

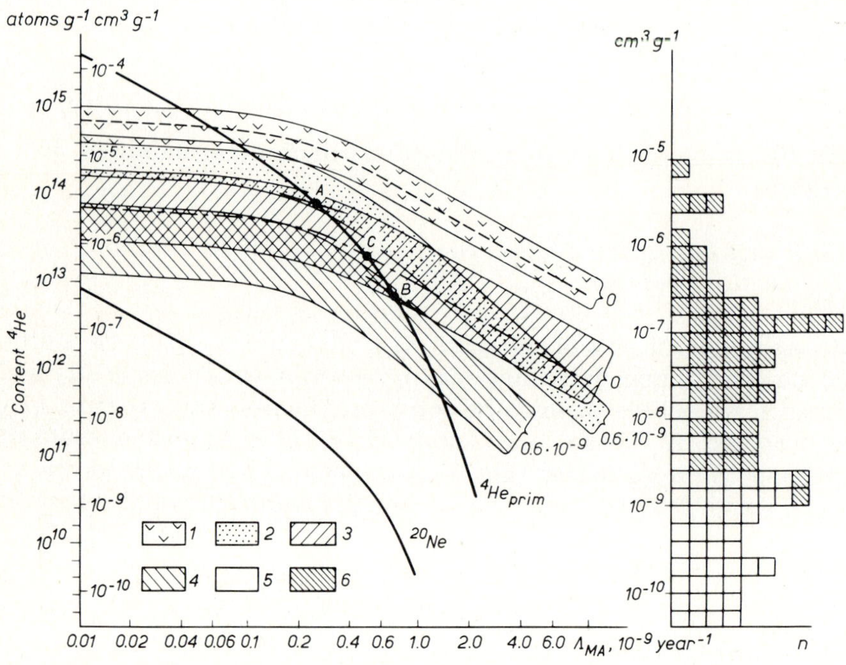

Fig. 11.2. Present-day contents of He$_{prim}$, He$_{rad}$ and Ne$_{prim}$ in the earth's mantle depending on the degassing constant Λ_{MA} (Tolstikhin, 1975b; Tolstikhin et al., 1975).
1—4 = areas of helium concentration: *1* = radiogenic helium, if Λ_{MC} is equal to 0; *2* = the same, if Λ_{MC} amounts to $0.6 \cdot 10^{-9}$ yr^{-1} (upper and lower limits of the areas correspond to uranium content of $3 \cdot 10^{-8}$ and $1 \cdot 10^{-8}$ g g^{-1}, respectively); *3* = primordial helium, if Λ_{MC} is equal to 0; *4* = the same, if Λ_{MC} amounts to $0.6 \cdot 10^{-9}$ yr^{-1}; the areas *3* and *4* correspond to observation 4 (Table 11.2). The dashed and dashed—dot lines represent concentrations of radiogenic and primordial ^{4}He when the uranium content is $2.25 \cdot 10^{-8}$ g g^{-1}, $\Lambda_{MC} = 0.12 \cdot 10^{-9}$ yr^{-1} and $(^3He/^4He)_M = 3 \cdot 10^{-5}$. Line ^{20}Ne indicates such ^{20}Ne content in the earth's mantle that observation 16 (Table 11.2) is satisfied for every value of Λ_{MA} — that is, the calculated neon content in the atmosphere equals its observed content. Line ^{4}He$_{prim}$ was drawn according to the relationship ^{20}He$_{prim}$ = 500 ^{20}Ne. The segments of lines intersecting line ^{4}He$_{prim}$ at points *A* and *C* are lines with maximum possible ranges of parameters U and Λ_{MC}; therefore, points *A*, *B* and *C* correspond to the maximum, optimum and minimum concentrations of ^{4}He$_{prim}$, respectively, satisfying observations 2, 4 and 16 (Table 11.2) at the same time. On the right a histogram of the radiogenic (6) and primordial (5) helium contents in samples of possible mantle origin is shown; *n* = number of cases.

rally consistent with observation 11, Table 11.2), we determine the two bound curves of the $^4He_{rad}$ content in the mantle after which we turn them into the boundaries of the ^4He primordial content. The latter can be calculated from the radiogenic content on the grounds of observations 1, 4 and 7 (Table 11.2):

$$(^3He/^4He)_{prim} = 2.4 \cdot 10^{-4}$$
$$(^3He/^4He)_{rad} = 2 \cdot 10^{-8} \tag{11.16}$$
$$(^3He/^4He)_M = (^3He_{prim} + {}^3He_{rad})/(^4He_{prim} + {}^4He_{rad}) = (2-4) \cdot 10^{-5}$$

From eq. 11.16 we get:

$$^4He_{prim} = (0.09 - 0.2) \cdot {}^4He_{rad} \tag{11.17}$$

Using formula 11.17 we can re-calculate the content of $^4He_{rad}$ (for example, areas *1* and *2*, Fig. 11.2) into the content of $^4He_{prim}$ (areas *3* and *4*, respectively). The segments of the boundaries of the $^4He_{prim}$ concentration in the mantle are shown in Fig. 11.2, near points *A* and *C*. Note that points lying between these segments yield values of $^4He_{prim}$ and Λ_{MA} which satisfy observations 1, 4, 7, and 11 (Table 11.2).

It is important that the content of primordial helium in the mantle can be calculated in a different, independent way from the content of neon in the earth's atmosphere. The relationship between the contents of primordial ^{20}Ne in the mantle and in the atmosphere is described by the equation:

$$^{20}Ne_M = {}^{20}Ne_0 \exp(-\Lambda_{MA} t) \tag{11.18}$$

Formula 11.18 is a solution of the equation analogous to eq. 11.6. From assumption 2 (p. 229) we know that:

$$^{20}Ne_0 = {}^{20}Ne_M + {}^{20}Ne_A$$

and substituting this sum into eq. 11.18 we obtain:

$$^{20}Ne_M = {}^{20}Ne_A/[\exp(\Lambda_{MA} t) - 1] \tag{11.19}$$

where $^{20}Ne_A = 1.49 \cdot 10^{-8}$ cm^3 g^{-1} and $t = 4.5$ b.y.

Eq. 11.19 gives concentrations of $^{20}Ne_M$ when observation 16 (Table 11.2) holds true for every value of degassing constant Λ_{MA}. These "concordant" concentrations are shown on Fig. 11.2 by curve ^{20}Ne. An analogous curve is drawn for primordial helium in such a way that the equality $^4He_{prim.M} = 500\ {}^{20}Ne_M$ is fulfilled (observation 2, Table 11.2); see also assumption 3 (p. 229). This curve ($^4He_{prim}$, Fig. 11.2) is also "concordant"; this implies that, if primordial helium had not escaped from the atmosphere, a relationship similar to eq. 11.19 would exist between its content in the atmosphere and in the mantle on the one hand and the degassing coefficient Λ_{MA} on the other.

Segment AC of the $^4\mathrm{He}_{prim}$ curve lying between the boundaries found earlier (Fig. 11.2) enables us to obtain the range of the U_0, Th_0, Λ_{MA} and Λ_{MC} values which may be considered as a solution of the model. The additional bound of the solution is determined by the thermal escape of helium isotopes from the upper atmosphere (Fig. 11.3). In our further discussion we will present solutions in a plot with coordinates Λ_{MA} vs Λ_{MC} while the contents of radioactive elements are shown as parameters. A typical area of solutions is shown in Fig. 11.3; the limits of the area are determined by the most effective restrictions of the model, namely the isotope composition of helium in the mantle, the concentrations of radioactive elements in the crust, the temperatures of the thermopause and in some cases (at low values of the degassing coefficient Λ_{MA}) the amount of primordial $^{36}\mathrm{Ar}$ in the atmosphere.

The areas are always located below the concordant line (Fig. 11.4) because the differentiation coefficients are lower that those of the degassing ones, $\Lambda_{MC} < \Lambda_{MA}$. This result means that volatile elements are more readily incorporated in magma than lithophile elements, such as K, Rb, U; this is in good agreement with the results of the U/He and K/Ar dating as well as

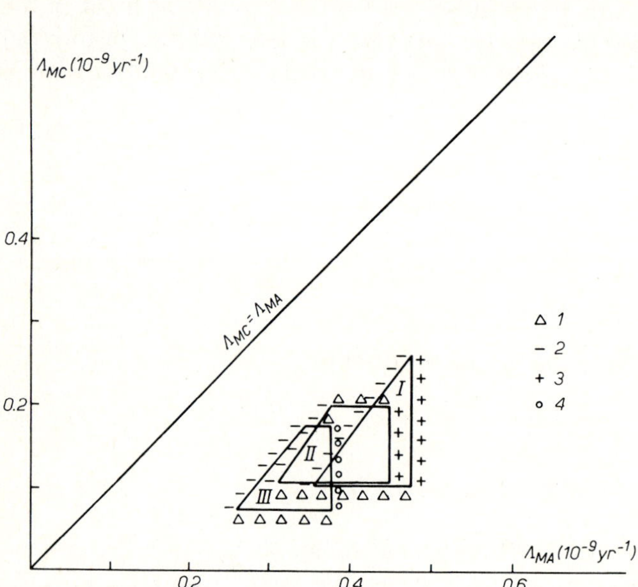

Fig. 11.3. Boundary conditions for the solutions of the model.
1 = minimum content of U in the crust; *2* = isotope composition of helium in the mantle; *3* = equality of the thermopause temperatures for $^3\mathrm{He}$ and $^4\mathrm{He}$ outflux; *4* = abundance of $^{36}\mathrm{Ar}$ in the atmosphere.

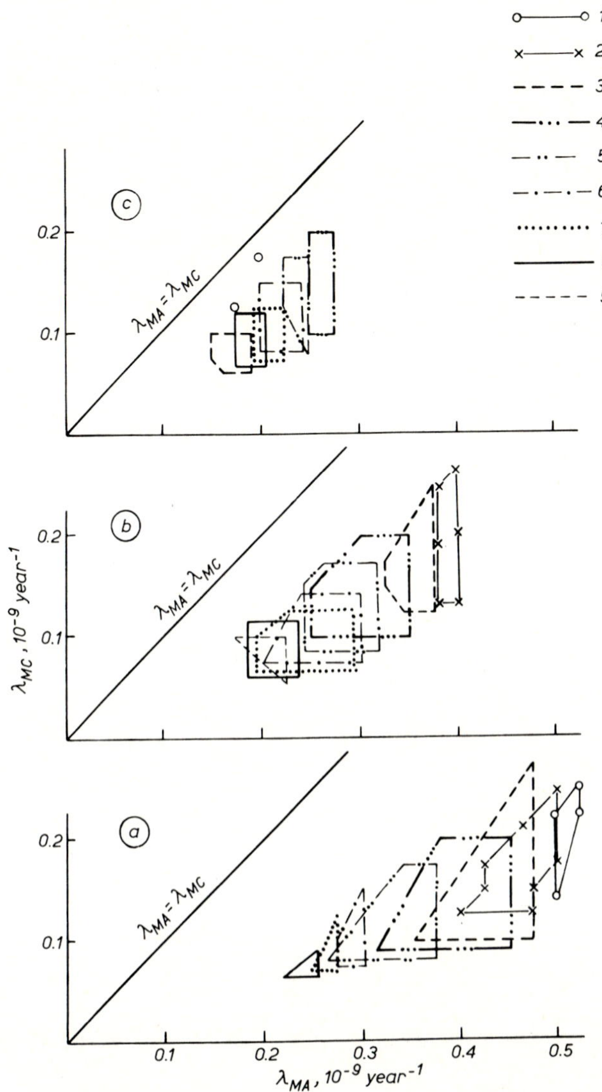

Fig. 11.4. Areas of solutions of the model presented in coordinates Λ_{MC} vs. Λ_{MA}. a. Early catastrophic degassing is absent. b. 25% of volatiles had been released during early degassing. c. 40% of volatiles had been released during early degassing. For the case when 50% of volatiles had been lost, the solutions are shown as dots.

$1—9$ = initial (at t = 0) uranium concentration (ppm) in the silicate earth, the K/U ratio being assumed as equal to 10^4 (by weight): 1 = 0.027, 2 = 0.028, 3 = 0.029, 4 = 0.033, 5 = 0.035, 6 = 0.038, 7 = 0.041, 8 = 0.043, 9 = 0.046.

with laboratory experiments. In the light of these data the coherent models of earth degassing do not seem convincing (Schwartzman, 1973a, b). According to the values of Λ_{MC} and Λ_{MA} obtained in our model (Table 11.5), the proportion of radioactive elements removed from the mantle into the crust ranges from 30 to 70%, whereas that of primordial noble gases released by the solid earth amounts to 50—90%.

On the other hand, the difference between the degassing and the differentiation coefficients is not very large. If we arrange the elements discussed here in an array according to their volatility, with uranium and helium as the end-members, other gases will crowd around helium rather than uranium, and their degassing coefficients will be quite similar to the helium ones. In other words, the results of model computing show a similar residence time of most volatile elements, especially noble gases, in the mantle, which is in striking contradiction with the data reported by Anufriev (1980) who suggested an improbable degassing coefficient of Xe = $0.00076 \cdot 10^{-9}$ yr^{-1}.

The maximum value of the mantle degassing coefficient, $\Lambda_{MA} = 0.5 \cdot 10^{-9}$ yr^{-1}, listed in Table 11.5 for the case excluding the possibility of early catastrophic degassing ($L = 0$), is somewhat lower than that obtained previously, $\Lambda_{MA} = 0.75 \cdot 10^{-9}$ yr^{-1} (Tolstikhin et al., 1975). This difference is due to constraints defined by the model of helium escape from the terrestrial atmosphere (see sections 10.3 and 10.4). Note that both models (those of degassing and dissipation) are in agreement.

The concentrations of the radioactive elements, U and K, derived from the model are consistent with available independent estimates (compare Tables 11.4 and 11.5). No solutions have been obtained with concentrations of uranium lying beyond the range presented in Fig. 11.4 and in Table 11.5.

TABLE 11.5

Solution of the model (compare with Table 11.3)

No.	Parameter	Dimensions	Values		
			$L = 0$	$L = 0.25$	$L = 0.40$
1	U_0	ppm	0.025—0.045	0.03—0.05	0.035—0.05
2	Th_0	ppm	$3\,U_0$	$3\,U_0$	$3\,U_0$
3	K	ppm	160—240	170—250	170—260
4	$^{36}Ar_0$	10^{-8} cm^3 g^{-1}	$1.9 \cdot {}^{20}Ne$	$1.9 \cdot {}^{20}Ne$	$1.9 \cdot {}^{20}Ne$
5	$^{20}Ne_0$	10^{-8} cm^3 g^{-1}	1.6—2.8	1.7—2.9	2.0—3.1
6	P_3	atoms g^{-1} yr^{-1}	0.5	0.5	0.5
7	Λ_{MA}	10^{-9} yr^{-1}	0.25—0.50	0.20—0.40	0.15—0.27
	Λ_{MC}	10^{-9} yr^{-1}	0.06—0.25	0.07—0.25	0.07—0.20
	Λ_{CA} (Ar)	10^{-9} yr^{-1}	0.4	0.4	0.4
	Λ_{CA} (He)	10^{-9} yr^{-1}	10	10	10
8	T_a	°K	1230—1370	1230—1360	1230—1360

The model concentrations of K in the earth are significantly lower than the chondritic concentrations, which implies either an achondritic abundance of this element in the earth protomatter or the extraction of K from the fractionating silicate earth soon after its accretion. It is interesting to note that the proportion between K and $^{36}Ar_0$ is practically constant, independent of the fractionation coefficients Λ_{MA}, Λ_{MC}, and Λ_{CA}, and equal to (5.3 ± 1.0) 10^3 g cm^{-3} (Fig. 11.5). This fact makes it possible to find out the potassium content in the terrestrial planets if the abundance of argon isotopes in the atmosphere is known. Using the data reported by Pollack and Black (1979) as well as those presented in Table 11.5, one can calculate 6—11 ppm and 35—60 ppm K for Mars and Venus, respectively. Abundance of this element in the planets may be taken as evidence for their degassing history: the tectonic activity of Mars might have ceased long ago, whereas the tectonic activity and juvenile outgassing of the earth are still going on.

Fig. 11.4 and Table 11.5 show how the solutions vary depending on the proportion of primordial noble gases released in the early catastrophic degassing event. From simple qualitative considerations it is clear that, if some amount of the gases had escaped during this event, then the degassing coefficients, Λ_{MA}, should decrease in order to preserve a greater amount of gases in the mantle, whereas the differentiation coefficients, Λ_{MC}, should increase in order to remove a greater amount of radioactive elements into the crust. This tendency illustrated in Fig. 11.4 is in agreement with the following considerations: while the proportion of gases initially released increases, the solution areas shift to the line determined by equality $\Lambda_{MA} = \Lambda_{MC}$. It is important that no solution has been obtained for the case of $L = 0.5$ or more, and the higher the values of L the smaller is the area of solutions. Hence, according to the model described, the ratio between the initially and the continuously released gases is not very high.

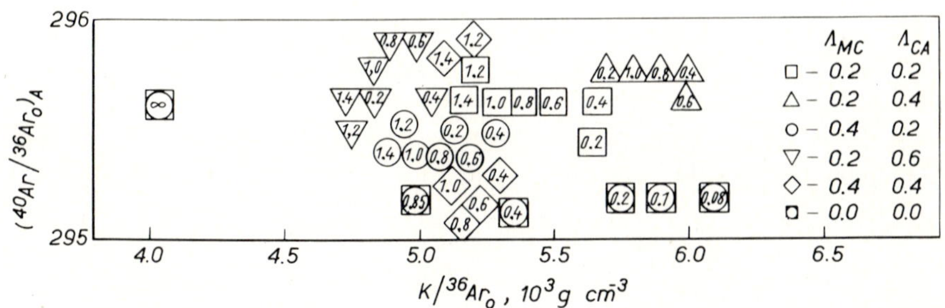

Fig. 11.5. The atmospheric ratio $Ar^{40}/Ar^{36} = 295.5$ is maintained at the relatively constant ratio of $K/^{36}Ar_0 = (5.3 \pm 1) \cdot 10^3$ g cm^{-3} (Tolstikhin et al., 1975). The effect of the degassing parameters is not great. The values of Λ_{MA} are given within the symbols; all values of Λ should be multiplied by 10^{-9} yr^{-1}; $L = 0$.

11.3.2. Radiogenic ^{4}He and ^{40}Ar in rocks and gases of the earth's crust

The K/U ratio in the total earth assumed in this model is equal to 10^4, which is close to the value of the K/U ratio in the crust; it follows that the values of the differentiation coefficient Λ_{MC} for these elements are also equal.

To compare the calculated results with observations 8 and 9 (Table 11.2), it is necessary to estimate the concentrations of radiogenic helium and argon in an average crustal rock and the rate of their escape into the atmosphere. The concentration of ^{40}Ar in the crust, calculated for the parameters $\Lambda_{MC} = 0.15 \cdot 10^{-9}$ yr^{-1}, $\Lambda_{CA} = 0.4 \cdot 10^{-9}$ yr^{-1} and K = 200 ppm, is $2.2 \cdot 10^{-4}$ cm^3 g^{-1}, the mass of the crust being taken as 0.56% of the mass of the mantle. The flow of ^{40}Ar from the crust is $0.34 \cdot 10^6$ atoms cm^{-2} s^{-1}. The losses of radiogenic helium and argon in rocks and minerals of the crust follow a pattern which differs greatly from that typical of the mantle: in the crust the gases migrate through cold rocks whereas in the mantle they are removed by hot silicate matter. This important difference is reflected in the values of the degassing coefficients. Despite the convincing evidence that the mantle—atmosphere degassing coefficients, Λ_{MA}, for all noble gases are the same (see section 11.3), the crust—atmosphere coefficients, Λ_{CA}, depend on the size of the radiogenic atoms, their mass and the initial energy of radioactive decay (fission). As all these parameters differ considerably with respect to radiogenic helium and argon, the value of Λ_{CA} (He) is much higher than Λ_{CA} (Ar). The latter was determined above as $0.4 \cdot 10^{-9}$ yr^{-1}. A rough comparison of the diffusion parameters of radiogenic ^{4}He and ^{40}Ar in minerals indicates that the degassing coefficients of these gases differ by a factor of 10—50. Thus, we may assume that the value of Λ_{CA} for radiogenic helium is about $10 \cdot 10^{-9}$ yr^{-1}. The arbitrary choice of this value is justified by the fact that when Λ_{CA} (He) $> \lambda_{238}$ the value of Λ_{CA} (He) only slightly influences the rate of ^{4}He flow. Using the selected parameters, we find that the ^{4}He content in the crust is $3 \cdot 10^{-5}$ cm^3 g^{-1} and the rate of its escape into the atmosphere is $1.1 \cdot 10^6$ atoms cm^{-2} s^{-1}. Now the calculated values of the ^{4}He/^{40}Ar ratio in rocks and gases of the earth's crust can be derived as follows:

$$(^4\text{He}/^{40}\text{Ar})_{\text{rock}} \approx 0.14$$
$$(^4\text{He}/^{40}\text{Ar})_{\text{gas}} \approx 3.2$$

which is in satisfactory agreement with observation 8 (see Table 11.2). It should be noted that observations 6 and 7 in our model are automatically satisfied. Thus, the calculated data agree with all observations 6—11 for the earth's crust.

11.3.3. Calculated fluxes of helium isotopes from the earth into the atmosphere

For optimal parameters of U = 0.016—0.018 ppm, K/U = 10^4, Λ_{MA} = 0.4 · 10^{-9} yr^{-1} and Λ_{MC} = 0.15 · 10^{-9} yr^{-1}, the calculated amounts of radiogenic and primordial isotopes of ^{4}He are 1.06 · 10^{42} and 0.175 · 10^{42} atoms, respectively; this sets the ^{4}He flux from the mantle into the atmosphere at 3.0 · 10^6 atoms cm^{-2} s^{-1}. Then the total ^{4}He flux from the solid earth (crust plus mantle) is 4.1 · 10^6 atoms cm^{-2} s^{-1}. The product of the ^{4}He amount in the mantle, the ^{3}He/^{4}He ratio in the mantle helium and the mantle—atmosphere degassing coefficient Λ_{MA} set the ^{3}He flux value from the mantle at 90 atoms cm^{-2} s^{-1}. Estimates of both ^{4}He and ^{3}He fluxes presented here (for comparison see Tables 10.1 and 10.2) are similar to those described by Tolstikhin et al., (1975).

Attention should be drawn to the fact that assuming a high ^{3}He/^{4}He ratio of 3 · 10^{-5} for the total mantle leads to high values of fluxes of both helium isotopes; the latter may be considered as the upper bound of terrestrial fluxes. A solution of the model has also been achieved with considerably lower values of the ^{3}He flux, about 40 atoms cm^{-2} s^{-1}.

We believe that further accumulation of experimental data in the field of aeronomy and isotope geochemistry as well as more powerful models of the earth degassing and differentiation will enable us to obtain more precise values for the fluxes of helium isotopes into the atmosphere.

CONCLUSIONS

Now we think it appropriate to make an attempt to summarize present-day knowledge of the origin and distribution of helium isotopes in nature. First we will mention major natural processes producing helium isotopes; then we will discuss some scientific and applied problems that can be resolved thanks to the progress in helium isotope geochemistry.

Nuclear fusion is the main process which produces both stable helium isotopes in nature; the constancy of the helium content in matter of various galaxies gives evidence that a large share of helium was produced at the enigmatic moment when the universe started to exist. Helium is widely distributed all though the solar system. The abundance of helium in the sun comes second after hydrogen; moreover, helium appears to be one of the major components of the giant planets. The abundance of primordial helium in stars is so high that nuclear reactions which occur in their interiors have little effect on the isotope ratio of $^3He/^4He \sim 10^{-4}$. Primordial helium is observed in meteorites, lunar soil, and, as has been recently established, it is preserved in deep earth's interior.

Another process which yields helium rich in the heavy isotope ^4He is the radioactive α-decay of heavy elements. This process occurs continuously and everywhere, including planets, asteroids and meteorites. Reactions stimulated by nuclear decay produce the light isotope ^3He. As a rule the proportion of radiogenic ^3He is very low as compared to its heavy counterpart; the $^3He/^4He$ ratio in radiogenic helium of typical materials of the solar system is about 10^{-8}. In the earth's crust and mantle radiogenic helium is continuously generated, and radiogenic ^4He is the most abundant isotope of helium in the solid earth and in the atmosphere.

The third process which produces helium mostly enriched in the light isotope is the interaction between high-energy galaxy rays and matter. An essential feature of the process is a decrease of its intensity with depth: even the flux of the most penetrable muon component is decreased by a factor of 10 when the depth is increased by 10^4 g cm^{-2} only. Spallogenic reactions yield helium characterized by the highest ratio of $^3He/^4He$, $\sim 10^{-1}$. The content of such helium enables us to estimate the so-called radiation or exposure age — that is, the time during which matter was exposed to cosmic radiation. The proportion of spallogenic helium in terrestrial helium is considerably lower than that of the primordial and the radiogenic types of helium. Nevertheless, the contribution of spallogenic helium in the ^{3}He amount in the earth's atmosphere is not negligible, and it should be taken into account when estimations of the ^{3}He influx into the atmosphere are made.

Now let us turn to the distribution of helium isotopes in the various shells of the earth.

Investigations of the last decades showed that the mantle preserves some proportions of primordial helium trapped by the earth as a microcomponent of proto-planetary matter. Mantle helium is characterized by a ^{3}He/^{4}He ratio of about $3 \cdot 10^{-5}$, and it seems to be the major reservoir of the light helium isotope in the earth. The outgassing of the mantle provides the major flux of the isotope in the atmosphere. The ^{3}He/^{4}He ratio in the mantle appears to be more or less homogeneous whereas it is variable in the crust; this may be considered as a token of the homogeneous distribution of elements in the mantle, even those which differ as much in their geochemical behaviour as helium and uranium.

Thus, mantle matter is always marked by helium enriched in the light isotope, and this marker enables us to distinguish mantle matter from that of the ancient crust. Because of the unique features of the helium isotope geochemistry, the ^{3}He/^{4}He ratio is, and will always be, the only unequivocal marker of juvenile volatile components.

When transferred into the crust, mantle matter releases volatile components which fill up the atmosphere. This degassing process on the one hand and enrichment of crustal materials in radioactive elements on the other lead to a sharp decrease in the ^{3}He/^{4}He ratio because of an accumulation of the radiogenic component during the subsequent crustal history of matter.

Rocks, gases and waters of the earth's crust contain mainly radiogenic helium characterized by a ^{3}He/^{4}He ratio of about $2 \cdot 10^{-8}$. Fluxes of radiogenic ^{4}He from the crust and the mantle into the atmosphere are similar.

Though the ^{3}He/^{4}He ratios in tectonically stable regions are more or less constant and close to the radiogenic value, considerable deviations provided by high lithium (beryllium) or uranium (thorium) contents in rocks are observed, the ^{3}He/^{4}He value amounting to 10^{-5} or decreasing to 10^{-10}, respectively.

In some peculiar regions of the earth the ^{3}He/^{4}He ratios can increase due to solar and spallogenic helium which appears on the earth's surface as a component of cosmic dust. In the dust of deep mid-ocean valleys as well as in that of polar and alpine glaciers, — that is, within the limits of areas which are not subject to contamination by terrigenic soils — the ^{3}He/^{4}He ratios reach $\sim 10^{-5}$. On the whole, the contribution of such helium appears to be negligible.

In young subsurface waters and in waters of lakes and ponds containing mostly dissolved air helium, the ^{3}He/^{4}He ratios can be several times higher than the atmospheric value due to an addition of radiogenic helium (^{3}He) produced by the β-decay of ^{3}H. The total terrestrial proportion between tritigenic and primordial ^{3}He is very low.

Helium of seawater consists of three components, dissolved air helium being the major one. Juvenile helium injected from the oceanic crust and mantle increases the ^{3}He/^{4}He ratio in seawater from $1.4 \cdot 10^{-6}$ to $1.8 \cdot 10^{-6}$, and in rare cases the ratio reaches 10^{-5}. An excess of ^{3}He and ^{4}He contents in water enables one to estimate the isotopes fluxes from the oceanic crust into the atmosphere. A certain excess of ^{3}He can also be provided by tritium decay.

Helium of the atmosphere, which is the major gas reservoir of the planet, is a mixture of light and heavy isotopes of various origins, the ^{3}He/^{4}He ratio amounting to $(1.39 \pm 0.01) \cdot 10^{-6}$. The heavy isotope ^{4}He is released from the crust and the mantle, its total flux constituting $(2-4) \cdot 10^6$ atoms cm^{-2} s^{-1}. The light isotope ^{3}He is mainly outgassed from the mantle, which is the major reservoir of this isotope on earth. According to model calculations the ^{3}He flux from the mantle can amount to $50-100$ atoms cm^{-2} s^{-1}. A considerable amount of ^{3}He appears to be introduced in the atmosphere as a component of solar wind. Both isotopes escape from the atmosphere into space. The most likely process of the dissipation is the thermal one; its nature explains why the outfluxes depend so strongly on the thermopause temperature and the masses of the escaping atoms.

Ions and atoms of helium occur in cosmic space in the region of terrestrial planets; the typical isotopic ratio of this helium is $\approx 3 \cdot 10^{-4}$; its source is the sun.

The distribution of helium isotopes in terrestrial reservoirs may help to resolve a number of fundamental problems in the earth sciences of which we will mention the five most significant ones.

(1) *The scenario of earth's accretion.* The fact of the preservation of primordial helium in the solid earth should be taken as a pre-condition for models of the earth's accretion and its early history. The models of high-temperature accretion and very intensive differentiation of the earth during its early stage of evolution must account for the preservation of the most volatile components within the planet.

(2) *The origin and history of terrestrial fluids.* As has been stated before, the ^{3}He/^{4}He ratio in the mantle is a unique tracer of juvenile volatiles. Thanks to this it is now possible to make certain whether there is a contribution of mantle volatiles in observed fluids, the crust or the mantle being their major source. Moreover, the ratio may quite regularly serve as a basis for a qualitative estimation of this contribution. Careful investigations will enable us to estimate the contribution of the crustal, mantle and atmospheric helium in the helium of a sample. In addition, there is a fair chance to estimate the contribution of chemically active gases from these three reservoirs.

(3) *The origin and evolution of terrestrial heat flow.* The direct correlation between the $^3He/^4He$ ratios in underground fluids and the heat flow enables us to distinguish the conductive and the convective components of the flow. The correlation gives grounds to conclude that the process of a convective transfer of heat and matter is the most important one. Taken together the data on heat flow and the $^3He/^4He$ ratio allows the following of deep fault zones through which the upwelling of juvenile volatiles and heat occurs or has occurred during the last half billion years.

(4) *The aeronomic problem of helium.* Recent degassing models have led to high values of the ^3He flux from the earth into the atmosphere. This result enables the making of a step forward in the study of helium escape from the terrestrial atmosphere.

It has been found that the values of the thermopause temperature computed for the influx of both helium isotopes, ^3He and ^4He, are very close to those inferred from contemporary models of the upper atmosphere.

(5) *Space components on the earth's surface.* High ratios of $^3He/^4He$ in space matter ensure their identification as well as a more or less precise estimation of their precipitation.

Isotopic investigations of terrestrial helium may be also helpful in resolving some applied problems. It makes possible direct prospecting of deposits of radioactive and some other elements, whose high concentrations lead to helium isotopic compositions unusual for a given region. When the $^3He/^4He$ ratio in helium released from a deposit substantially deviates from that typical of the country rocks, it is a sign of possible enrichment of the deposit by radioactive (low ratios) or lithium (high ratios) elements.

Measurements of helium isotopic compositions throw some light on the origin of volatile components as well as their deposits. Available data show that more than 95% of the oil and gas resources originated in crustal sedimentary rocks, and there is no evidence that any of their components were derived from the mantle. The $^3He/^4He$ ratio in helium of volatile components injected through the earth's surface is in some cases the only tool that helps to identify the components and decide whether they are derived from deep or surface reservoirs; such an approach is especially helpful when shelves, as well as swamp and lake areas, are explored. The isotope analyses of noble gases from microinclusions in ore minerals and rocks similarly enable us to identify mantle, crustal and atmospheric components and widens our knowledge of the sources of ore deposits and some peculiarities of their origin.

Some applied problems of physical limnology and hydrogeology can be solved with the help of the ^3He—^4He method of determinating helium residence time in waters.

The distribution of helium isotopes in seawater may be considered as a key for mapping ocean streams and estimating their velocity in regions where the water is marked by helium injected through the ocean floor and characterized by a specific proportion of ^{3}He and ^{4}He.

Thirty five years ago, when isotope geochemistry of helium was in its cradle yet, Aldrich and Nier (1948) called it "a new and fascinating field of investigation"; now the reader can see for himself how accurate this statement has proved to be.

REFERENCES

Afrosimov, V.V., Gordeev, Yu.S., Lavrov, V.M. and Nikulin, V.K., 1971. Inelastic transitions and scattering in the collisions of Na and K ions with atoms of noble gases. In: *VII, ICPEAC, Amsterdam*, p. 143 (abstract).

Afrosimov, V.V., Gordeev, Yu.S. and Lavrov, V.M., 1975. The differential cross-sections of scattering in the collisions of He^+—Ar^+. *Zh. Eksp. Teor. Fiz.*, 68: 1715—1723 (in Russian).

Akimov, A.P., Berzina, I.G., Gurvich, M.Yu. and Lutz, B.G., 1968. Uranium content in eclogite inclusion from kimberlite pipes. *Dokl. Akad. Nauk SSSR*, 181: 1245—1248 (in Russian).

Alaerts, L., Lewis, R.S., Matsuda, J. and Anders, E., 1980. Isotopic anomalies of noble gases in meteorites and their origin. 6. Presolar components in the Murchison C2 chondrite. *Geochim. Cosmochim. Acta*, 44: 189—210.

Aldrich, L.T. and Nier, A.O., 1946. The abundance of ^3He in atmosphere and well helium. *Phys. Rev.*, 70: 983.

Aldrich, L.T. and Nier, A.O., 1948. The occurrence of ^3He in natural sources of helium. *Phys. Rev.*, 74: 1590—1593.

Alekseevsky, N.E., Dubrovin, N.B. and Koretsky, G.A., 1966. A small-size mass spectrometer with high resolution and heterogeneous magnetic field for analysis of light gases. *Dokl. Akad. Nauk SSSR*, 166: 1088—1090 (in Russian).

Alekseichuk, B.K., Anufriev, G.S., Afonina, G.I., Bronshtein, A.M., Efis, Yu.M., Zagulin, V.A., Kleshkov, E.M., Mamyrin, B.A., Nenarokomova, V.T., Pavlenko, V.A. and Rafalson, A.E., 1979. Mass spectrometer MI 9302 for analyses of small amounts of inert and chemically active gases. *Prib. Tekh. Eksp.*, 4: 206—210 (in Russian).

Alexander, E.C. and Schwartzman, D.W., 1976. Argon isotopic evolution of upper mantle. *Nature*, 259: 104—108.

Alexandrov, M.L., Gall', L.N. and Pliss, N.S., 1974. Study of ion scattering in mass spectrometers by a static simulation technique. *Zh. Tekh. Fiz.*, 44: 1302—1305 (in Russian).

Alimova, I.A., Boltenkov, B.S., Gartmanov, V.N., Mamyrin, B.A. and Shustrov, B.N., 1966. Determination of small amounts of helium in solids. *Geokhimiya*, 9: 1044—1048 (in Russian).

Alimova, I.A., Boltenkov, B.S., Gartmanov, V.N., Kocharov, G.E., Mamyrin, B.A. and Naidenov, V.O., 1970. *An Equipment for Determination of Small Amounts of ^3H, ^3He and ^{37}Ar and Its Possible Usage for Astrophysical Investigations*. A.F. Ioffe Fiz.-Tekh. Inst. Akad. Nauk SSSR, Leningrad, 12 pp. (in Russian; preprint).

Alpert, Ya.L., 1972. *Propagation of Electromagnetic Waves in the Ionosphere*. Nauka, Moscow, 563 pp. (in Russian).

Alvarez, L.W. and Cornog, R., 1939a. ^3He in helium. *Phys. Rev.*, 56: 379.

Alvarez, L.W. and Cornog, R., 1939b. Helium and hydrogen of mass 3. *Phys. Rev.*, 56: 613.

250

Anders, E., 1964. Origin, age and composition of meteorites. *Space Sci. Rev.*, 3: 583—714.

Anders, E., 1971. Meteorites and the early solar system. *Annu. Rev. Astron. Astrophys.*, 9: 1—34.

Anders, E., 1981. Noble gases in meteorites: evidence for presolar matter and superheavy elements. *Proc. R. Soc. Lond., Ser. A.*, 374: 207—238.

Anders, E., Heymann, D., and Mazor, E., 1970. Isotope composition of primordial helium in carbonaceous chondrites. *Geochim. Cosmochim. Acta*, 34: 127—132.

Anonymous, 1974. *Tectonic Basement Map of the USSR Area, Scale 1 : 5,000,000*. Akad. Nauk SSSR, MG SSSR, Moscow.

Anonymous, 1976. *Mass Spectrometer MI 9302: Description of the Instrument*. Akad. Nauk SSSR, Leningrad, 4 pp. (in Russian).

Anufriev, G.S., 1980. On the regularities of the earth outgassing. *Dokl. Akad. Nauk SSSR*, 250: 672—674 (in Russian).

Anufriev, G.S., Kamensky, I.L. and Pavlov, V.P., 1976. Anomalous isotope composition of neon in hot springs of modern volcanic zones. *Dokl. Akad. Nauk SSSR*, 231: 1454—1457 (in Russian).

Anufriev, G.S., Gartmanov, V.N., Mamyrin, B.A. and Pavlov, B.P., 1977. A vacuum extraction system with low background. *Prib. Tekh. Eksp.*, 1: 248—250 (in Russian).

Anufriev, G.S., Krylov, A.Ya. and Boltenkov, B.S., 1978. Relict noble gases in rocks of the ocean floor. *Dokl. Akad. Nauk. SSSR*, 241: 1424—1427 (in Russian).

Anufriev, G.S., Afonina, G.I., Mamyrin, B.A., Nenarokomova, V.T. and Rafalson, A.E., 1979. Operation of the mass spectrometer MI 9302 in a static pumping mode. *Prib. Tekh. Eksp.*, 4: 211—213 (in Russian).

Arrhenius, G. and Alfvén, H., 1971. Fractionation and condensation in space. *Earth Planet. Sci. Lett.*, 10: 253—267.

Ashkinadze, G.Sh., Gorokhovsky, B.M., Drubetskoy, E.R., Sprintson, V.D. and Tolstikhin, I.N., 1976. An equipment for determination of small amounts of argon by the isotope delution technique. In: Yu.A. Shukolyukov and I.M. Morozova (Editors), *Development of Nuclear Geochronology Methods*. Nauka, Leningrad, pp. 60—77 (in Russian).

Avetis'yanz, A.A., Makarenko, F.A. and Sergienko, S.I., 1975. Heat flow in the Transcaucasian region. *Dokl. Akad. Nauk SSSR*, 222: 665—668 (in Russian).

Axford, W.I., 1968. The polar wind and the terrestrial helium budget. *J. Geophys. Res.*, 73: 6855—6859.

Banks, P.M. and Holzer, T.E., 1969. High-latitude plasma transport: the polar wind. *J. Geophys. Res.*, 74: 6317—6332.

Banks, P.M. and Kockarts, G., 1973. *Aeronomy, A*. Academic Press, New York, N.Y., 430 pp.

Barer, R., 1957. *Diffusion in Solids*. Inostr. Lit., Moscow, 265 pp. (in Russian).

Barnes, R.O. and Bieri, R.H., 1976. Helium flux through marine sediments of the northeast Pacific Ocean. *Earth Planet. Sci. Lett.*, 28: 331—337.

Baskov, E.A., Vetshtein, V.E., Surikov, S.N., Tolstikhin, I.N., Malyuk, T.A. and Mishnina, T.A., 1973. Isotope composition of H, O, C, Ar and He in thermal waters and gases of Kuril—Kamchatka volcanic zone as a criterion of their origin. *Geokhimiya*, 2: 180—190 (in Russian).

Basu, A.R., Ray, S.L., Saha, A. K. and Sarkar, S.N., 1981. Eastern-Indian 3800 mln year-old crust and early mantle differentiation. *Science*, 212: 1502—1506.

Bates, D.R. and MacDowell, J., 1959. Atmospheric helium. *J. Atmos. Terr. Phys.*, 16: 393—394.

Bates, D.R. and Patterson, T.R.L., 1962. Helium ions in the upper atmosphere. *Planet. Space Sci.*, 9: 599—605.

Bennet, G.A. and Manuel, O.K., 1970. Xenon in natural gases. *Geochim. Cosmochim. Acta*, 34: 593—610.

Bernatowicz, T.J. and Podosek, F.A., 1978. Nuclear components in the atmosphere. In: E.C. Alexander and M. Ozima (Editors), *Terrestrial Rare Gases*. Advances in Earth and Planetary Sciences, 3. Cent. Acad. Publ., Tokyo, pp. 99—135.

Bieri, R.H. and Koide, M., 1972. Dissolved noble gases in the East Equatorial and Southeast Pacific. *J. Geophys. Res.*, 77: 1667—1676.

Bieri, R.H., Koide, M. and Goldberg, E.D., 1964. Noble gases in sea-water. *Science*, 146: 1035—1037

Bieri, R.H., Koide, M. and Goldberg, E.D., 1967. Geophysical implications of the excess helium found in Pacific waters. *J. Geophys. Res.*, 72: 2497—2511.

Bieri, R. H., Koide, M. and Goldberg, E. D., 1968. Noble gas contents of marine waters. *Earth Planet. Sci. Lett.*, 4: 329—341.

Bieri, R. H., Koide, M., Martell, E. A. and Scholz, T. G., 1971. Noble gases in the atmosphere between 43 and 63 kilometers. *J. Geophys. Res.*, 75: 6731—6735.

Birch, F., 1958. Differentiation in the mantle. *Bull. Geol. Soc. Am.*, 69: 483—485.

Bitterberg, W., Brutchhausen, K., Offerman, D. and Zahn, U., 1970. Lower thermosphere composition and density above Sardinia in October, 1967. *J. Geophys. Res.*, 75: 5528—5534.

Biutner, E.K., 1958. Gas dissipation from planetary atmospheres. I. The dissipation of isothermal ideal gas in a central field of gravitation. *Sov. Astron.*, 2: 528—537.

Biutner, E.K., 1959. Gas dissipation from planetary atmospheres. II. The total velocity of gas dissipation from a planetary atmosphere. The problem of terrestrial helium. *Sov. Astron.*, 3: 92—101.

Black, D.C., 1972. On the origin of trapped helium, neon and argon isotopic variations in meteorites. *Geochim. Cosmochim. Acta*, 36: 347—395.

Black, D.C. and Pepin, R.O., 1969. Trapped neon in meteorites. *Earth Planet. Sci. Lett.*, 6: 395—405.

Black, L.P., Gale, N.H., Moorbath, S., Pankhurst, R.J. and McGregor, V.R., 1971. Isotope dating of very early Precambrian amphibolite facies gneisses from the Godthaab district, West Greenland. *Earth Planet. Sci. Lett.*, 12: 245—259.

Blaut, E.M., 1965. *Dynamische Massenspektrometer*, F. Viveg, Braunschweig, 167 pp.

Bogard, D.D., Clarke, R.S., Keith, J.E. and Reynolds, M.A., 1971. Noble gases and radionuclides in Lost City and other recently fallen meteorites. *J. Geophys. Res.*, 76: 4076—4083.

Bogolyubov, A.N., Glebovskaya, V.S., Gorblyuk, V.G., Dobraykov, A.P., Matveeva, E.S., Puchkov, G.P., Tafeev, G.P. and Tuktarova, A.B., 1975. *Water—Helium Survey*. Nedra, Leningrad, 71 pp. (in Russian).

Boltenkov, B.S., 1973. *Measurements of Low Amounts of Helium and Their Application to Experimental Physics*. Dissertation, A.F. Ioffe Fiz.-Tekh. Inst., Akad. Nauk SSSR, Leningrad, 18 pp. (in Russian; abstract).

Boltenkov, B.S., Gartmanov, V.N., Il'yasov, E.I., Kocharov, G.E., Mamyrin, B.A., Surkov, Yu.A. and Khabarin, L.V., 1974. Investigation of the isotope composition of helium in a fine fraction of the lunar soil from "Lunar-16" mission. In: A.P. Vinogradov (Editor), *Lunar Soil From Mare Fertility*. Nauka, Moscow, pp. 365—369 (in Russian).

Bondarenko, V.N., 1970. *Statistic Solutions of Some Geological Problems*. Nedra, Moscow, 246 pp. (in Russian).

Brass, G., 1976. The variation of marine $^{87}Sr/^{86}Sr$ ratio during Phanerozoic time: interpretation using a flux model. *Geochim. Cosmochim. Acta*, 40: 721—730.

Brown, J.F., Harper, C.T. and Odom, A.L., 1974. Petrogenic implications of argon isotopic evolution in the upper mantle. *Nature*, 250: 130—133.

Bugaev, E.V., Kotov, Yu.D. and Rosental, I.L., 1970. *Cosmic Muons and Neutrinos*. Atomizdat, Moscow, 320 pp. (in Russian).

Bühler, F., Axford, W.I., Chivers, H.J.A. and Marti, K., 1976. Helium isotopes in an aurora. *J. Geophys. Res.*, 81: 111—115.

Burnett, D.S., Lippolt, H.J. and Wasserburg, G.J., 1966. The relative isotopic abundance of ^{40}K in terrestrial and meteoritic samples. *J. Geophys. Res.*, 71: 1249—1269.

Cameron, A.G.W., 1973. Abundance of the elements in the solar system. *Space Sci. Rev.*, 15: 121—146.

Cercignani, C., 1975. *Theory and Application of the Boltzman Equation.* Scottish Academic Press, Edinburg, London, 459 pp.

Chen, J.H. and Wasserburg, G.J., 1981. The isotopic composition of uranium and lead in Allende inclusions and meteoritic phosphates. *Earth Planet. Sci. Lett.*, 52: 1—15.

Cherdyntsev, V.V. and Shitov, Yu.V., 1967. Excess ^{36}Ar in volcanic and hydrothermal gases of the USSR. *Geokhimiya*, 5: 618—621 (in Russian).

Cherdyntsev, V.V., Shitov, Yu.V. and Lizarskaya, I.V., 1967. Isotope composition of argon from volcanic gases of the USSR. *Dokl. Akad. Nauk SSSR*, 172: 1180—1181 (in Russian).

Chernook, S.V. (Editor), 1964. *International Tectonic Map of Europe, Scale 1:2,500,000.* Akad. Nauk SSSR, GUGK SSSR, Moscow (in Russian).

Clarke, S.P., Turekian, K.K. and Grossman, L., 1976. Model of the early history of the earth. In: Yu. Robertson (Editor), *The Nature of the Solid Earth.* Mir, Moscow, pp. 9—23 (in Russian).

Clarke, W.B. and Kugler, G., 1973. Dissolved helium in ground water. A possible method for uranium and thorium prospecting. *Econ. Geol.*, 68: 243—251.

Clarke, W.B., Beg, M.A. and Craig, H., 1969. Excess ^3He in the sea: evidence for terrestrial primordial helium. *Earth Planet. Sci. Lett.*, 6: 213—220.

Clarke, W.B., Beg, M.A. and Craig, H., 1970. Excess ^3He at the North Pacific Geosecs Station. *J. Geophys. Res.*, 75: 7676—7678.

Clarke, W.B., Jenkins, W.J. and Top, Z., 1976. Determination of tritium by mass spectrometric measurement of ^3He. *Int. J. Appl. Radiat. Isot.*, 27: 515—522.

Clarke, W.B., Top, Z., Beavan, A.D. and Ganghi, S.S., 1977. Dissolved helium in lakes: uranium prospecting in the Precambrian terrain of Central Labrador. *Econ. Geol.*, 72: 233—242.

Clayton, R.N., 1978. Isotopic anomalies in the early solar system. *Annu. Rev. Nucl. Sci.*, 28: 501—522.

Clayton, R.N. and Mayeda, T., 1978. Genetic relations between iron and stony meteorites. *Earth Planet. Sci. Lett.*, 40: 168—174.

Clayton, R.N., Onuma, N. and Mayeda, T., 1976. A classification of meteorites based on oxygen isotopes. *Earth Planet. Sci. Lett.*, 30: 10—18.

Coffey, D., Lorentz, D.C. and Smith, F.T., 1969. Collision spectroscopy. II. Inelastic scattering of He by Ne. *Phys. Rev.*, 1871: 201—220.

Cook, G.A. (Editor), 1961. *Argon, Helium and Rare Gases, 1,2.* Wiley—Interscience, New York, N.Y., 452 pp. (Vol. 1), 442 (Vol. 2).

Coon, J.H., 1949. ^3He isotopic abundance. *Phys. Rev.*, 75: 1355—1357.

Costa, V., Ferreira, M.P., Macedo, R. and Reynolds, J.H., 1975. Rare gas dating I: A demountable metal system with low blanks. *Earth Planet. Sci. Lett.*, 25: 131—141.

Craig, H. and Clarke, W.B., 1970. Oceanic ^3He: contribution from cosmogenic tritium. *Earth Planet. Sci. Lett.*, 9: 45—48.

Craig, H. and Weiss, R.F., 1971. Dissolved gas saturation anomalies and excess helium in the ocean. *Earth Planet. Sci. Lett.*, 10: 289—296.

Craig, H. and Lupton, J.E., 1976. Primordial neon, helium and hydrogen in oceanic basalts. *Earth Planet. Sci. Lett.*, 31: 369—385.

Craig, H. and Lupton, J.E., 1978. Helium isotope variations: evidence for mantle plumes at Yellowstone, Kilauea and Ethiopian rift valley. *Trans. Am. Geophys. Union*, 59: 1194.

Craig, H., Weiss, R.F. and Clarke, W.B., 1967. Dissolved gases in the equatorial and south Pacific ocean. *J. Geophys. Res.*, 72: 6165—6181.

Craig, H., Clarke, W.B. and Beg, M.A., 1975. Excess ^{3}He in deep water on the East Pacific Rise. *Earth Planet. Sci. Lett.*, 26: 125—132.

Craig, H., Lupton, J.E. and Horibe, Y., 1978a. A mantle helium component in circum-Pacific volcanic gases: Hakone, the Marianas and Mt. Lassen. In: E.C. Alexander and M. Ozima (Editors), *Terrestrial Rare Gases*. Advances in Earth and Planetary Sciences, 3. Cent. Acad. Publ., Tokyo, pp. 3—16.

Craig, H., Lupton, J.E., Welhan, J.A. and Poreda, R., 1978b. Helium isotope ratios in Yellowstone and Lassen park volcanic gases. *Geophys. Res. Lett.*, 5: 897—900.

Damon, P.E. and Kulp, J.L., 1957. Determination of radiogenic helium in zircon by stable isotope delution technique. *Trans. Am. Geophys. Union*, 38: 945—953.

Damon, P.E. and Kulp, J.L., 1958a. Inert gases and the evolution of the atmosphere. *Geochim. Cosmochim Acta*, 13: 280—292.

Damon, P.E. and Kulp, J.L., 1958b. Excess helium and argon in beryl and other minerals. *Am. Mineral.*, 43: 433—459.

De Paolo, D.J. and Wasserburg, G.J., 1976. Inferences about magma sources and mantle structure from variations of ^{143}Nd/^{144}Nd. *Geophys. Res. Lett.*, 3: 743—746.

Devirts, A.L., Kamensky, I.L. and Tolstikhin, I.N., 1971. Helium isotopes and tritium in volcanic hot springs. *Dokl. Akad. Nauk SSSR*, 197: 450—452 (in Russian).

Drubetskoy, E.R., Verkhovsky, A.B., Tolstikhin, I.N. and Kazakov, I.K., 1977. Age data and helium and argon isotopes in rocks of West Sangelen. In: *Geologicheskaya interpretatsiya geokhronologicheskikh dannykh*. Tezisy dokladov, Irkutsk, p. 34 (in Russian).

Drubetskoy, E.R., Loginov, Yu.V., Lopatin, B.G. and Tolstikhin, I.N., 1979. On the isotope composition of argon in the earth mantle. *Geokhimiya*, 8: 1247—1250 (in Russian).

Dymond, J. and Hogan, L., 1973. Noble gas abundance patterns in deep-sea basalts — primordial gases from the mantle. *Earth Planet. Sci. Lett.*, 20: 131—139.

Dymond, J. and Hogan, L., 1978. Factors controlling the noble gas abundance patterns of deep-sea basalts. *Earth Planet. Sci. Lett.*, 38: 117—128.

Eberhardt, P., 1974. A neon-E-rich phase in the Orgueil carbonaceous chondrite. *Earth Planet. Sci. Lett.*, 24: 182—187.

Eberhardt, P., 1978. A neon-E-rich phase in Orgueil: results of stepwise heating experiments. In: *9th Proc. Lunar Planet. Sci. Conf.*, pp. 1027—1051.

Eberhardt, P., Geiss, J. and Grogler, N., 1966. Distribution of rare gases in the pyroxene and feldspar of the Khor Temiki meteorite. *Earth Planet. Sci. Lett.*, 1: 7—12.

Eberhardt, P., Geiss, J., Grogler, N., Krähenbühl, U., Schwaller, H., Schwarzmüller, J. and Stettler, A., 1970. Trapped solar wind noble gases, exposure age and K/Ar age in Apollo 11 lunar fine material. *Science*, 167: 558—560.

Eberhardt, P., Jungck, M.H.A., Meier, F.O. and Niederer, F., 1979. Presolar grains in Orgueil: evidence from neon-E. *Astrophys. J.*, 234: 169—171.

Ebert, G., 1968. *Brief Reference book of Physics*. Fiz.-Mat. Giz., Moscow, 552 pp. (in Russian).

Ehrenberg, H.Fr., 1953. Isotopenanalysen an Blei aus Mineralen. *Z. Phys.*, 134: 317—333.

Ermolin, G.M., 1957. On a technique of quantitative separation of helium—neon mixtures. *Tr. Radievogo Inst. V.G. Khlopina, Akad. Nauk SSSR, Leningrad*, 6: 119—138 (in Russian).

Eugster, O., Grögler, N., Mendia, M.D., Eberhardt, P. and Geiss, J., 1973. Trapped solar wind noble gases and exposure age of Lunar 16 lunar fines. *Geochim. Cosmochim. Acta*, 37: 1991—2005.

Fairbank, H.A., Lane, C.T., Aldrich, L.T. and Nier, A.O., 1947. The concentration of ^{3}He in the liquid and vapor phase of ^{4}He. *Phys. Rev.*, 71: 911—913.

254

Fairhall, A.W., 1969. Concerning the source of the excess ^{3}He in the sea. *Earth Planet. Sci. Lett.*, 7: 249—250.

Fanale, F.P., 1971. A case for catastrophic early degassing of the earth. *Chem. Geol.*, 8: 79—105.

Fanale, F.P. and Cannon, W.A., 1972. Origin of planetary primordial rare gas: the possible role of adsorption. *Geochim. Cosmochim. Acta*, 36: 319—329.

Faure, G. and Powell, J.L., 1972. *Strontium Isotope Geology*. Springer-Verlag, Berlin, 188 pp.

Fisher, D.E., 1970. Search for ^{3}He in deep-sea basalts. *Earth Planet. Sci. Lett.*, 8: 77—78.

Fisher, D.E., 1972. Uranium content and radiogenic ages of hypersthene, bronzite, amphoterite and carbonaceous chondrites. *Geochim. Cosmochim. Acta*, 36: 15—33.

Fisher, D.E., 1975. Trapped helium and argon and the formation of the atmosphere by degassing. *Nature*, 256: 113—114.

Fisher, D.E., 1978. Terrestrial potassium and argon abundance as limits to models of atmospheric evolution. In: E.C. Alexander and M. Ozima (Editors), *Terrestrial Rare Gases*. Advances in Earth and Planetary Sciences, 3. Cent Acad. Publ., Tokyo, pp. 173—184.

Fluss, M.J., Dadley, N.D. and Malewicki, R.L., 1972. Tritium and α-particle yields in fast and thermal neutron fission of ^{235}U. *Phys. Rev.*, C, 6: 2252—2259.

Flynn, K.K., Srinivasan, B., Manuel, O.K. and Glendenin, L.E., 1972. Distribution of mass and charge in spontaneous fission of ^{244}Cm. *Phys. Rev.*, C, 6: 2211—2214.

Funkhouser, J.G. and Naughton, J.J., 1968. Radiogenic He and Ar in ultramafic inclusions from Hawaii. *J. Geophys. Res.*, 73: 4601—4607.

Galimov, E.M., 1968. *Geochemistry of Carbon Stable Isotopes*. Nedra, Moscow, 224 pp. (in Russian).

Galimov, E.M., Migdasov, A.A. and Ronov, A.B., 1975. Variation in isotope composition of carbonate and organic carbon from sedimentary rocks in the earth history. *Geokhimiya*, 3: 323—342 (in Russian).

Gall', R.N., 1969. Ion-optical scheme of the mass spectrometer with energy focusing for analysis of impurities. *Zh. Tekh. Fiz.*, 39: 360—364 (in Russian).

Gartmanov, V.N., 1974. *Foil Method for the Investigation of Elemental and Isotopic Compositions of Corpuscle Fluxes on the Space-Crafts "Zond", "Soyuz" and "Kosmos"*. Dissertation, A.F. Ioffe Fiz.-Tekh. Inst. Akad. Nauk SSSR, Leningrad, 21 pp (in Russian; abstract).

Gast, P.W., 1968. Upper mantle chemistry and the evolution of the earth's crust. In: R.A. Phiney (Editor), *The History of the Earth's Crust*. Princeton University Press, Princeton, N.J., pp. 15—27.

Gast, P.W., 1976. Chemical composition of the earth, the moon and chondrite meteorites. In: Yu. Robertson (Editor), *The Nature of Solid Earth*. Mir, Moscow, pp. 23—41 (in Russian).

Geiss, J., 1976. Noble gas isotopes and deuterium in the solar system. In: G. Reeves (Editor), *The Origin of the Solar System*. Mir. Moscow, pp. 321—343 (in Russian).

Geiss, J., Eberhardt, P., Bühler, F., Meister, J. and Signer, P., 1970. Apollo-11 and Apollo-12 solar wind composition experiments: fluxes of He and Ne isotopes. *J. Geophys. Res. Space Phys.*, 75: 5972—5979.

Gerling, E.K., 1939. Helium diffusion as a criterion of the possibility of age determination by the helium—uranium method. *Dokl. Akad. Nauk SSSR*, 24: 572—575 (in Russian).

Gerling, E.K., 1957. Migration of helium from rocks and minerals. *Tr. Radievogo Inst. V.G. Khlopina, Akad. Nauk SSSR, Leningrad*, 6: 64—87 (in Russian).

Gerling, E.K., 1961. *Recent Advances in the Argon Method of Age Determination and Its Usage in Geology*. Izd. Akad. Nauk SSSR, Moscow, Leningrad, 130 pp. (in Russian).

Gerling, E.K. and Levsky, L.K., 1956. On the origin of inert gases in stone meteorites. *Geokhimiya*, 7: 59—64 (in Russian).

Gerling, E.K., Shukolyukov, Yu.A., Koltsova, T.V., Matveeva, I.I. and Yakovleva, S.S., 1963. The determination of the earth age by ancient minerals and rocks. In: A.P. Vinogradov (Editor), *Chemistry of the Earth's Crust*. Akad. Nauk SSSR, Moscow, pp. 366—374 (in Russian).

Gerling, E.K., Tolstikhin, I.N., Shukolyukov, Yu.A., Nesmelova, Z.N. and Azbel, I.Ya., 1967a. Argon isotopes and helium in terrestrial carbon—hydrogen gases. *Geokhimiya*, 5: 608—617 (in Russian).

Gerling, E.K., Morozova, I.M., Shukolyukov, Yu.A., Maslennikov, V.A., Levchenkov, O.A., Matveeva, I.I. and Vaskovsky, D.P., 1967b. On the origin of excess ^{40}Ar in chlorite. *Geokhimiya*, 10: 1035—1043 (in Russian).

Gerling, E.K., Morozova, I. M. and Sprintson, V.D., 1968. On the nature of excess argon in some minerals. In: A.P. Vinogradov (Editor), *The Problems of Geochemistry and Cosmology, Tr. XXIII Mezh. Geol. Kongr.*, Nauka Moscow, pp. 76—81 (in Russian).

Gerling, E.K., Mamyrin, B.A., Tolstikhin, I.N. and Yakovleva, S.S., 1971. Isotope composition of helium in rocks. *Geokhimiya*, 5: 608—617.

Gerling, E.K., Tolstikhin, I.N., Mamyrin, B.A., Anufriev, G.S., Kamensky, I.L. and Prasolov, E.M., 1972. Recent investigations in isotope geochemistry of helium. In: A.I. Tugarinov (Editor), *Tr. I Mezh. Geokhim. Kongr., 1*, Nauka, Moscow, pp. 200—216 (in Russian).

Gerling, E.K., Tolstikhin, I.N., Drubetskoy, E.R., Levkovsky, R.Z., Sharkov, E.V. and Kozakov, I.K., 1976. He and Ar isotopes in rock-forming minerals. *Geokhimiya*, 11: 1603—1610 (in Russian).

Ginzburg, A.I. and Panteleev, A.I., 1971. On the problem of excess argon in beryls. *Geokhimiya*, 10: 1218—1227 (in Russian).

Gorokhov, I.M., 1964. Determination of Rb—Sr whole-rock age of Korosten granites and Dnieper migmatites and metabasites from the Ukraine. *Geokhimiya*, 8: 744—753 (in Russian).

Gorshkov, G.V., Zyabkin, V.A., Lyatkovskaya, N.M. and Zvetkov, Yu.S., 1966. *Natural Neutron Background of the Atmosphere and the Earth's Crust*. Atomizdat, Moscow, 410 pp. (in Russian).

Hall, L.G., Hines, C.K. and Sley, I.E., 1959. A high speed cycloidal mass spectrometer. In: J.D. Waldron (Editor), *Advances in Mass Spectrometry*. Pergamon Press, London, pp. 266—271.

Hamano, Y. and Ozima, M., 1978. Earth—atmosphere evolution model based on Ar isotopic data. In: E.C. Alexander and M. Ozima (Editors), *Terrestrial Rare Gases*. Advances in Earth and Planetary Sciences. Cent. Acad. Publ., Tokyo, pp. 155—172.

Hampel, W., Takagi, J., Sakamoto, K. and Tanaka, S., 1975. Measurement of muon-induced ^{26}Al in terrestrial silicate rock. *J. Geophys. Res.*, 80: 3757—3760.

Harper, C.T. and Schamel, S., 1971. Note on the isotope composition of argon in quartz veins. *Earth Planet. Sci. Lett.*, 12: 129—134.

Haskin, L.A., Frey, F.A., Schmidt, R.A. and Smith, R.H., 1966. *Meteoritic, Solar and Terrestrial Rare—Earth Distributions*. Pergamon Press, London, 187 pp.

Heier, K.S., 1963. U, Th, K in eclogitic rocks. *Geochim. Cosmochim. Acta*, 27: 849—860.

Herzog, G.F., 1973. Variability of ^{3}He and ^{21}Ne production rates in ordinary chondrites. *Geochim. Cosmochim. Acta*, 37: 2125—2133.

Herzog, G.F. and Anders, E., 1974. Primordial noble gases in separated meteoritic minerals. II. *Earth Planet. Sci. Lett.*, 24: 173—181.

Heymann, D. and Mazor, E., 1967. Light-dark structure and rare gas content of the carbonaceous chondrite Nogoya. *J. Geophys. Res.*, 72: 2704—2707.

Hickman, D.R. and Nier, A.O., 1972. Measurements of the neutral composition of the lower thermosphere above Fort Churchill by rocket-borne mass spectrometer. *J. Geophys. Res.*, 77: 2880—2887.

Hoffman, D.C., Lawrence, F.O., Mewherter, J.L. and Rourke, F.M., 1971. Detection of ^{244}Pu in Nature. *Nature*, 234: 132—134.

Hoffman, J.H. and Nier, A.O., 1958. Production of helium in iron meteorites by the action of cosmic rays. *Phys. Rev.*, 112: 2112—2117.

Holzer, T.E. and Axford, W.I., 1971. Interaction between interstellar helium and the solar wind. *J. Geophys. Res.*, 76: 6965—6970.

Hudson, D.J., 1964. *Statistics. Lectures on Elementary Statistics and Probability*. Geneva, 300 pp.

Hurley, P.M., 1968. Absolute abundance and distribution of Rb, K and Sr in the earth. *Geochim. Cosmochim. Acta*, 32: 273—284.

Hurley, P.M. and Rand, J.K., 1969. Pre-drift continental nuclei. *Science*, 164: 1229—1242.

Ionov, N.I. and Karataev, V.I., 1962. Double-stage mass spectrometer for analysis of small impurities. *Prib. Tekh. Eksp.*, 3: 119—122 (in Russian).

Ionov, N.I., Karataev, V.I., Mamyrin, B.A. and Shustrov, B.N., 1961. Double-stage magnetic mass-spectrometer. Inventor's certificate No. 139031. *Bull. Izobr.*, 12: 34 (in Russian).

Izakov, M.N., 1979. Noble gases in the atmosphere of Venus, Earth and Mars and the problem of the atmospheres' origin. *Kosm. Issled.*, XVII: 602—610.

Jacchia, L.G., 1977. Thermosphere temperature, density and composition: new models. *Spec. Rep. Smithson. Astrophys. Observ.*, No. 375: 106 pp.

Jacobsen, S.B. and Wasserburg, G.J., 1979. The mean age of mantle and crustal reservoirs. *J. Geophys. Res.*, 84: 7411—7427.

Jeans, J.H., 1925. *The Dynamical Theory of Gases*. Cambridge University Press, Cambridge, 342 pp.

Jeffery, P.M. and Anders, E., 1970. Primordial noble gases in separated meteoritic minerals. *Geochim. Cosmochim. Acta*, 34: 1175—1198.

Jenkins, W.J. and Clarke, W.B., 1976. The distribution of ^{3}He in the Western Atlantic Ocean. *Deep-Sea Res.*, 23: 481—494.

Jenkins, W.J., Beg, M.A., Clarke, W.B., Wangersky, P.J. and Craig, H., 1972. Excess ^{3}He in the Atlantic Ocean. *Earth Planet. Sci. Lett.*, 16: 122—126.

Jenkins, W.J., Edmond, J.M. and Corliss, J.B., 1978. Excess ^{3}He and ^{4}He in Galapagos submarine hydrothermal waters. *Nature*, 272: 156—158.

Johnson, H.E. and Axford, W.I., 1969. Production and loss of ^{3}He in the earth's atmosphere. *J. Geophys. Res.*, 74: 2433—2438.

Jungck, M.H.A. and Eberhardt, P., 1979. Neon-E in Orgueil density separates. *Meteoritics*, 14: 439—441.

Kalbitzer, S., Kiko, J. and Zähringer, S., 1969. The diffusion of ^{3}He and ^{4}He in LiF. *Z. Naturforsch., Teil B*, 24a: 1996—2000.

Kamensky, I.L., 1970. *Investigations of Isotope Composition of Terrestrial Helium*. Dissertation, Lensoveta Institite of Technology, Leningrad, 130 pp. (in Russian).

Kamensky, I.L., Yakutseny, V.P., Mamyrin, B.A., Anufriev, G.S. and Tolstikhin, I.N., 1971. Helium isotopes in nature. *Geokhimiya*, 8: 914—931 (in Russian).

Kamensky, I.L., Prasolov, E.M. and Tikhomirov, V.V., 1974. On the juvenile components in natural gases of Sakhalin. *Geokhimiya*, 8: 1220—1225 (in Russian).

Kamensky, I.L., Lobkov, V.A., Prasolov, E.M., Beskrovny, N.S., Kudryavtseva, E.I., Anufriev, G.S. and Pavlov, V.P., 1976. Components of the upper mantle of the earth in gases of Kamchatka. *Geokhimiya*, 5: 682—695 (in Russian).

Kaneoka, I., 1974. Investigation of excess argon in ultramafic rocks of the Kola Peninsula by ^{40}Ar/^{39}Ar method. *Earth Planet. Sci. Lett.*, 22: 145—156.

Kaneoka, I., 1980. Rare gases isotopes and mass fractionation: an indicator of gas transport into or from a magma. *Earth Planet. Sci. Lett.*, 48: 284—292.

Kaneoka, I. and Takaoka, N., 1978. Excess ^{129}Xe and high ^{3}He/^{4}He ratios in olivine phenocrysts of Kapuho lava and xenolithic dunites from Hawaii. *Earth Planet. Sci. Lett.*, 39: 382—386.

Kaneoka, I. and Takaoka, N., 1980. Rare gas isotopes in Hawaiian ultramafic nodules and volcanic rocks: constraint on generic relationships. *Science*, 208: 1366—1368.

Kaneoka, I., Takaoka, N. and Aoki, K.I., 1977. Rare gases in a phlogopite nodule and a phlogopite-bearing peridotite in South African kimberlites. *Earth Planet. Sci. Lett.*, 36: 181—186.

Kaneoka, I., Takaoka, N. and Aoki, K.I., 1978. Rare gases in mantle-derived rocks and minerals. In: E.C. Alexander and M. Ozima (Editors), *Terrestrial Rare Gases*. Advances in Earth and Planetary Sciences, 3. Cent. Acad. Publ., Tokyo, pp. 71—83.

Kasprzak, W.T., Krankowsky, D. and Nier, A.O., 1968. A study of day—night variations in the neutral composition of the lower thermosphere. *J. Geophys. Res.*, 73: 6765—6782.

Kaula, W.M., 1979. Thermal evolution of earth and moon growing by planetesimal impacts. *J. Geophys. Res.*, B84: 999—1008.

Keil, K. and Fredriksson, K., 1964. The Fe, Mg and Ca distribution the coexisting olivines and rhombic pyroxenes in chondrites. *J. Geophys. Res.*, 69: 3487—3515.

Kel'man, V.M. and Yavor, S.Ya., 1968. *Electronic Optics*, Nauka, Leningrad, 487 pp. (in Russian).

Kel'man, V.M., Nazarenko, L.M. and Yakushev, E.M., 1972. Theory of symmetric prism mass spectrometer. *Zh. Tekh. Fiz.*, 42: 1963—1968 (in Russian).

Kel'man, V.M., Nazarenko, L.M. and Yakushev, E.M., 1976. Symmetric prism mass spectrometer with high resolution. *Zh. Tekh. Fiz.*, 46: 1700—1706 (in Russian).

Kel'man, V.M., Karetskaya, C.P., Fedulina, L.V. and Yakushev, E.M., 1979. *Electron-Optical Elements of Prism Mass Spectrometers of Charged Particles*. Nauka, Alma-Ata, 195 pp. (in Russian).

Kerridge, J.F. and Vedder, J.F., 1972. Accretionary processes in the early Solar system: an experimental approach. *Science*, 177: 49—51.

Khabarin, L.V., 1975. *Helium Isotopes in the Upper Shells of the Earth*. Dissertation, A.F. Ioffe Fiz.-Tekh. Inst., Akad. Nauk SSSR, Leningrad, 20 pp. (in Russian; abstract).

Khlopin, V.G. and Gerling, E.K., 1948. New results in the geochemistry of noble gases. *Dokl. Akad. Nauk SSSR*, 61: 297—299 (in Russian).

Kirsten, T., 1976. *Time and the Solar System*. Max-Plank-Institute für Kernphysik, Heidelberg, 152 pp.

Kirsten, T. and Gentner, W., 1966. K—Ar Alterbestimmungen an Ultrabasiten des Baltischen Schildes. *Z. Naturforsch.*, 21a: 119—122.

Klyavin, O.V., Mamyrin, B.A., Khabarin, L.V., Chernov, Yu.M. and Yudenich, V.S., 1976. Helium penetration in LiF crystals by their deformation in liquid ^{4}He and ^{3}He. *Zh. Tekh. Fiz.*, 46: 1281—1285 (in Russian).

Kockarts, G., 1973. Helium in the terrestrial atmosphere. *Space Sci. Rev.*, 14: 723—757.

Kockarts, G. and Nicolet, M., 1962. Le problème aeronomique de l'helium et de l'hydrogène neutres. *Ann. Géophys.*, 18: 269—282.

Komarov, A.N. and Ilupin, I.P., 1978. Fission-track ages of Yakutian kimberlites. *Geokhimiya*, 7: 1004—1013 (in Russian).

Komarov, A.N. and Zitkov, A.C., 1973. Uranium in ultramafic xenoliths from basalts. *Izv. Akad. Nauk SSSR, Ser. Geol.*, 10: 79—85 (in Russian).

König, H., Wänke, H., Bein, G.S., Rakestraw, N.W. and Suess, H.E., 1964. Helium, neon and argon in the oceans. *Deep-Sea Res.*, 11: 243—251.

Kononov, V.I. and Polyak, B.G., 1977. Geothermal activity. In: V.V. Belousov and G.B. Udintsev (Editors). *Iceland and the Middle-Ocean Rise (Deep Structure, Seismicity, Geothermy)*. Nauka, Moscow, pp. 8—82 (in Russian).

Kononov, V.I., Mamyrin, B.A., Polyak, B.G. and Khabarin, L.V., 1974. The helium isotopes in the Icelandic hydrothermae. *Dokl. Akad. Nauk SSSR*, 217: 172—175 (in Russian).

Krähenbühl, U., Morgan, J.W., Ganapathy, R. and Anders, E., 1973. Abundance of 17 trace elements in carbonaceous chondrites. *Geochim. Cosmochim. Acta*, 37: 1353—1370.

Krall, N. and Trivelpiece, A., 1973. *Principles of Plasma Physics*. McGraw-Hill, New York, N.Y., 674 pp.

Krankowsky, D., Kasprzak, W.T. and Nier, A.O., 1968. Mass spectrometric studies of the composition of the lower thermosphere during summer 1967. *J. Geophys. Res.*, 73: 7291—7306.

Krummenacher, D., 1970. Isotope composition of argon in modern surface volcanic rocks. *Earth Planet. Sci. Lett.*, 8: 109—117.

Krylov, A.Ya., Mamyrin, B.A., Khabarin, L.V., Mazina, T.I. and Silin, Yu,I., 1974. Helium isotopes in basement rocks of the ocean floor. *Geokhimiya*, 8: 1221—1226 (in Russian).

Kugler, G. and Clarke, W.B., 1972. Mass spectrometric measurements of ^{3}H, ^{3}He and ^{4}He produced in thermal neutron ternary fission of ^{235}U: evidence for short-range ^{4}He. *Phys. Rev., C*, 5: 551—560.

Kukharenko, A.A., Il'insky, G.A., Ivanova, T.N., Galakhov, A.V., Kozyreva, I.V., Gel'man, E.M., Broneman-Starynkevich, I.D., Stolyarova, I.N., Skrizanskaya, V.I., Ryzkova, P.I. and Milental, B.N., 1968. Bulk composition of the Khibiny alkaline massif. *Zap. Vses. Mineral. O.-va.*, XCVII: 133—149 (in Russian).

Kung, C.-C. and Clayton, R.N., 1978. Nitrogen abundances and isotopic compositions in stony meteorites. *Earth Planet. Sci. Lett.*, 38: 421—435.

Kunz, W. and Schintlmeister, I., 1965. *Tabellen der Atomekerne, Teil II, Kernreaktionen*. Akademie-Verlag, Berlin, 1022 pp.

Kuroda, P.K., 1969. ^{244}Pu in the early Solar system. *Nature*, 221: 726—729.

Kurz, M.D. and Jenkins, W.J., 1981. The distribution of helium in oceanic basalt glasses. *Earth Planet. Sci. Lett.*, 53: 41—54.

Kutas, R.I. and Gordienko, V.V., 1972. The Karpatian heat field and some problems of geothermy. *Tr. Mosk. O. va. Ispyt. Prir., Otd. Geol.*, 46: 75—80 (in Russian).

Kuz'min, A.F., 1976. *The Separating and Focussing Properties of Quazihomogeneous and Quazistationary Crossed Electric and Magnetic Fields*. Dissertation, A.F. Ioffe Fiz.-Tekh. Inst., Akad. Nauk SSSR, Leningrad, 19 pp. (in Russian; abstract).

Lambert, I.B. and Heier, K.S., 1968. Estimates of the crustal abundance of Th, U and K. *Chem. Geol.*, 3: 233—238.

Lancet, M.S. and Anders, E., 1973. Solubilities of noble gases in magnetite: implications for planetary gases in meteorites. *Geochim. Cosmochim. Acta*, 37: 1371—1388.

Larimer, J.W., 1971. Composition of the earth: chondritic or achondritic. *Geochim. Cosmochim. Acta*, 35: 769—771.

Lee, W.H.K. and Uyeda, S., 1965. Review of heat flow data. In: W.H.K. Lee (Editor), *Terrestrial Heat Flow*. Am. Geophys. Union, Washington, D.C., pp. 87—190.

Leutwein, F.G. and Kaplan, G., 1963. Quelques recherches sur l'aptitude de certains cristaux de néoformation à capture argon radiogénique. *C. R. Acad. Sci. Paris*, 257: 1313—1317.

Leventhal, J.S. and Libby, W.F., 1968. Tritium geophysics from 1961—1962 nuclear tests. *J. Geophys. Res.*, 73: 2715—2720.

Levsky, L.K., 1963. Diffusion of helium from meteorites. *Geokhimiya*, 6: 544—548 (in Russian).

Levsky, L.K., 1973. *Noble Gas Isotopes in Meteorites and Some Problems of Nucleogenesis*. Dissertation, V.I. Vernadsky Inst. Geokhim. Analit. Khim. Akad. Nauk. SSSR, Moscow, 47 pp. (in Russian; abstract).

Lewis, R.S., Gros, J. and Anders, E., 1979. Isotopic anomalies of noble gases in meteorites and their origin. 2. Separated minerals from Allende. *J. Geophys. Res.*, 82: 779—792.

Lindenfeld, M.J. and Shizgal, B., 1979. Non-Maxwellian effects associated with thermal escape of planetary atmosphere. *Planet. Space Sci.*, 27: 739—752.

Lippolt, H.J., 1966. Die Zusammensetzung des Überschussargon in Schwarzwälder Flussspaten. *Z. Naturforsch.*, 21a: 1162—1167.

Lippolt, H.J. and Gentner, W., 1963. K—Ar dating of some limestones and fluorites. In: *Radioactive Dating.* Int. Atom. Energy Agency Symp., Athens, 1962, pp. 239—244.

Lomonosov, I.S., Mamyrin, B.A., Prasolov, E.M. and Tolstikhin, I.N., 1976. Isotope composition of helium and argon in hot springs of the Baikal rift zone. *Geokhimiya*, 11: 1743—1746 (in Russian).

Lupton, J.E., 1973. Direct accretion of ^{3}He and ^{3}H from cosmic rays. *J. Geophys. Res.*, 78: 8330—8337.

Lupton, J.E., 1976. The ^{3}He distribution in deep water over the mid-Atlantic ridge. *Earth Planet. Sci. Lett.*, 32: 371—374.

Lupton, J.E., 1979. Helium-3 in the Quaymas basin: evidence for injection of mantle volatiles in the Gulf of California. *J. Geophys. Res.*, 84: 7446—7452.

Lupton, J.E. and Craig, H., 1975. Excess ^{3}He in oceanic basalts: evidence for terrestrial primordial helium. *Earth Planet. Sci. Lett.*, 26: 133—139.

Lupton, J.E., Weiss, R.F. and Craig, H., 1977a. Mantle helium in the Red Sea brines. *Nature*, 266: 2440—2446.

Lupton, J.E., Weiss, R.F. and Craig, H., 1977b. Mantle helium in hydrothermal plumes in the Galapagos rift. *Nature*, 267: 603—604.

Lupton, J.E., Klinkhammer, G.P., Normark, W.R., Haymon, R.H., MacDonald, K.C., Weiss, R.F. and Craig, H., 1980. Helium-3 and manganese at the 21° N East Pacific rise hydrothermal site. *Earth Planet. Sci. Lett.*, 50: 115—127.

Lysak, S.V. and Zorin, Yu.A., 1976. *Geothermal Field of the Baikal Rift Zone.* Nauka, Moscow, 92 pp. (in Russian).

MacDaniel, E., 1964. *Collision Phenomena in Ionized Gases.* Wiley, New York, N.Y., 775 pp.

MacDonald, G.J.F., 1959. Chondrites and the chemical composition of the earth. In: P.H. Abelson (Editor), *Research in Geochemistry.* Wiley, New York, N.Y., pp. 476—494.

MacDonald, G.J.F., 1963. The escape of helium from the earth atmosphere. *Rev. Geophys.*, 1: 305—349.

MacDougall, J.D. and Kothari, B.K., 1976. Formation chronology for C2 meteorites. *Earth Planet. Sci. Lett.*, 33: 36—44.

Makarenko, F.A., Smirnov, Ya.B. and Sergienko, S.I., 1968. Deep heat flow and tectonic structure of the Cis-Caucasus. *Dokl. Akad. Nauk SSSR*, 183: 901—904 (in Russian).

Mal'kov, B.A. and Gustomesov, B.A., 1975. Jurassic belemnites in the kimberlite pipe "Naked" on Olekma rise, northern Yakutia. *Izv. Akad. Nauk SSSR, Ser. Geol.*, 11: 137—140 (in Russian).

Mamchur, G.P., Matvienko, A.D. and Yarynyg, O.A., 1975. On the origin of chambered pegmatities of Volyn in the light of δ^{13}C data. *Geol. Zh.*, 35: 91—97 (in Russian).

Mamyrin, B.A., 1966. *Investigation of Time-of-Flight Ion Separation.* Dissertation, A.F. Ioffe Fiz.-Tekh. Inst., Akad. Nauk SSSR, Leningrad, 30 pp. (in Russian; abstract).

Mamyrin, B.A., 1972. Dynamic mass spectrometry. In: *Tr. Vses. Konf. Mass Spectrometrii.* Akad. Nauk SSSR, Leningrad, pp. 43—52 (in Russian).

Mamyrin, B.A. and Khabarin, L.V., 1977. Manifold metal furnace for inert gas extraction. *Prib. Tekh. Eksp.*, 2: 232—233 (in Russian).

Mamyrin, B.A. and Shustrov, B.N., 1957. Mass spectrometer with resolution power of several thousands. *Zh. Tekh. Fiz.*, 27: 1347—1356 (in Russian).

Mamyrin, B.A. and Shustrov, B.N., 1962. A high resolution mass spectrometer with two-stage time-of-flight separation of ions. *Prib. Tekh. Exper.*, 5: 135—141 (in Russian).

Mamyrin, B.A. and Tolstikhin, I.N., 1981. *The Helium Isotopes in Nature.* Energoizdat, Moscow, 222 pp. (in Russian).

Mamyrin, B.A., Tolstikhin, I.N., Anufriev, G.S. and Kamensky, I.L., 1969a. Anomalous isotopic composition of helium in volcanic gases. *Dokl. Akad. Nauk SSSR*, 184: 1197—1199 (in Russian).

Mamyrin, B.A., Tolstikhin, I.N., Anufriev, G.S. and Kamensky, I.L., 1969b. The usage of a magnetic resonance mass spectrometer for isotopic analyses of primordial helium. *Geokhimiya*, 5: 595—602 (in Russian).

Mamyrin, B.A., Anufriev, G.S., Kamensky, I.L. and Tolstikhin, I.N., 1970a. Isotope composition of atmospheric helium. *Dokl. Akad. Nauk SSSR*, 195: 111 (in Russian).

Mamyrin, B.A., Anufriev, G.S., Kamensky, I.L. and Tolstikhin, I.N., 1970b. The determination of atmospheric helium isotope composition. *Geokhimiya*, 6: 721—730 (in Russian).

Mamyrin, B.A., Anufriev, G.S., Kamensky, I.L. and Tolstikhin, I.N., 1970c. Isotope composition of helium in the atmosphere. *Dokl. Akad. Nauk SSSR*, 195: 188—189 (in Russian).

Mamyrin, B.A., Shustrov, B.N., Anufriev, G.S., Boltenkov, B.S., Zagulin, V.A., Kamensky, I.L., Tolstikhin, I.N. and Khabarin, L.V., 1972a. Measurements of isotope composition of helium using the magnetic resonance mass spectrometer. *Zh. Tekh. Fiz.*, 42: 2577—2582 (in Russian).

Mamyrin, B.A., Shustrov, B.N., Anufriev, G.S., Boltenkov, B.S., Kamensky, I.L., Tolstikhin, I.N. and Khabarin, L.V., 1972b. Magnetic resonance mass spectrometer for terrestrial helium isotope analysis. *Prib. Tekh. Eksp.*; 6: 148—150 (in Russian).

Mamyrin, B.A., Tolstikhin, I.N., Anufriev, G.S. and Kamensky, I.L., 1972c. Isotope composition of helium in Icelandic hot springs. *Geokhimiya*, 11: 1396 (in Russian).

Mamyrin, B.A., Gerasimovsky, V.I. and Khabarin, L.V., 1974. Helium isotopes in rocks of East Africa and Iceland rift zones. *Geokhimiya*, 5: 693—700 (in Russian).

Mamyrin, B.A., Khabarin, L.V. and Yudenich, V.S., 1978. Extremely high ratios of ^{3}He/^{4}He in technogenic metals and semiconductors. *Dokl. Akad. Nauk SSSR*, 241: 1054—1957 (in Russian).

Manuel, O.K., 1978. A comparison of terrestrial and meteoritic noble gases. In: E.C. Alexander and M. Ozima (Editors), *Terrestrial Rare Gases*. Advances in Earth and Planetary Sciences, 3. Cent. Acad. Publ., Tokyo, pp. 85—92.

Mason, B., 1962. *Meteorites*. Wiley, New York, N.Y., 274 pp.

Masurenkov, Yu.P., 1971. Heat flow and the depth of the magma source under Elbrus Volcano. *Byull. Vulkanol. Stn. SO, Akad. Nauk*, 47: 79—82 (in Russian).

Matsubayashi, O., Nagao, K. and Takaoka, N., 1978. Magmatic isotope anomaly of He found in natural gases along the Japanese Islands. *Bull. Volc. Soc. Jap.*, 24: 21—22.

Matsuo, S., Suzuki, M. and Mizutai, Y., 1978. Nitrogen to argon ratio in volcanic gases. In: E.C. Alexander and M. Ozima (Editors), *Terrestrial Rare Gases*. Advances in Earth and Planetary Sciences, 3. Cent. Acad. Publ., Tokyo, pp. 17—25.

Matveeva, E.S., Tolstikhin, I.N. and Yakutseny, V.P., 1978. Helium isotope criterion to study the origin of gases and to recognize neotectonic zones (in the Caucasus region). *Geokhimiya*, 3: 307—317 (in Russian).

Mazor, E., Heymann, D. and Anders, E., 1970. Noble gases in carbonaceous chondrites. *Geochim. Cosmochin. Acta*, 34: 781—824.

McCulloch, M.T. and Wasserburg, G.J., 1978. More anomalies from Allende meteorite: samarium. *Geophys. Res. Lett.*, 5: 599—602.

McSween, H.Y., 1979. Are carbonaceous chondrites primitive or processes? A review. *Rev. Geophys. Space Phys.*, 17: 1059—1078.

Meier, F.O., Jungck, M.H.A. and Eberhardt, P., 1980. Evidence for pure neon-22 in Orgueil and Murchison. *Lunar Planet. Sci.*, XI: 723—725.

Melton, C.E. and Giardini, A.A., 1980. The isotope composition of argon included in an Arkansas diamond and its significance. *Geophys. Res. Lett.*, 7: 461—464.

Miller, R.L. and Kahn, J.S., 1962. *Statistical Analysis in the Geological Sciences.* Wiley, New York, N.Y., 450 pp.

Mitra, S., 1952. *The Upper Atmosphere.* Asiatic Soc., Calcutta, 713 pp.

Moor, C.A. and Esfandiary, B., 1971. Geochemistry and geology of helium. *Adv. Geophys.,* 15: 2—54.

Moorbath, S., 1977. Ages, isotopes and evolution of the Precambrian crust. *Chem. Geol.,* 20: 151—187.

Moorbath, S., O'Nions, R.K., Pankhurst, R.J., Gale, N.H. and McGregor, V.R., 1972. Further Rb/Sr age determinations of very early Precambrian rocks of the Godthaab District, West Greenland. *Nature, Phys. Sci.,* 240: 78—82.

Morozova, I.M. and Ashkinadze, G.Sh., 1971. *Migration of Noble Gas Atoms in Minerals.* Nauka, Leningrad, 120 pp. (in Russian).

Morrison, P. and Beard, O.B., 1949. ^{3}He and the origin of terrestrial helium. *Phys. Rev.,* 75: 1332.

Morrison, P. and Pine, J., 1955. Radiogenic origin of the helium isotopes in rocks. *Ann. N.Y. Acad. Sci.,* 62: 69—92.

Müller, D. and Hartman, G., 1969. A mass spectrometric investigation of the lower thermosphere above Fort Churchill with special emphasis on the helium content. *J. Geophys. Res.,* 74: 1287—1293.

Nagao, K., Takaoka, N. and Matsubayashi, O., 1979. Isotopic anomalies of rare gases in the Nigorikawa geothermal area, Hokkaido, Japan. *Earth Planet. Sci. Lett.,* 44: 82—90.

Nagao, K., Takaoka, N., Matsuo, S., Mizutani, Y. and Matsubayashi, O., 1980a. Change in rare gas isotope composition of the fumarolic gases from the Showa-shinzan Volcano. *Geochem. J.,* 14: 139—143.

Nagao, K., Takaoka, N., Wakita, H., Matsuo, S. and Fujii, N., 1980b. Isotopic compositions of rare gases in the Matsushiro earthquake fault region. *Geochem. J.,* 14: 63—69.

Nagao, K., Takaoka, N. and Matsubayashi, O., 1981. Rare gas isotope compositions in natural gases of Japan. *Earth Planet. Sci. Lett.,* 53: 175—188.

Naidenov, B.M., Bogolepov, V.G., Polyvyanny, E.Ya. and Zakharchenko, A.I., 1972. Argon isotopes in mineral-bearing fluids of pegmatites. *Geokhimiya,* 6: 734—737 (in Russian).

Naughton, J.J., Funkhouser, J.B. and Barnes, J.L., 1966. Fluid inclusions in K—Ar age anomalies and related inert gas studies. *Trans. Am. Geophys. Union,* 47: 197.

Naughton, J.J., Lee, J.H., Keeling, D., Finlayson, J.B. and Doroty, G., 1973. Helium flux from the earth's mantle as estimated from Hawaiian fumarolic degassing. *Science,* 180: 55—57.

Nesmelova, Z.N. (Editor), 1969. *Sampling and Analysis of Terrestrial Gases.* Nedra, Leningrad, 85 pp. (in Russian).

Nicolet, M., 1957. The aeronomic problem of helium. *Ann. Géophys.,* 13: 1—21.

Nicolet, M., 1964. *Aeronomy.* Mir, Moscow, 298 pp. (in Russian).

O'Nions, R.K., Hamilton, P.J. and Evensen, N.M., 1977. Variations in ^{143}Nd/^{144}Nd and ^{87}Sr/^{86}Sr ratios in oceanic basalts. *Earth Planet. Sci. Lett.,* 34: 13—22.

O'Nions, R.K., Evensen, N.M., and Hamilton, P.J., 1979. Geochemical modelling of mantle differentiation and crustal growth. *J. Geophys. Res.,* 84: 6091—6101.

Orowan, E., 1969. Density of the moon and nucleation of planets. *Nature,* 222: 867.

Ozima, M., 1973. Evolution of the atmosphere: continuous or catastrophic? *Rock Mag. Paleogeophys.,* 1: 111—113.

Ozima, M., 1975. Ar isotopes and earth—atmosphere evolution models. *Geochim. Cosmochim. Acta,* 39: 1127—1133.

Ozima, M. and Kudo, K., 1972. Excess argon in submarine basalts and the earth—atmosphere evolution model. *Nature, Phys. Sci.,* 239: 23—24.

Paneth, F.A., Reasbeck, P. and Mayne, K.I., 1952. Helium 3 content and age of meteorites. *Geochim. Cosmochim. Acta,* 2: 300—303.

Paneth, F.A., Reasbeck, P. and Mayne, K.I., 1953. Production by cosmic rays of ^{3}He in meteorites. *Nature*, 172: 200—201.

Pepin, R.O., 1967. Trapped neon in meteorites. *Earth Planet. Sci. Lett.*, 2: 13—18.

Pepin, R.O. and Signer, P., 1965. Primordial rare gases in meteorites. *Science*, 149: 253—265.

Phinney, D., Tennyson, J. and Frick, U., 1978. Xenon in CO_2 well gas revisited. *J. Geophys. Res.*, 83: 2312—2319.

Podosek, F.A., 1978. Isotopic structures in solar system materials. *Annu. Rev. Astron, Astrophys.*, 16: 293—334.

Pollack, J. and Black, D.C., 1979. Implications of the gas compositional measurements of Pioneer Venus for the origin of planetary atmospheres. *Science*, 205: 56—59.

Polyak, B.G. and Smirnov, Ya.B., 1966. Continental heat flow. *Dokl. Akad. Nauk SSSR*, 168: 170—172 (in Russian).

Polyak, B.G. and Smirnov, Ya.B., 1968. A relation between heat flow and tectonic structure of the continents. *Geotektonika*, 4: 3—20 (in Russian).

Polyak, B.G. and Tolstikhin, I.N., 1983. *Geotectonics, Heat Flux and Helium Isotopes: Tripple Relationship*. Kola Dep. USSR Acad. Sci., Apatite, 52 pp.

Polyak, B.G., Kononov, V.I., Tolstikhin, I.N., Mamyrin, B.A. and Khabarin, L.V., 1976. The helium isotopes in thermal fluids. In: A.I. Johnson (Editor), *Thermal and Chemical Problems of Thermal Waters. Int. Assoc. Hydrol. Sci., Publ.*, 119: 15—29.

Polyak, B.G., Prasolov, E.M., Buachidze, G.I., Kononov, V.I., Mamyrin, B.A., Surovtseva, I.I., Khabarin, L.V. and Yudenich, V.S., 1979a. Isotope composition of He and Ar in fluids of the Alpine—Apennine region and its relation with volcanic activity. *Dokl. Akad. Nauk SSSR*, 247: 1220—1225 (in Russian).

Polyak, B.G., Tolstikhin, I.N. and Yakutseny, V.P., 1979b. Isotope composition of helium and heat flow — geochemical and geophysical aspects of tectogenesis. *Geotektonika*, 5: 3—23 (in Russian).

Polyak, B.G., Kononov, V.I., Mamyrin, B.A., Menyailov, I.A., Meshik, A.P., Prasolov, E.M., Khabarin, L.V. and Yudenich, V.S., 1980. Relation between isotope composition of helium and strontium in manifestations of volcanic and hydrothermal activity. In: *VIII Vses. Simp. Stab. Isotopam v Geokhimii*. GEOKHI Akad. Nauk SSSR, Moscow, pp. 13—17 (in Russian; abstract).

Prasolov, E.M., 1972. *Abundance and Genesis of Argon and Helium Isotopes in Microinclusions in Minerals*. Dissertation, Lensoveta Institute of Technology, Leningrad, 140 pp. (in Russian).

Prasolov, E.M., 1976. Excess Ar in gas-fluid inclusions in minerals and rocks. In: Yu.A. Shukolyukov and I.M. Morozova (Editors), *Razvitie i primenenie metodov yadernoy geokhronologii*. Nauka, Leningrad, pp. 153—176 (in Russian).

Prasolov, E.M. and Lobkov, V.A., 1977. On the origin and migration of methane (in the light of carbon isotope composition). *Geokhimiya*, 1: 122—135 (in Russian).

Prasolov, E.M. and Tolstikhin, I.N., 1969. Isotope composition of helium and argon from microinclusions in amphibole. *Geokhimiya*, 2: 231—234 (in Russian).

Prasolov, E.M. and Tolstikhin, I.N., 1970. Isotope composition of helium and argon in microinclusions in quartz. *Dokl. Akad. Nauk SSSR*, 191: 653—655 (in Russian).

Prasolov, E.M. and Tolstikhin, I.N., 1972. On the origin of ^{3}He in microinclusions of honeycomb quartz from Volyn. *Geokhimiya*, 6: 727—730 (in Russian).

Prasolov, E.M., Verkhovsky, A.B. and Polyak, B.G., 1982. Atmospheric noble gases in modern hydrothermae (theoretical calculations and experimental results). *Geokhimiya*, 12: 1691—1704 (in Russian).

Pudovkin, M.I., Tolstikhin, I.N. and Golovchanskaya, I.V., 1981. Recent achievements in helium isotope dissipation research. *Geochem. J.*, 15: 51—61.

Raitt, W.J., Schunk, R.W. and Banks, P.M., 1978. Quantitative calculations of helium ion escape fluxes from the polar ionosphere. *J. Geophys. Res.*, 83: 5617—5623.

Rajan, R.S., 1974. On the irradiation history and origin of gas-rich meteorites. *Geochim. Cosmochim. Acta*, 38: 777—788.

Rama, S.N.J., Hart, S.R. and Roedder, E., 1965. Excess radiogenic argon in fluid inclusions. *J. Geophys. Res.*, 70: 509—511.

Rees, C.E. and Thode, H.G., 1977. A ^{33}S anomaly in the Allende meteorite. *Geochim. Cosmochim. Acta*, 41: 1679—1682.

Revelle, R. and Suess, H.E., 1962. Interchange of properties between sea and atmosphere. In: M.N. Hill (Editor), *The Sea, 1*. Interscience, New York, N.Y., pp. 315—327.

Reynolds, J.H. and Turner, G., 1964. Rare gases in the chondrite Renazzo. *J. Geophys. Res.*, 69: 3263—3275.

Reynolds, J.H., Frick, U., Neil, J.M. and Phinney, D.L., 1978. Rare-gas-rich separates from carbonaceous chondrites. *Geochim. Cosmochim. Acta*, 42: 1775—1797.

Rik, G.R. and Shukolyukov, Yu.A., 1954. Isotope composition of potassium from meteorites. *Dokl. Akad. Nauk SSSR*, 94: 667—669 (in Russian).

Ringwood, A.E., 1966. Chemical evolution of terrestrial planets. *Geochim. Cosmochim. Acta*, 30: 41—104.

Rodionov, D.A., 1964. *Distribution Functions of Elements and Minerals in Erupted Rocks*. Nauka, Moscow, 102 pp. (in Russian).

Russell, R.D. and Ozima, M., 1971. The potassium—rubidium ratio of the earth. *Geochim. Cosmochim. Acta*, 35: 679—687.

Russell, R.D. and Reynolds, J.H., 1965. The age of the earth. In: N.I. Khitarov (Editor), *Geochemical Problems*. Nauka, Moscow, pp. 37—49 (in Russian).

Safronov, V.S., 1969. *Evolution of the Solar Nebula and Accumulation of the Earth and Planets*. Nauka, Moscow, 244 pp. (in Russian).

Safronov, V.S. and Kozlovskaya, S.V., 1977. Heating of the earth due to impacts of planetesimals in the course of its accumulation. *Fiz. Zemli*, 10: 3—13 (in Russian).

Saito, K., Basu, A.R. and Alexander, E.C., 1978. Planetary-type noble gases in an upper mantle-derived amphibole. *Earth Planet. Sci. Lett.*, 39: 274—280.

Schwartzman, D.W., 1973a. Argon degassing models of the earth. *Nature, Phys. Sci.*, 245: 20—21.

Schwartzman, D.W., 1973b. Ar degassing and the origin of the sialic crust. *Geochim. Cosmochim. Acta*, 37: 2479—2495.

Sheldon, W.R. and Kern, J.W., 1972. Atmospheric helium and geomagnetic field reversals. *J. Geophys. Res.*, 77: 6194—6201.

Shewmon, P., 1966. *Diffusion in Solids*. Metallurgiya, Moscow, 195 pp. (in Russian).

Shillibeer, H.A. and Russell, R.D., 1955. The argon-40 content of the earth atmosphere and the age of the earth. *Geochim. Cosmochim. Acta*, 8: 16—21.

Shukolyukov, Yu.A., 1970. *Nuclear Fission of Uranium in Nature*. Atomizdat, Moscow, 270 pp. (in Russian).

Shukolyukov, Yu.A. and Levsky, L.K., 1972. *Geochemistry and Cosmochemistry of Noble Gases*. Atomizdat, Moscow, 335 pp. (in Russian).

Shukolyukov, Yu.A. and Tolstikhin, I.N., 1965. Xe and Ar isotopes in the ancient rocks of the earth. *Geokhimiya*, 10: 1179—1185 (in Russian).

Shukolyukov, Yu.A., Ashkinadze, G.Sh., Verkhovsky, A.B. and Sharif-Zade, V.B., 1974. Geochemistry of neon isotopes. In: E.K. Gerling and Yu.A. Shukolyukov (Editors), *Geokhimiya radiogennykh i radioaktivnykh isotopov*. Nauka, Leningrad, pp. 5—45 (in Russian).

Shukolyukov, Yu.A., Kapusta, Ya.S. and Verkhovsky, A.B., 1979. Methods of mass spectrometric analysis of xenon produced by nuclear fission. *At. Energ.*, 47: 389—391 (in Russian).

Shustrov, B.N., 1960. A new circuit of an impulse magnetic mass spectrometer with high resolution. *Zh. Tekh. Fiz.*, 30: 860—864 (in Russian).

Signer, P. and Suess, H.E., 1963. Rare gases in the Sun, in the atmosphere and in meteorites. In: J. Geiss and E.D. Goldberg (Editors). *Earth Sciences and Meteorites*. North-Holland, Amsterdam, pp. 241—272.

Smelov, S.B., Vinogradov, V.I., Kononov, V.I. and Polyak, B.G., 1975. Isotope composition of argon in thermal fluids of Iceland. *Dokl. Akad. Nauk SSSR*, 222: 429—432 (in Russian).

Smirnov, Ya.B., 1972. Terrestrial heat flow and problems of geosyncline energy. In: P.N. Kropotkin (Editor), *Energetika geologicheskikh i geofizicheskikh processov*. Nauka, Moscow, pp. 52—74 (in Russian).

Smirnov, Ya.B. (Editor), 1982. *Map of Heat Flow of the USSR and Adjacent Regions, Scale 1:10,000,000, 1982*. GUGK SM SSSR, Moscow.

Smith, R.B., Shuey, R.T., Pelton, J.R. and Bailey, J.P., 1977. Yellowstone hot spot: contemporary tectonics and crustal properties from earthquake and aeronomic data. *J.Geophys. Res.*, 82: 3665—3676.

Smyslov, A.A., 1969. Heat regime of the earth crust and underlying masses. In: I.I. Borovikov and A.I. Semenov (Editors), *Geologicheskoe stroenie SSSR, V.* Nedra, Moscow, pp. 261—278 (in Russian).

Smyslov, A.A., Moiseenko, U.I. and Chadovich, T.Z., 1979. *Heat Regime and Radioactivity of the Earth*, Nedra, Leningrad, 191 pp. (in Russian).

Sokolov, V.A., 1966. *Geochemistry of Volatiles of the Earth Crust and the Atmosphere*. Nedra, Moscow, 300 pp. (in Russian).

Srinivasan, B., Alexander, E.C. and Manuel, O.K., 1969. Xenon and krypton from the spontaneous fission of Californium-252. *Phys. Rev.*, 179: 1166—1169.

Srinivasan, B., Gros, J. and Anders, E., 1977. Noble gases in separated meteoritic minerals: Murchison (C2), Ornans (C3), Karoonda (C5) and Abee (E4). *J. Geophys. Res.*, 82: 762—778.

Stickland, A.L. (Editor), 1972. *Cospar International Reference Atmosphere 1972*. Akademie-Verlag, Berlin, 450 pp.

Takagi, J., 1970. Rare gas anomalies and intense muon fluxes in the past. *Nature*, 227: 362—363.

Takagi, J., Sakamoto, K. and Tanaka, S., 1967. Terrestrial Xe anomaly and explosion of our Galaxy. *J. Geophys. Res.*, 72: 2267—2270.

Takaoka, N. and Ozima, M., 1978a. Rare gas isotopic compositions in diamonds. *Nature*, 271: 45—46.

Takaoka, N. and Ozima, M., 1978b. Rare gas isotopic compositions in diamonds. In: E.C. Alexander and M. Ozima (Editors), *Terrestrial Rare Gases*. Advances in Earth and Planetary Sciences, 3. Cent. Acad. Publ., Tokyo, pp. 65—70.

Taylor, S.R., 1979. Geochemical constraints on planetary compositions. In: *Abstr. Interdisciplinary Symposia XVII General Assembly of IUGS, Canberra, A.C.T.*, p. 212.

Telesnin, R.V., 1973. *Molecular Physics*. Vysshaya shkola, Moscow, 360 pp. (in Russian).

Tolstikhin, I.N., 1975a. *Helium Isotopes in Nature*. Dissertation, V.I. Vernadsky Inst. Geokhim. Analit. Khim., Moscow, 51 pp. (in Russian; abstract).

Tolstikhin, I.N., 1975b. Helium isotopes in the earth's interior and in the atmosphere: a degassing model of the earth. *Earth Planet. Sci. Lett.*, 26: 88—96.

Tolstikhin, I.N., 1978. A review: some recent advances in isotope geochemistry of light rare gases. In: E.C. Alexander and M. Ozima (Editors), *Terrestrial Rare Gases*. Advancies in Earth and Planetary Sciences, 3. Cent. Acad. Publ., Tokyo, pp. 33—62.

Tolstikhin, I.N., 1979. Mantle, crust and atmospheric volatiles in cold and thermal fluids. In: *Abstr. Inter-disciplinary Symposia, XVII General Assembly of IUGS, Canberra, A.C.T.*, p. 01/09.

Tolstikhin, I.N., 1980a. The problem of the earth accretion and degassing in the light of recent isotopic data. *Geokhimiya*, 3: 335—349 (in Russian).

Tolstikhin, I.N., 1980b. *Origin and History of Volatile Components of the Terrestrial Planets*. KF Akad. Nauk SSSR, Apatite, 54 pp.

Tolstikhin, I.N. and Drubetskoy, E.R., 1975. The ^{3}He/^{4}He and (^{4}He/^{40}Ar)$_{rad}$ isotope ratios for the earth's crust. *Geochem. Int.*, 12: 133—145.

Tolstikhin, I.N. and Drubetskoy, E.R., 1977. The helium isotopes in rocks and minerals of the earth's crust. In: Yu.A. Shukolyukov (Editor), *Problemy datirovaniya dokembriyskikh obrazovaniy*. Nauka, Leningrad, pp. 172—197 (in Russian).

Tolstikhin, I.N. and Kamensky, I.L., 1969. On the possibility of age determination of underground water by the tritium—helium-3 method. *Geokhimiya*, 8: 1027—1029 (in Russian).

Tolstikhin, I.N. and Kamensky, I.L., 1970. Sampler for liquids. Inventor's certificate No. 268344. *Byull. Izobr.*, 14: 8 (in Russian).

Tolstikhin, I.N. and Khabarin, L.V., 1976. Light noble gas isotopes in meteorites, in the earth and in the atmosphere, I. He, Ne and Ar in meteorites (a review). In: Yu.A. Shukolyukov and I.M. Morozova (Editors), *Razvitie i primenenie metodov yadernoy geokhronologii*. Nauka, Leningrad, pp. 78—102 (in Russian).

Tolstikhin, I.N. and Prasolov, E.M., 1971. Study of noble gases from gas-liquid inclusions in rocks and minerals. In: N.P. Ermakov and L.N. Khetchikov (Editors), *Investigations of Solutions and Melts by Analysis of Microinclusions in Minerals. Tr. Sib. Nauchno-Issled. Inst. Geol. Geofiz. Mineral. Ser.*, XIV: 86—97 (in Russian).

Tolstikhin, I.N., Shukolyukov, Yu.A. and Voitov, G.I., 1967. On the isotope composition of Xe and Kr in carbon-hydrogene gases of the Khibiny massiv. *Zap. Leningr. Gorn. Inst.*, LI: 111—113 (in Russian).

Tolstikhin, I.N., Kamensky, I.L. and Mamyrin, B.A., 1969. Isotopic criterion for the investigation of the origin of terrestrial helium. *Geokhimiya*, 2: 201—204 (in Russian).

Tolstikhin, I.N., Mamyrin, B.A., Baskov, E.A., Kamensky, I.L. and Surikov, S.N., 1972a. Helium isotopes in gases of hot springs of the Kuril—Kamchatka volcanic zone. In: A.I. Tugarinov (Editor), *Ocherki sovremennoy geokhimii i analiticheskoy khimii*. Nauka, Moscow, pp. 405—411 (in Russian).

Tolstikhin, I.N., Mamyrin, B.A. and Khabarin, L.V., 1972b. Anomalous isotope composition of helium in some xenoliths. *Geochim. Int.*, 9: 407—409.

Tolstikhin, I.N., Mamyrin, B.A., Khabarin, L.V. and Erlikh, E.N., 1974a. Isotope composition of helium in ultrabasic xenoliths from volcanic rocks of Kamchatka. *Earth Planet. Sci. Lett.*, 22: 75—84.

Tolstikhin, I.N., Prasolov, E.M., Khabarin, L.V. and Azbel, I.Ya., 1974b. The estimation of the helium diffusion coefficient in crystalline quartz. In: Yu.A. Shukolyukov and E.K. Gerling (Editors), *Geokhimiya radioaktivnykh i radiogennykh isotopov*. Nauka, Leningrad, pp. 79—90 (in Russian).

Tolstikhin, I.N., Azbel, I.Ya. and Khabarin, L.V., 1975. Isotopes of light nobles gases in the mantle, crust and the atmosphere. *Geochem. Int.*, 12: 10—20.

Tolstikhin, I.N., Drubetskoy, E.R., Erlikh, E.N. and Mamyrin, B.A., 1976. On the origin of felsic volcanic rocks of Kamchatka. *Geochem. Int.*, 13: 32—36.

Tolstikhin, I.N., Drubetskoy, E.R., Mitrofanov, F.P. and Kozakov, I.K., 1977a. The helium isotopes in rocks of the Sangilen massif. *Geokhimiya*, 4: 495—501 (in Russian).

Tolkstikhin, I.N., Verkhovsky, A.B. and Drubetskoy, E.R., 1977b. The helium isotope abundance in the "nondissipated" atmosphere. *Geokhimiya*, 8: 1107—1110 (in Russian).

Tolstikhin, I.N., Drubetskoy, E.R. and Sharas'kin, A.Ya., 1978. On the isotope composition of Ar in the earth mantle. *Geokhimiya*, 4: 514—520 (in Russian).

Top, Z. and Clarke, W.B., 1981. Dissolved helium isotopes and tritium in lakes: further results for uranium prospecting in Central Labrador. *Econ. Geol.*, 7: 2018—2031.

Top, Z., Clarke, W.B., Eismont, W.C. and Jones, E.P., 1980. Radiogenic helium in Baffin Bay bottom water. *J. Mar. Res.*, 38: 435—452.

Torgersen, T. and Clarke, W.B., 1978. Excess ^{4}He in Teggau lake: possibilities for a uranium ore body. *Science*, 199: 769—771.

Torgersen, T. and Jenkins, W.J., 1979. Comparative helium isotopes for the Snake River Plain—Yellowstone system. *Trans. Am. Geophys. Union*, 60: 945.

Torgersen, T., Top, Z., Clarke, W.B., Jenkins, W.J. and Broecher, W.S., 1977. A new method for physical limnology — tritium—helium-3 age results for lakes Erie, Huron, Ontario. *Limnol. Oceanogr.*, 22: 181—193.

Torgersen, T., Clarke, W.B. and Jenkins, W.J., 1979. The tritium/helium 3 method in hydrology. In: *Isotope Hydrology 1978, II.* International Atomic Energy Agency, Vienna, pp. 917—930.

Torgersen, T., Hammond, D.E., Clarke, W.B. and Pend, T.-H., 1981. Fayetteville, Green lake, New York: ^{3}H—^{3}He water mass ages and secondary chemical structure. *Limnol. Oceanogr.*, 26: 110—122.

Trofimov, A.V., 1949. Isotope composition of S in meteorites and terrestrial materials. *Dokl. Akad. Nauk SSSR*, 66: 181—184 (in Russian).

Tugarinov, A.I., Osipov, Yu.G. and Reutin, Yu.V., 1975. On the helium flux through fault zones and endogenic ore deposits. *Geokhimiya*, 11: 1615—1625 (in Russian).

Turekian, K.K., 1959. The terrestrial economy of He and Ar. *Geochim. Cosmochim. Acta*, 17: 37—42.

Turekian, K.K., 1964. Degassing of argon and helium from the earth. In: P.J. Rancazio and A.G.W. Cameron (Editors), *The Origin and Evolution of Atmospheres and Oceans.* Wiley, New York, N.Y., pp. 74—82.

Turekian, K.K. and Wedepohl, K.H., 1961. Distribution of the elements in the earth's crust. *Geol. Soc. Am. Bull.*, 72: 175—192.

Urey, H.C. and Craig, H., 1953. The composition of some meteorites and on the origin of the meteorites. *Geochim. Cosmochim. Acta*, 4: 36—82.

Van Schmus, W.R. and Wood, J.A., 1967. A chemical-petrologic classification for the condritic meteorites. *Geochim. Cosmochim. Acta*, 31: 747—766.

Veizer, J. and Compston, W., 1976. ^{87}Sr/^{86}Sr in Precambrian carbonates as an index of crustal evolution. *Geochim. Cosmochim. Acta*, 40: 905—914.

Verkhovsky, A.B. and Shukolyukov, Yu.A., 1975. On the possible occurrence of primordial earth neon in sudberite of the Monchegorsk Pluton. *Dokl. Akad. Nauk SSSR*, 224: 685—688 (in Russian).

Verkhovsky, A.B., Shukolyukov, Yu.A. and Ashkinadze, G.Sh., 1977. Radiogenic helium, neon and argon in terrestrial gases. In: Yu.A. Shukolyukov (Editor), *Problemy datirovaniya dokembriyskikh obrazovaniy.* Nauka, Leningrad, pp. 152—170 (in Russian).

Vinogradov, A.P., 1959. *Chemical Evolution of the Earth.* V.I. Vernadsky Memorial Symp., I. Akad. Nauk SSSR, Moscow, 43 pp. (in Russian).

Vinogradov, A.P., 1961. On the origin of the earth crust. *Geokhimiya*, 1: 3—29 (in Russian).

Vinogradov, A.P. and Zadorozhny, I.K., 1972. Rare gases in the regolith from the sea of Fertility. *Space Res.*, XII (1): 23—31.

Vinogradov, A.P., Dontsova, E.I. and Chupakhin, M.S., 1958. Isotope composition of oxygen in intrusive rocks and meteorites. *Geokhimiya*, 3: 187—190 (in Russian).

Vinogradov, A.P., Devirts, A.L. and Dobkina, E.I., 1968. Recent concentrations of tritium in natural waters. *Geokhimiya*, 10: 1147—1162 (in Russian).

Vinogradov, V.I., Kononov, V.I. and Polyak, B.G., 1974. Sulphur isotope composition in thermal manifestations of Iceland. *Dokl. Akad. Nauk SSSR*, 217: 1149—1152 (in Russian).

Voronov, A.N., Prasolov, E.M. and Tikhomirov, V.V., 1974. The relations between radiogenic isotopes of helium and argon in bed gases. *Geokhimiya*, 12: 1842—1855 (in Russian).

Vostrov, G.A. and Bol'shakov, O.I., 1966. Permeability of glasses for helium. *Prib. Tekh. Eksp.*, 2: 112—116.

Wakita, H., Nagasawa, H., Uyeda, S. and Kuno, H., 1967. Uranium, thorium and potassium contents of possible mantle materials. *Geochem. J.*, 1: 183—198.

Wakita, H., Fujii, N., Matsuo, S., Notsu, K., Nagao, K. and Takaoka, N., 1978. "Helium spots": caused by diapiric magma from upper mantle. *Science,* 200: 430—432.

Wasserburg, G.J. and Burnett, D.S., 1969. The status of isotopical age of iron and stone meteorites. In: P.M. Millman (Editor), *Meteorite Research.* Reidel, Dordrecht, pp. 467—479.

Wasserburg, G.J., Mazor, E. and Zartman, R.E., 1963. Isotopic and chemical composition of some terrestrial natural gases. In: J. Geiss and E.D. Goldberg (Editors), *Earth Science and Meteoritics.* North-Holland, Amsterdam, pp. 219—240.

Wasserburg, G.J., MacDonald, G.J.F., Hoyle, F. and Fowler, W.A., 1964. Relative contribution of U, Th and K to heat productions in the earth. *Science,* 143: 465—467.

Wasserburg, G.J., Huneke, J.C. and Burnett, D.S., 1969. Correlation between fission tracks and fission type xenon in meteoritic whitlickite. *J. Geophys. Res.,* 74: 4221—4232.

Wasserburg, G.J., Papanastassiou, D.A., Tera, F. and Huneke, J.C., 1977. The accumulation and bulk composition of the moon: outline of a lunar chronology. *Philos. Trans. R. Soc. Lond., Ser. A,* 285: 7—22.

Wasson, J.T., 1969. Primordial rare gases in the atmosphere of the earth. *Nature,* 223: 163.

Weidenschilling, S.J., 1976. Accretion of terrestrial planets. II. *Icarus,* 27: 161—171.

Weiss, R.F., 1970. Helium isotope effect in solution in water and sea water. *Science,* 168: 247—248.

Weiss, R.F., 1971. Solubility of helium and neon in water and seawater. *J. Chem. Eng. Data,* 16: 235—241.

Weiss, R.F., 1977. Hydrothermal manganese in the deep sea: scavenging residence time and $Mn/^3He$ relationships. *Earth Planet. Sci. Lett.,* 37: 257—262.

Welhan, J.A. and Craig, H., 1979. Methane and hydrogen in East Pacific rise hydrothermal fluids. *Geophys. Res. Lett.,* 6: 829—831.

Welhan, J.A., Lupton, J.E. and Craig, H., 1978a. Helium isotope ratios in Southern California fault zones. *Trans. Am. Geophys. Union,* 59: 12.

Welhan, J.A., Poreda, R., Lupton, J.E. and Craig, H., 1978b. Gas chemistry and helium isotopes at Cerro Prieto. In: *Proc. First Symp. Cerro Prieto Geothermal Field, Baja California, Mexico.* U.S. Natl. Techn. Serv., Springfield, pp. 113—118.

Wetherill, G.W., 1975. Radiometric chronology of the early Solar system. *Annu. Rev. Nucl. Sci.,* 16: 283—328.

White, F.A. and Forman, L., 1967. Four-stage mass spectrometer of large radius. *Rev. Sci. Instr.,* 38: 355—359.

Wilson, H.W., 1963. Two-stages mass spectrometers at Aldermaston. In: R.M. Elliot (Editor), *Advances in Mass Spectrometry,* 2. Pergamon Press, London, pp. 206—211.

Yakutseny, V.P., 1968. *The Geology of Helium.* Nedra, Leningrad, 232 pp. (in Russian).

Yanshin, A.L. (Editor), 1966. *Tectonic Map of Europe and Asia, Scale 1:5,000,000.* GIN Akad. Nauk, GUGK SM SSSR, MGiON SSSR, Moscow.

Zähringer, J., 1968. Rare gases in stony meteorites. *Geochim. Cosmochim. Acta,* 32: 209—237.

Zartman, R.E., Wasserburg, G.J. and Reynolds, J.H., 1961. Helium, argon and carbon in some natural gases. *J. Geophys. Res.,* 66: 277—306.

Zverev, B.P., Simakhin, Yu.F. and Dutov, A.G., 1972. The determination of lithium in solids by the reactions $^6Li(n,\alpha)^3H$. *At. Ener.,* 32: 39—42 (in Russian).

SUBJECT INDEX

272

—, step-wise heating, 11, 85, 87, 89, 108, 113, 155

Pacific island arc, 174
Pacific Ocean, 106, 112, 113, 200
Pamir, 164, 168
Paramushir Island, 172
Pechora syneclise, 167
pegmatite, 156, 157
penetrability, 1, 2
penetration, 40, 41
peridotite, 109
permeability, 2
phenocryst, 110, 118, 130, 132
pitchblende, 153
plutonium, 99
Po Basin, 171
potash salt, 143
potassium, 76—78, 142, 156, 231, 236, 238, 239
prospecting index, 191, 192
purification, 3, 16—18, 31, 40, 49, 72

quartz, 2, 3, 120, 144, 156, 161

radioactive decay, 75, 97—99, 104, 119, 130, 140, 157, 193, 208, 211, 228, 240, 244, 245
— —: α-decay, 98, 244
— —: β-decay, 105, 245
Red Sea, 106, 165, 201
resolution, 33, 34, 38, 40, 42, 44, 51, 53—58, 61—65, 110
— of double-stage mass spectrometer, 38, 40, 51, 53, 54
— of $^4He^+$—$^{12}C^{3+}$ mass multiplet, 34
— of $^3He^+$—$^3H^+$ mass multiplet, 42
— of $^3He^+$—$(H_3^+ + HD^+)$ mass multiplet, 34
— of magnetic resonance mass spectrometer, 56, 58, 61—63, 65, 69
— of one-stage mass spectrometer, 48, 53
resonance capture of neutrons, 153
Rioni trough, 170
Rostov Height, 167
Ruanda, 115
rubidium, 236

Sakhalin Island, 165, 172, 184
sampling, 5, 6
saturation, 84, 191, 194, 196, 199

scattering, 35, 37, 39, 49, 63
Scythian plate, 164, 168, 169
sensitivity, 33, 37, 38, 40, 44, 45, 48, 53—56, 61—63, 65, 69, 71, 72, 74
— of double-stage mass spectrometer, 40, 54
— of magnetic resonance mass spectrometer, 56, 61—63, 65, 69, 71, 72, 74
— of one-stage mass spectrometer, 48, 53
— threshold, 45
separation, 17, 28, 38, 42, 51, 63, 64, 69, 72, 148, 150
— of ions in magnetic resonance mass spectrometer, 63, 64, 69, 72
— of ions in static mass spectrometer, 38, 51
— of minerals, 148, 150
— of neon and helium, 17, 28
shape of peak, 39, 63, 69
Siberian Platform, 164, 167, 168
Sicily, 171, 172
Sochi, 29
soda glass, 2
solar system, 76, 79, 94, 244
— wind, 84, 89, 92, 203, 211
solubility, 184
solution, 193
sorption, 41
South Africa, 111, 112
South Okhotsk Deep, 172
Spitsbergen, 108
spodumene, 153, 154
standard mixtures, 23, 26, 28
Stavropol Uplift, 168
step-wise heating, *see* outgassing
strontium, 77, 145
Sukhumi, 29
sulphur, 76, 106, 181, 184

Terek—Kuma trough, 170
thermal fluids (*see also* hot springs), 97, 104, 107, 118, 133, 145, 180, 198
— springs, 73
thermocline, 188, 189
thermocouple, 14, 19
thermopause, 213, 214, 219, 222, 223, 233, 236, 245
tholeiites (*see also* basalt), 104, 133
thorium, 98, 99, 103, 142, 146, 148, 150, 151, 156, 192, 244
— content in minerals, 148, 150, 156